AF714571

The Diverse World of PDEs

Geometry and Mathematical Physics

CONTEMPORARY MATHEMATICS

788

The Diverse World of PDEs

Geometry and Mathematical Physics

Alexandre Vinogradov Memorial Conference
Diffieties, Cohomological Physics, and Other Animals
December 13–17, 2021
Independent Unversity of Moscow
and Moscow State University, Moscow, Russia

I. S. Krasil'shchik
A. B. Sossinsky
A. M. Verbovetsky
Editors

AMS AMERICAN MATHEMATICAL SOCIETY

EDITORIAL COMMITTEE

Michael Loss, Managing Editor

John Etnyre Angela Gibney Catherine Yan

2020 *Mathematics Subject Classification*. Primary 32-XX, 35-XX, 37-XX, 47-XX, 53-XX, 58-XX, 76-XX, 82-XX, 83-XX.

Library of Congress Cataloging-in-Publication Data

Names: Alexandre Vinogradov Memorial Conference on Diffieties, Cohomological Physics, and Other Animals (2021 : Moscow, Russia), author. | Krasil'shchik, I. S. (Iosif Semenovich), editor. | Sosinskiĭ, A. B. (Alekseĭ Bronislavovich), editor. | Verbovetsky, Alexander, editor. | Vinogradov, A. M. (Aleksandr Mikhaĭlovich), honoree

Title: The diverse world of PDEs : geometry and mathematical physics / I.S. Krasil'shchik, A.B. Sossinsky, A.M. Verbovetsky, editors.

Description: Providence, Rhode Island : American Mathematical Society, [2023] | Series: Contemporary mathematics, 0271-4132 ; 788 | "Alexandre Vinogradov Memorial Conference on Diffieties, Cohomological Physics, and Other Animals, December 13–17, 2021, Independent Unversity of Moscow and Moscow State University, Moscow, Russia"– title page. | Includes bibliographical references.

Identifiers: LCCN 2023006519 | ISBN 9781470471477 (paperback) | ISBN 9781470474089 (ebook)

Subjects: LCSH: Differential equations, Partial–Congresses. — Differential equations, Nonlinear–Congresses. | Geometry, Differential–Congresses. | Mathematical physics–Congresses. | AMS: Several complex variables and analytic spaces. | Partial differential equations. | Dynamical systems and ergodic theory. | Operator theory. | Differential geometry. | Global analysis, analysis on manifolds. | Fluid mechanics. | Statistical mechanics, structure of matter. | Relativity and gravitational theory.

Classification: LCC QA377 .A5552 2023 | DDC 515/.353–dc23/eng20230626

LC record available at https://lccn.loc.gov/2023006519

Color graphic policy. Any graphics created in color will be rendered in grayscale for the printed version unless color printing is authorized by the Publisher. In general, color graphics will appear in color in the online version.

Copying and reprinting. Individual readers of this publication, and nonprofit libraries acting for them, are permitted to make fair use of the material, such as to copy select pages for use in teaching or research. Permission is granted to quote brief passages from this publication in reviews, provided the customary acknowledgment of the source is given.

Republication, systematic copying, or multiple reproduction of any material in this publication is permitted only under license from the American Mathematical Society. Requests for permission to reuse portions of AMS publication content are handled by the Copyright Clearance Center. For more information, please visit `www.ams.org/publications/pubpermissions`.

Send requests for translation rights and licensed reprints to `reprint-permission@ams.org`.

© 2023 by the American Mathematical Society. All rights reserved.
The American Mathematical Society retains all rights
except those granted to the United States Government.
Printed in the United States of America.

∞ The paper used in this book is acid-free and falls within the guidelines
established to ensure permanence and durability.
Visit the AMS home page at `https://www.ams.org/`

10 9 8 7 6 5 4 3 2 1 23 22 21 20 19 18

To the memory of Alexandre Vinogradov (1938–2019)

Contents

The Editors Preface

A remarkable Soviet/Russian/Italian mathematician Alexandre M. Vinogradov passed away on September 20, 2019 (see the obituary in Russian Math. Surveys **75** (2020), no. 2, 369–375). Independent University of Moscow in collaboration with the Faculty of Mechanics and Mathematics of Moscow Lomonosov State University organized an international conference *Diffieties, Cohomological Physics, and Other Animals* in memory of Vinogradov (`https://gdeq.org/amv_conf`). The Conference took place in the period December 13–17, 2021 and comprised a very wide spectrum of topics that more or less reflect the wideness of Vinogradov's mathematical interests. The current volume of Contemporary Mathematics contains the articles that deal with relations between nonlinear PDEs and geometry and mathematical physics.

I. Krasil′shchik, A. Sossinsky, A. Verbovetsky
Independent University of Moscow

Contemporary Mathematics
Volume **788**, 2023
https://doi.org/10.1090/conm/788/15817

Geometric analysis of metric Legendre foliated cocycles on contact manifolds via SODE structure

Fatemeh Ahangari

Abstract. This paper is predominantly devoted to thoroughgoing investigation of the Legendre foliated cocycles which are totally compatible with second order ordinary differential equation fields (SODE). For this goal, the spencer theory of formal integrability, relative 1-forms, Bott connection and the Vrăceanu connection, are applied as the prominent tools in order to designate the sufficient conditions for the metric concomitant to the analyzed SODE structure, to extend to a bundle-like metric for the lifted Legendre foliated cocycle on the tangent space of an arbitrary contact manifold. Significantly, the geometric structure of the constructed metric foliated cocycles on the tangent bundle are extensively scrutinized by virtue of the two local invariants corresponding to Legendre foliations comprising a symmetric 2-form and a symmetric 3-form characterizing on the tangent bundle of the foliation. In particular, it is focused on the essence of the metric Legendre foliated cocycles adapted to SODE structure on two specific class of contact metric manifolds including the Sasakian manifolds and (k, μ)-spaces. Furthermore, the pivotal circumstances regarding the existence and the geometric features of the conjugate Legendre foliated cocycles on the tangent space of the contact metric manifolds which are entirely compatible with the ambient SODE structure are conclusively explored.

1. Introduction

In recent years, there has been a great deal of concentration toward the field of contact geometry, predominantly on account of its noteworthy applications in modelling of physical phenomena specifically in optics and time-depending mechanical systems. A contact manifold as proposed by J. W. Gray [**10**] is inherently an odd-dimensional differential manifold which carries a contact structure. This contact structure is definitely correlated to pseudogroups of contact transformations. In the subject area of geometry and mathematical physics, the scrutinization of Lagrangian foliations on symplectic manifolds is of particular significance. More recent is the survey of the theory of Legendre foliations as a viable class of foliations on contact manifolds. The prominent point is that there exists a close association

2020 *Mathematics Subject Classification.* Primary 53D10, 53C12; Secondary 53C60, 53C22.
Key words and phrases. Legendre foliations, projective metrizability, Bott connection, Vrăceanu connection, contact metric (k, μ)-spaces.

©2023 American Mathematical Society

among the geometry of Legendre foliations and the geometry of Lagrangian foliations of symplectic manifolds. Literally, Legendre foliations on contact manifolds are intrinsically odd-dimensional counterpart of Lagrangian foliations on symplectic manifolds. In their groundbreaking papers of 90's, M. Y. Pang [**18**] and P. Libermann [**15**] carried out the first comprehensive systematic investigations into geometric structure of Legendre foliations on contact manifolds which are relatively recent in this context and explicitly combines the theory of contact manifolds with foliation theory. The overall goal of this paper is to analyze the metric Legendre foliated cocycles which are totally adapted to second order ordinary differential equations (SODE).

Over time, an extensive literature has developed on qualitative analysis of systems of (non-) autonomous second (higher) order ordinary (partial) differential equations fields through some associated geometric structures. Second order ordinary differential equations (SODE) are of special significance mainly due to their extensive applications in various domains of mathematics, science and engineering. The inverse problem of the calculus of variations consists in determining whether or not a given system of differential equations is inspired by a variational problem. Necessary and sufficient conditions for this problem, known as the Helmholtz conditions, were explicitly acquired only for SODE's in both autonomous and non-autonomous case. Hence, a remarkable type of SODE is the one which can be deduced from a variational principle. In their seminal paper of 2009, I. Bucataru and M. F. Dahl [**8**], by utilizing the Frölicher-Nijenhuis formalism exhaustively obtained the global Helmholtz conditions expressed in terms of a semi-basic 1-form, that designate identically when an autonomous second order ordinary differential equations field is a Lagrangian vector field. Moreover, a significant case of the inverse problem of the calculus of variations refers to homogenous differential equation fields and comprises two noteworthy problems: projective and Finsler metrizability. For both problems, in [**8**] Helmholtz conditions are dervided in terms of a homogenous semi-basic 1-form.

The projective metrizability problem for a homogenous system of SODE, which can be characterized explicitly by virtue of a spray S, explores a Finsler metric F whose concomitant geodesics coincide with the geodesics of the spray S, up to an orientation preserving reparametrization. This was successfully established as described by I. Bucataru and Z. Muzsnay [**9**], by virtue of reformulating the projective metrizability problem for a spray in terms of a first order partial differential operator $\mathcal{P}_1$ and a set of concomitant algebraic conditions on semi-basic 1-forms. Thereupon, the formal integrability of $\mathcal{P}_1$ is thoroughly discussed on account of two sufficient conditions provided by Cartan-Kähler Theorem. Indeed, it is explicitly demonstrated that the symbol of $\mathcal{P}_1$ is involutive and accordingly one of the two conditions is always fulfilled. While scrutinizing the second condition, it is illustrated that there exists only one obstruction into the formal integrability of $\mathcal{P}_1$ which is identically due to the curvature tensor of the induced nonlinear connection. Therefore, whenever the curvature obstruction is satisfied, the projective metrizability problem reduces to investigation of the corresponding algebraic conditions, which as illustrated in [**9**] are always fulfilled in the analytic case.

In this research, a thoroughgoing investigation of the Legendre foliated cocycles which are compatible with a given SODE structure is proposed. For this prominent motive, we have applied the spencer theory of formal integrability, relative 1-forms, Bott connection and the Vrăceanu connection, as the powerful tools in order to characterize the sufficient conditions for the metric concomitant to the analyzed SODE structure, to extend to a bundle-like metric for the for the lifted Legendre foliated cocycle on the tangent space of an arbitrary contact manifold. Significantly, we have precisely analyzed the geometric structure of the resulted metric foliated cocycles on the tangent bundle by virtue of the two local invariants existing on the tangent bundle of the foliation introduced by M. Y. Pang [**18**], including the fundamental quadratic form $\mathbf{\Pi}$ which is a symmetric 2-form and the invariant $\mathbf{G}$ which is a symmetric 3-form. In particular, we have studied the metric Legendre foliated cocycles adapted to SODE structure on two noteworthy class of contact metric manifolds comprising the Sasakian manifolds and (k, μ)-spaces [**17**]. Moreover, the essence and the geometric properties of the conjugate Legendre foliated cocycles on the tangent space of contact metric manifolds which are compatible with the concomitant SODE structure are fundamentally analyzed.

The major framework of this paper is as follows: In section 2, we recall first some fundamental aspects of the Frölicher-Nijenhuis formalism and apply it to the geometric objects that live canonically on the tangent space such as vertical distribution, Liouville vector field and semi-basic forms. Thereupon, by virtue of this geometric setting, the reformulation of the projective metrizability problem in terms of the first-order partial differential operator P_1 which acts on semi-basic 1-forms is briefly discussed. In section 2, we have presented some preliminaries and qualitative information regarding the geometric structure of the Legendre foliations on contact manifolds. Section 3 is dedicated to comprehensive study of the metric Legendre foliated cocycles which are compatible with SODE structure. Some concluding points are declared at the end of the paper.

2. Frölicher-Nijenhuis formalism and projective metrizability problem

Suppose that M be an m-dimensional arbitrary manifold and (TM, π, M) with local coordinates (x^i, y^i) indicates its corresponding tangent bundle. Furthermore, the distribution VTM indicates the concomitant vertical subbundle. As it is well known, the inverse problem of Lagrangian mechanic is profoundly derived from the fact that whether or not a given system of second order ordinary differential equations (SODEs) can be categorized as the Euler-Lagrange equations of some Lagrangian. This can be explicitly described as follows: What are the principal circumstances under which the solutions of a system of second order differential equations (SODE)

$$\frac{d^2x^i}{dt^2} + 2G^i\big(x, \dot{x}\big) = 0, \quad i \in \Big\{1, \dots, m\Big\}, \tag{2.1}$$

can be construed from a variational principle? In other words, for the optional Lagrangian function L, can be considered as the solutions of the following system of differential equations which is designated by the Euler-Lagrange equations:

$$\frac{d}{dt}\Big(\frac{\partial L}{\partial \dot{x}^i}\Big)\frac{\partial L}{\partial x^i} = 0, \quad i \in \Big\{1, \dots, m\Big\},$$

Indeed, one privileged point of view regarding the problem mentioned above, applies the viable Helmholtz conditions which are predominantly considered as the necessary and sufficient conditions for the indispensable essence of a specified multiplier matrix $g_{ij}(x,\dot{x})$ such that the following identity holds:

$$g_{ij}(x,\dot{x})\left(\frac{d^2x^j}{dt^2}+2G^i(x,\dot{x})\right)=\frac{d}{dt}\left(\frac{\partial \mathrm{L}}{\partial \dot{x}^i}\right)-\frac{\partial \mathrm{L}}{\partial x^i}, \tag{2.2}$$

for appointive Lagrangian function $L(x,\dot{x})$. It is worth mentioning that the multiplier matrix g_{ij} absolutely constitutes a symmetric $\binom{0}{2}$-type tensor field g along the tangent bundle projection. Geometric formulation of Helmholtz conditions in terms of g_{ij} has been extensively scrutinized from various standpoints in the last few years.

In this section, according to [**8, 9, 11**], we present some preliminaries and fundamental concepts regarding Frölicher-Nijenhuis theory on $TM\backslash\{0\}$ together with Helmholtz conditions for projective metrizability and concomitantly their formal integrability.

Suppose that A is a vector valued l-form on $TM\backslash\{0\}$ and α is a k-form on $TM\backslash\{0\}$ where $k\geq 1$ and $l\geq 0$. Then the inner product of A and α is the $(k+1-l)$-form $i_A\alpha$ which is expressed by:

$$i_A\alpha(X_1,\dots,X_{k+l-1})$$
$$=\tfrac{1}{l!\left(k-1\right)!}\textstyle\sum_{\sigma\in S_{k+l-1}}\mathrm{sign}(\sigma)\alpha\left(A\left(X_{\sigma(1)},\dots,X_{\sigma(l)}\right),X_{\sigma\left(l+1\right)},\dots,X_{\sigma\left(k+l-1\right)}\right),$$

where $X_1,\dots,X_{k+l-1}\in\mathfrak{X}(TM\backslash\{0\})$ and S_p denotes the permutation group of the elements $1,\dots,p$. In particular, when $l=0$ then A is a vector field on $TM\backslash\{0\}$ and $i_A\alpha$ is the usual inner product of k-form α concerning a vector field A. Besides, whenever $l=1$ then A is a $\binom{1}{1}$-type tensor field and $i_A\alpha$ is the following k-form:

$$i_A\alpha(X_1,\dots,X_k)=\sum_{i=1}^{k}\alpha\left(X_1,\dots,AX_i,\dots,X_k\right).$$

Suppose that A is a vector valued l-form on $TM\backslash\{0\}$. Then for $k,l\geq 0$ the exterior derivative with reference to A is this specified map:

$$\begin{aligned}&d_A:\Lambda^k(TM\backslash\{0\})\longrightarrow\Lambda^{k+1}(TM\backslash\{0\}),\\&d_A=i_A\circ d-(-1)^{l-1}d\circ i_A,\end{aligned}\tag{2.3}$$

where the C^∞ module of k-forms is represented by $\Lambda^k(M)$. Furthermore, a k-form ω on $TM\backslash\{0\}$ is designated d_A-closed if $d_A\omega=0$ and d_A-exact whenever there is at least one $\Theta\in\Lambda^{k-1}\left(TM\backslash\{0\}\right)$ such that $\omega=d_A\Theta$. Specifically, when $A\in\mathfrak{X}(TM\backslash\{0\})$ and $k\geq 0$, we acquire $d_A=\mathcal{L}_A$, where $\mathcal{L}_A$ is the common Lie derivative $\mathcal{L}_A:\Lambda^k(TM\backslash\{0\})\longrightarrow\Lambda^k(TM\backslash\{0\})$. In this case, relation (2.3) is exactly the Cartan's formula. For the two vector valued forms A and B on $TM\backslash\{0\}$ of degrees respectively $l\geq 0$ and $k\geq 0$ the Frölicher-Nijenhuis bracket of A and B can be regarded explicitly as a vector valued $(k+l)$-form $[A,B]$ on $TM\backslash\{0\}$ which is uniquely determined by:

$$d_{[A,B]}=d_A\circ d_B-(-1)^{kl}d_B\circ d_A,$$

when A and B are vector fields, then the Frölicher-Nijenhuis bracket $[A,B]$ is identically the usual Lie bracket $[A,B]=\mathcal{L}_A B$. Moreover, for a vector field $X\in\mathfrak{X}(TM\backslash\{0\})$ and a $\binom{1}{1}$-type tensor field A on $TM\backslash\{0\}$, the Frölicher-Nijenhuis bracket $[X,A]=\mathcal{L}_X A$ is the $\binom{1}{1}$-type tensor field on $TM\backslash\{0\}$ which is expressed by:

$$\mathcal{L}_X A=\mathcal{L}_X\circ A-A\circ\mathcal{L}_X.$$

Furthermore, for $\binom{1}{1}$-type tensor fields A and B the next commutation formula on $\Lambda^k(TM\backslash\{0\}), k\geq 0$ holds:

$$\begin{aligned}&(a): i_A d_B-d_B i_A=d_{B\circ A}-i_{[A,B]},\\ &(b): \mathcal{L}_X i_A-i_A\mathcal{L}_X=i_{[X,A]},\\ &(c): i_X d_A+d_A i_X=\mathcal{L}_{AX}-i_{[X,A]}.\end{aligned}\tag{2.4}$$

It is worth mentioning that formula (2.4.c) is referred as the generalized Cartan's formula.

In the local coordinate system, the tangent structure $\mathcal{J}$ is fundamentally demonstrated by $\mathcal{J}=\frac{\partial}{\partial y^i}\otimes dx^i$ and the vector field $\mathbb{C}\in\mathfrak{X}(TM)$ defined by $\mathbb{C}=y^i\frac{\partial}{\partial y^i}$ is identified as the Liouville vector field. In addition, a $k-$form ω is called semi-basic if $\omega(X_1,X_2,\ldots,X_k)=0$ whenever one of the vector fields X_i is vertical for $i\in\{1,\ldots,k\}$. Moreover, the module of semi-basic k-forms is denoted by $\mathrm{Sec}(\Lambda^k T_V^*)$. Also, a vector valued k-form A on $TM\backslash\{0\}$ is said to be semi-basic if it admits values in the vertical bundle and specifically when one of the vectors X_i, $i\in\{1,\ldots,k\}$ is vertical the following relation holds: $A(X_1,X_2,\ldots,X_k)=0$.

Hence according to Frölicher Nijenhuis theory a semispray (spray) on M is a vector field $S\in\mathfrak{X}(TM\backslash\{0\})$ such that $\mathcal{J}S=\mathbb{C}$ (and $[\mathbb{C},S]=S$). Now consider the almost tangent structure $\Gamma=-\mathcal{L}_S\mathcal{J}=h-v$ where h and v are the horizontal and vertical projectors determined by S respectively. Then the Jacobi endomorphism (or Douglas tensor) Φ is characterized as the following $(1,1)$-type tensor field

$$\Phi=vo\mathcal{L}_S h=-vo\mathcal{L}_S v=R^i_j\frac{\partial}{\partial y^i}\otimes dx^j.$$

The dynamical covariant derivative ∇ is defined by:

$$\begin{aligned}&\nabla X=h[S,hX]+v[S,vX],\quad \forall\ X\in\mathfrak{X}(TM\backslash\{0\})\\ &\nabla=\mathcal{L}_S+ho\mathcal{L}_S h+vo\mathcal{L}_S v=\mathcal{L}_S+\Psi.\end{aligned}$$

Taking into account that ∇ is a zero-degree derivation on $\Lambda^k(TM\backslash\{0\})$ it can be distinctively decomposed into the sum of a Lie derivation $\mathcal{L}_S$ and an algebraic derivation i_Ψ as follows: $\nabla=\mathcal{L}_S-i_\Psi$. According to [**8**] the following significant relations hold:

$$\begin{aligned}&(a):\ \nabla S=0,\quad \nabla\mathbb{C}=0,\quad \nabla i_S=i_S\nabla,\quad \nabla i_{\mathbb{C}}=i_{\mathbb{C}}\nabla;\\ &(b):\ \nabla h=0,\quad \nabla v=0,\quad \nabla\mathcal{J}=0,\quad \nabla\mathbb{F}=0;\\ &(c):\ d\nabla-\nabla d=d_\Psi,\quad \nabla i_h=i_h\nabla=0,\quad \nabla i_{\mathcal{J}}-i_{\mathcal{J}}\nabla=0.\end{aligned}$$

The significant point is that system (2.1) can be explicitly characterized via the semispray S which is locally expressed as follows:

$$S=y^i\frac{\partial}{\partial x^i}-2G^i(x,y)\frac{\partial}{\partial y^i}.\tag{2.5}$$

Moreover, system (2.2) can be identically reformulated as

$$\mathcal{L}_S\Big(\frac{\partial L}{\partial \dot{x}^i}dx^i\Big) = dL, \tag{2.6}$$

where $\mathcal{L}_S$ denotes the Lie derivative with respect to semispray S. Conclusively, the Helmholtz conditions for the multiplier matrix g_{ij} are unequivocally determined by the following crucial identities:

$$g_{ij} = g_{ji}, \qquad \frac{\partial g_{ij}}{\partial y^k} = \frac{\partial g_{ik}}{\partial y^j}, \tag{2.7}$$

$$\nabla g_{ij} = 0, \qquad g_{ik}R^k_j = g_{jk}R^k_i. \tag{2.8}$$

It is noticeable that conditions (2.7) are necessary and sufficient conditions for the existence of a Lagrange function which has as Hessian the matrix multiplier g_{ij}. Moreover, the conditions (2.8) represent the compatibility among the multiplier matrix and the analyzed SODE and pertaining geometric constructions such as: The Douglas tensor (Jacobi endomorphism) Φ and the dynamical covariant derivative.

So a semispray S on M is denoted by a Lagrangian vector field if there is at least one $\mathrm{L} \in C^\infty(TM\backslash\{0\})$ such that $\mathcal{L}_S d_{\mathcal{J}}\mathrm{L} = d\mathrm{L}$. It is worth mentioning that by virtue of Frölicher-Nijenhuis Formalism the Helmholtz conditions can be identically reformulated in terms of some regular, linear partial differential operators. Accordingly, the formal integrability of these PDE operators are precisely addressed by utilizing Spencer theory as a powerful tool. Consequently, according to [**9**] we can state the following theorem:

THEOREM 2.1. *A semispray S is a Lagrangian vector field if and only if there exists a semi-basic 1-form Θ on $TM\backslash\{0\}$ that satisfies the following reformulations of Helmholtz conditions:*

$$d_h\Theta = 0, \quad d_{\mathcal{J}}\Theta = 0, \quad \nabla d\Theta = 0, \quad d_\Phi\Theta = 0. \tag{2.9}$$

Consequently, the semispray S is a Lagrangian vector field if and only if the following partial differential operator

$$\mathcal{P}_{\mathrm{L}} = \big(d_{\mathcal{J}}, d_h, d_\Phi, \nabla d\big) : \ \mathrm{Sec}(T_v^*) \longrightarrow \mathrm{Sec}\,\big(\oplus^{(4)}\,\Lambda^2 T_v^*\big) \tag{2.10}$$

is formally integrable.

(a): Furthermore, Two sprays S_1 and S_2 are *projectively equivalent* if their geodesics coincide up to an orientation preserving re-parametrization.

(b): A spray S is called *projectively metrizable* if it is projectively equivalent to the geodesic spray of a Finsler function.

Literally, according to [**9**], two sprays S_1 and S_2 are projectively equivalent if and only if there exists a 1-homogenous function $\mathscr{P} \in C^\infty\big(TM \setminus \{0\}\big)$ such that $S_2 = S_1 - 2\mathscr{P}\mathbb{C}$.

Overall, considering above discussion a spray S is projectively metrizable if there exists a 1-homogeneous function $F \in C^\infty(TM\backslash\{0\})$ such that $\mathcal{L}_S d_{\mathcal{J}} F = dF$; Moreover a spray S is Finsler metrizable if there exists a 2-homogeneous function $\mathrm{L} \in C^\infty(TM)\backslash\{0\}$ such that $\mathcal{L}_S d_{\mathcal{J}}\mathrm{L} = \mathrm{L}$.

Consequently, the Lagrangian function L will be determined as the potential of the homogeneous semi-basic 1-form. Thereupon, for the semi-basic 1-form Θ on $TM\backslash\{0\}$, the following two specific cases are totally analyzed [**8**]:

Case (1): If Θ is 0-homogeneous and satisfies the following two Helmholtz conditions: $d_{\mathcal{J}}\Theta = 0$ and $d_h\Theta = 0$, then the corresponding potential $i_S\Theta$ is a 1-homogeneous function which projectively metricizes the spray S.

Case (2): Provided that Θ is 1-homogeneous and complies with these three Helmholtz conditions: $d_{\mathcal{J}}\Theta = 0$, $d_h\Theta = 0$ and $\nabla d\Theta = 0$, then its associated potential $i_S\Theta$ is a 2-homogeneous function that Finsler metricizes the spray S.

As a result, reckon with [**8**] we explicitly educe the subsequent assertions:

(i): A spray S is projectively metrizable if there exists a 1-homogemeous function $F \in C^\infty\big(TM\backslash\{0\}\big)$ such that $\mathcal{L}_S d_{\mathcal{J}}F = dF$.

(ii): A spray S is Finsler metrizable if there exists a 2-homogemeous function $L \in C^\infty\big(TM\backslash\{0\}\big)$ such that $\mathcal{L}_S d_{\mathcal{J}}L = dL$.

Summing up the points asserted above, it is deduced that: taking into account the fact that $d_\Phi\Theta = 0$ is identically the same as $g_{ik}R^k_j = g_{jk}R^k_i$, one of the Helmholtz conditions (2.7) and (2.8) is obviously redundant. Besides, if L is 2-homogeneous we have: $2L = i_S\Theta$ which explicitly implies that $d_{\mathcal{J}}\Theta = 0$ if and only if $\Theta = d_{\mathcal{J}}L$. In addition, whenever $d_{\mathcal{J}}\Theta = 0$, the two conditions $d_h\Theta = 0$ and $\nabla d\Theta = 0$ are thoroughly equivalent to $\mathcal{L}_S d_{\mathcal{J}}L = dL \longleftrightarrow i_S dd_{\mathcal{J}}L = -dL \longleftrightarrow d_hL = 0$.

3. Legendre foliations on contact manifolds

In this section, taking into account [**2, 6, 18**], we recall some predominant concepts regrading the geometric structure of Legendre foliations on contact manifolds.

Let M be a real $(2n+1)$-dimensional smooth manifold which carries a 1-form η satisfying $\eta \wedge (d\eta)^n \neq 0$ everywhere on M, where the exponent represents the nth exterior power. Then $\big(M, \eta\big)$ is called a *contact manifold* with the *contact form* η. Then a global vector field ξ, called the *characteristic vector field* or *Reeb vector field* on the contact manifold $\big(M, \eta\big)$, is defined on M by these conditions: $i_\xi\eta = 1$ and $i_\xi d\eta = 0$. A contact manifold $\big(M, \eta\big)$ admits a natural $2n$-dimensional distribution $\mathcal{H}$ which is defined by the kernel of η. In other words, $\mathcal{H}$ is simply the subbundle of TM on which $\eta = 0$. To be more precise we can write:

$$\Gamma(\mathcal{H}) = \Big\{X \in \Gamma(TM) :\ \eta(X) = 0\Big\} \tag{3.1}$$

The distribution $\mathcal{H}$ is defined by the *contact distribution* on $\big(M, \eta\big)$.

In the following, we want to relate contact manifolds with the notion of the contact metric manifolds. Let (M, g) be a real $(2n+1)$-dimensional Riemannian manifold endowed with a tensor field φ of the type $\binom{1}{1}$, a 1-form η and a vector field ξ. Then $\big(M, g, \varphi, \xi, \eta\big)$ is denoted by a *contact metric manifold* if for any $X, Y \in \Gamma(TM)$, the following tensor fields satisfy:

$$\begin{cases} (\mathbf{a}) : \varphi^2 = -I + \eta \otimes \xi, \\ (\mathbf{b}) : \eta(X) = g\big(X, \xi\big), \\ (\mathbf{c}) : g\big(X, \varphi Y\big) = d\eta\big(X, Y\big). \end{cases} \tag{3.2}$$

Taking into account (3.2) we can easily check that the following important identities hold for any $X, Y \in \Gamma(TM)$.

$$(3.3)\qquad \begin{cases} (\mathbf{d}) : \eta(\xi) = 1 \quad , \quad (\mathbf{e}) : \varphi(\xi) = 0 \\ (\mathbf{f}) : \eta(\varphi X) = 0 \quad , \quad (\mathbf{g}) : g(X, \varphi Y) + g(Y, \varphi X) = 0, \\ (\mathbf{h}) : g(\varphi X, \varphi Y) = g(X, Y) - \eta(X)\eta(Y). \end{cases}$$

It is clear that a contact metric manifold is a contact manifold. In [**6**] it is proved that the converse is also true i.e. any contact manifold (M, η) admits a contact metric structure (g, φ, ξ, η). According to relation **(c)** of (3.2) the 2-form Ω on M can be defined by:

$$(3.4)\qquad \Omega(X, Y) = g(X, \varphi Y) \quad , \quad \forall\, X, Y \in \Gamma(TM),$$

and it is called the *fundamental 2-form* associated to the contact metric structure (g, φ, ξ, η). Since Ω is non-degenerate and we have $d\eta = 0$ on $\Gamma(\mathcal{H})^3$, it is deduced that Ω defines a symplectic structure on the contact distribution. Furthermore, we intend to analyze the integrability of the contact distribution $\mathcal{H}$. To be specific, it can obviously be explicated that the contact condition $\eta \wedge (d\eta)^n \neq 0$ by declaring that the distribution $\mathcal{H}$ is as far from being integrable as possible.

Taking into account (3.2.b) it can be inferred that the contact distribution $\mathcal{H}$ identically coincides with the complementary orthogonal distribution to the characteristic distribution span$\{\xi\}$. Now, assume that $\mathcal{H}$ is integrable. Hence, for any $X, Y \in \Gamma(\mathcal{H})$ we have $[X, Y] \in \Gamma(\mathcal{H})$ i.e. $\eta([X, Y]) = 0$. Consequently, for any $X, Y \in \Gamma(\mathcal{H})$ we have: $d\eta(X, Y) = 0$. On the other hand, from (3.2.c) and (3.3.e) we have:

$$(3.5)\qquad d\eta(X, \xi) = 0, \quad \forall\ X \in \Gamma(TM).$$

So, it is resulted that $d\eta = 0$ on M, which is totally impossible since M is a contact manifold. Ultimately, it is concluded that the contact distribution $\mathcal{H}$ on a contact manifold (M, η) is not an integrable distribution.

Moreover, For any $X, Y \in \Gamma(TM)$, the *exterior derivative* of η is expressed by the following formula:

$$(3.6)\qquad d\eta(X, Y) = \frac{1}{2}\Big(X(\eta(Y)) - Y(\eta(X)) - \eta([X, Y])\Big),$$

Hence, by applying (3.5), (3.6), (3.3.d) and (3.2.b) it is deduced that:

$$(3.7)\qquad \forall\, X \in \Gamma(\mathcal{H}) \qquad \eta([X, \xi]) = 0$$

Now, assume that P is a k-dimensional integral manifold of the contact distribution $\mathcal{H}$. Then according to (3.6) we have:

$$(3.8)\qquad \forall\, X, Y \in \Gamma(TN) \qquad d\eta(X, Y) = 0$$

Consequently, due to (3.2.c) we have $g(X, \varphi Y) = 0$, which means that $\varphi(TP) \subset TP^{\perp}$. Thus P is an anti-invariant submanifold of $(M, g, \varphi, \xi, \eta)$. Moreover, it is normal to the characteristic vector field ξ. Considering the point that φ is an automorphism of $\Gamma(\mathcal{H})$, it can be inferred that $p < n+1$. Subsequently, it is proved that the maximum dimension of an integral manifold of the contact distribution $\mathcal{H}$ is $k = n$. In addition, the existence of the integral manifolds of the maximum dimension can be resulted via the Darboux's theorem. According to this theorem, for an arbitrary $(n+1)$-dimensional contact manifold (M,η), about each point there exists local coordinates $(x^1,\dots,x^n,y^1,\dots,y^n,z)$ such that the 1-form η on M has the following expression:

$$\eta = dz - \sum_{i=1}^{n} y^i dx^i \tag{3.9}$$

As a consequence, due to (3.9) the n-dimensional integral manifold of the contact distribution $\mathcal{H}$ is defined by: $x^i = \text{const.}$, $z = \text{const.}$, $i \in \{1,\cdots,n\}$.

Afterwards, according to [**2**] we can state the following theorem:

THEOREM 3.1. *Let (M,η) be a $(2n+1)$-dimensional contact manifold. Then the maximal dimension of any integrable subbundle of the contact distribution $\mathcal{H}$ is n.*

From another point of view, taking into account [**18**], a contact distribution $\mathcal{H}$ on M is a $2n$-dimensional subbundle of TM which has the property that any local trivialization η of the line bundle $\mathbf{E}^* = \text{Ann}(\mathcal{H}) \subset T^*M$ (i.e. nonzero section of $\mathbf{E}^*_{|\mathcal{U}} \longrightarrow \mathcal{U}$ defined on a neighborhood $\mathcal{U} \subset M$) satisfies the condition $\eta \wedge (d\eta)^n \neq 0$. The pair $(M,\mathcal{H})$ is designated as a *contact manifold in the wider sense.* Specifically, whenever $\mathbf{E}^*$ is trivial as a vector bundle, $\mathbf{E}^*$ has a globally defined nonzero section η. Thereupon, η is identically a contact 1-form and a contact manifold is characterized to be a pair (M,η). Unequivocally, a contact manifold is also a contact manifold in the wider sense with the contact distribution designated by $\text{Ann}(\eta)$. Meanwhile, the contact 1-form identically specifies a vector field ξ on M, denoted by the characteristic vector field, by virtue of the conditions $i(\xi)\eta = 1$ and $i(\xi)d\eta = 0$. The characteristic vector field yields a natural decomposition $T^*M = \mathbf{E}^* \oplus \text{Ann}(\xi)$.

A *Legendre foliation (resp. in the wider sense)* $\mathcal{F}$ of a contact manifold (M,η) is a foliation by n-dimensional integral submanifolds of η (resp. of $\mathcal{H}$). Two legendre foliations (resp. in the wider sense) $\mathcal{F}_1$ and $\mathcal{F}_2$ are considered to be *equivalent* if there is a diffeomorphism $\mathscr{J} : M_1 \longrightarrow M_2$ satisfying these two main conditions:

$$\mathscr{J}^*\eta_2 = \eta_1 \;\; (\text{resp. } \mathscr{J}^*E_2^* = E_1^*), \quad \mathscr{J}^*\mathcal{F}_2 = \mathcal{F}_1,$$

where $\mathscr{J}^*\mathcal{F}_2$ is the foliation of M_1 whose leaves are the inverse images under $\mathscr{J}$ of leaves of $\mathcal{F}_2$.

Given a foliation $\mathcal{F}$, the tangent bundle $\pi_{\mathbf{L}} : \mathbf{L} \longrightarrow M$ to $\mathcal{F}$ is the subbundle of TM comprising the tangent vectors to the foliation. In an analogous manner, the conormal bundle $\pi_{\mathbf{N}} : \mathbf{N}^* \longrightarrow M$ to $\mathcal{F}$ is the subbundle of T^*M consisting of covectors normal to $\mathcal{F}$. A *tangential vector field* is a section of $\mathbf{L}$, and a *relative 1-form* is a section of $\mathbf{L}^*$. In the case when $\mathcal{F}$ is a Legendre foliation with contant 1-form η, the conormal bundle has a canonical splitting:

$$\mathbf{N}^* = \mathbf{E}^* \oplus \mathbf{Q}^*$$

where $\pi_{\mathbf{Q}} : \mathbf{Q}^* \longrightarrow M$ is the subbundle of $\mathbf{N}^*$ defined by: $\mathbf{Q}^* = \mathbf{N}^* \cap \text{Ann}(\xi)$. In addition, there exists a canonical vector bundle isomorphism $\mathscr{X} : \mathbf{L} \longrightarrow \mathbf{Q}^*$ characterized as follows: For $X \in \mathbf{L}$, let $\mathscr{X}(X) = i(X)d\eta \in T^*M$. But, $i(X)d\eta \in \Gamma(\mathbf{Q}^*)$ since $\{i(X)d\eta\}(\xi) = 0$ and $\{i(X)d\eta\}(Y) = -\eta([X,Y]) = 0$ for all $Y \in \Gamma(\mathbf{L})$. Noticeably, $\mathscr{X}$ is an injection because $d\eta$ is nondegenerate on $\text{Ann}(\xi)$, and, since $\mathbf{L}$, $\mathbf{Q}^*$ are of the same dimension, $\mathscr{X}$ is an isomorphism.

A *symplectification* is an approved procedure of achieving a symplectic manifold from a given contact manifold. Let $(M, \mathcal{H})$ be a contact manifold in the wider sense. Then the line bundle $\mathbf{E}^* \subset T^*M$, as a $(2n+2)$-dimensional manifold, carries the canonical symplectic 2-form $-d\widetilde{\eta}$ where $\widetilde{\eta}$ is the 1-form on $\mathbf{E}^*$ defined by the formula

$$\widetilde{\eta}_v(X) = v(\pi_{\mathbf{E}^*}X)$$

for $v \in \mathbf{E}^*$ and $X \in T_v\mathbf{E}^*$. Moreover, for each Legendre foliation $\mathcal{F}$ of $(M, \mathcal{H})$, $\pi^*_{\mathbf{E}}\mathcal{F}$ is a Lagrangian foliation on $\mathbf{E}^*$. Additionally, the leaves of $\pi^*_{\mathbf{E}}\mathcal{F}$ are integral submanifolds of $\widetilde{\eta}$. (refer to [**18**] for further details).

Generally, an arbitrary $(2n+2)$-dimensional manifold with 1-form $\widetilde{\eta}$ such that $-d\widetilde{\eta}$ is a symplectic 2-form is denoted by an *exact symplectic* manifold.

In [**18**] M. Pang proposed a viable theorem in the context of local equivalence of Lagrangian foliations of exact symplectic manifolds which is absolutely similar to Weinstein's result on Lagrangian foliations of symplectic manifolds [**21**]. For this noteworthy motive, by considering $\mathcal{W}$ as a neighborhood of a point in an exact symplectic manifold $(\widetilde{M}, \widetilde{\eta})$ equipped with the Lagrangian foliation $\widetilde{F}$, Pang constructed the bundle map:

$$\mathcal{H} : \widetilde{\mathbf{N}}^* \longrightarrow T^*(\mathcal{W}/\widetilde{\mathcal{F}})$$

characterized as follows: Let $\widetilde{\mathbf{N}}^*_{|\mathcal{W}} \longrightarrow \mathcal{W}$ be the conormal bundle to $\widetilde{\mathcal{F}}$ restricted to $\mathcal{W}$. The derivative of the projection map $\pi_{\mathcal{W}}$ defines a bundle map $\pi_{\mathcal{W}^*}$ between $\widetilde{\mathbf{N}}$ and the tangent bundle $T(\mathcal{W}/\widetilde{\mathcal{F}})$ with $\pi_{\mathcal{W}^*} : \widetilde{\mathbf{N}}_p \longrightarrow T_{\pi_{\mathcal{W}}(p)}(\mathcal{W}/\widetilde{\mathcal{F}})$ is an isomorphism for all $p \in \mathcal{W}$. The bundle map $\mathcal{H}$ at $p \in \mathcal{W}$ is determined by means of the equation $\mathcal{H}_p = \left(\pi^*_{\mathcal{W}}\right)^{-1}$ with $\mathcal{H}_p$ a vector space isomorphism. Conclusively, $\mathscr{H}$ is identified via the composition $\mathcal{H} \circ \widetilde{\eta}$ where $\widetilde{\eta}$ is considered as a map from $\mathcal{W}$ to $\widetilde{\mathbf{N}}^*$ in such a way that the following diagram commutes:

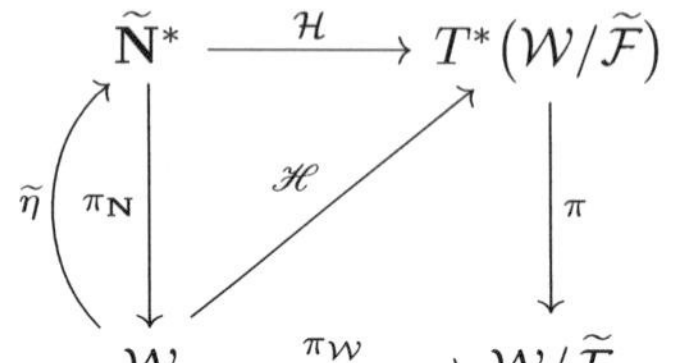

Accordingly, the following fundamental theorem is deduced [**18**]:

THEOREM 3.2. *Let $\mathcal{W}$ be a neighborhood of a point in an exact symplectic manifold $(\widetilde{M}, \widetilde{\eta})$ with Lagrangian foliation $\widetilde{F}$. Without loss of generality, we may assume that the leaf space $\mathcal{W}/\widetilde{\mathcal{F}}$ of the restriction of $\widetilde{\mathcal{F}}$ to $\mathcal{W}$ is a manifold. Then there is a unique local diffeomorphism:*

$$\mathscr{H} : \mathcal{W} \longrightarrow T^*(\mathcal{W}/\widetilde{\mathcal{F}})$$

such that

(1): *$\mathscr{H}^*\zeta = \widetilde{\eta}$ where ζ is the canonical 1-form on $T^*(\mathcal{W}/\widetilde{\mathcal{F}})$,*

(2): *$\pi \circ \mathscr{H} = \pi_{\mathcal{W}}$ where π and $\pi_{\mathcal{W}} : \mathcal{W} \longrightarrow \mathcal{W}/\widetilde{\mathcal{F}}$ are projection morphisms (i.e. each leaf of the foliation of $\mathcal{W}$ is mapped into the fiber of $T^*(\mathcal{W}/\widetilde{\mathcal{F}})$.*

The noteworthy point is that in order to classify Legendre foliations, two local invariants are introduced: the fundamental quadratic form $\mathbf{\Pi}$ which is a symmetric 2-form on the tangent bundle to $\mathcal{F}$ and the invariant $\mathbf{G}$, a symmetric 3-form on the tangent bundle to $\mathcal{F}$.

Suppose that X, Y, Z are tangential vector fields to the Legendre foliation $\mathcal{F}$. Then the quantities $\mathbf{\Pi}$ and $\mathbf{G}$ can be invariantly determined in terms of Lie derivatives as follows [**18**]:

$$
\begin{aligned}
&(\mathbf{a}) : \mathbf{\Pi}(X,Y) = -\Big\{\mathcal{L}_X\mathcal{L}_Y\eta\Big\}(\xi),\\
(3.10)\qquad &(\mathbf{b}) : \mathbf{G}(X,Y,Z) = \tfrac{1}{2}\Big\{X\big(\Pi(Y,Z)\big) + Y\big(\Pi(X,Z)\big) + Z\big(\Pi(X,Y)\big)\\
&\qquad\qquad + \big(\mathcal{L}_Y\mathcal{L}_X\mathcal{L}_Z\eta + \mathcal{L}_Z\mathcal{L}_X\mathcal{L}_Y\eta\big)(\xi)\Big\}.
\end{aligned}
$$

A *relative k-form* is a section of $\bigwedge^k \mathbf{L}^*$. Let $\mathbf{\Omega}^k_{\mathcal{F}}(M)$ represent the set of all relative k-forms. The relative exterior derivative $d_{\mathcal{F}} : \mathbf{\Omega}^k_{\mathcal{F}}(M) \longrightarrow \mathbf{\Omega}^{k+1}_{\mathcal{F}}(M)$ is structurally characterized as follows. The remarkable point is that the vector bundle derived via restricting $\mathbf{L}^*$ to a leaf $\mathscr{L} \in \mathcal{F}$ is straightforwardly the cotangent bundle $T^*\mathscr{L}$ of $\mathscr{L}$. If ν is a relative form, then the relative exterior derivative $d_{\mathcal{F}}\nu$ of ν is designated explicitly to be the ordinary exterior differential of ν restricted to each leaf $\mathscr{L}$. Literally, if $\imath^* : \mathbf{\Omega}^k(M) \longrightarrow \mathbf{\Omega}^k_{\mathcal{F}}(M)$ indicates the pull back of a k-form on M to a relative k-form, then the relative exterior derivative $d_{\mathcal{F}}$ is totally described by means of the following commutative diagram [**18**]:

$$
\begin{array}{ccc}
\mathbf{\Omega}^k(M) & \xrightarrow{\ \imath^*\ } & \mathbf{\Omega}^k_{\mathcal{F}}(M)\\
\Big\downarrow d & & \Big\downarrow d_{\mathcal{F}}\\
\mathbf{\Omega}^{k+1}(M) & \xrightarrow{\ \imath^*\ } & \mathbf{\Omega}^{k+1}_{\mathcal{F}}(M)
\end{array}
$$

For an arbitrary vector bundle $\mathbf{B} \longrightarrow M$, a *partial connection* $\nabla^{\mathbf{B}}$ on $\mathbf{B}$ is an $\mathbb{R}$-linear map:

$$\nabla^{\mathbf{B}} : \Gamma(\mathbf{B}) \longrightarrow \Gamma\big(\mathbf{L}^* \otimes \mathbf{B}\big)$$

satisfying the equation $\nabla^{\mathbf{B}}\big(f\nu\big) = f\nabla^{\mathbf{B}}\nu + d_{\mathcal{F}}f \otimes \nu$, for $f \in C^\infty(M)$, and $\nu \in \Gamma(\mathbf{B})$. The curvature R is a $\mathrm{Hom}(\mathbf{B},\mathbf{B})$-valued relative 2-form defined by $R\big(X,Y\big)\nu = \big[\nabla^{\mathbf{B}}_X, \nabla^{\mathbf{B}}Y\big]\nu - \nabla^{\mathbf{B}}_{[X,Y]}\nu$, where X and Y are tangential vector fields and $\big[\ ,\ \big]$ represents the commutator.

If $\nu = (\nu^\alpha)$ is a local frame for $\mathbf{B}$, there exists canonically a matrix valued 1-form ϖ called relative partial connection form, and a relative 2-form $\mathbf{\Omega}$ called the curvature matrix characterized by virtue of the following conditions:

$$\nabla^{\mathbf{B}}\nu = -\varpi \otimes \nu, \qquad R\,\nu = -\mathbf{\Omega} \otimes \nu.$$

Accordingly, this predominant identity $\mathbf{\Omega} = d_{\mathcal{F}}\varpi + \varpi \wedge \varpi$ is concluded.

The *Bott partial connection* on $\mathbf{N}^*$ is a partial connection determined as follows:

$$\nabla^{\mathbf{N}}_X\nu = \mathcal{L}_X\nu = i(X)d\nu$$

where X and ν are tangential vector field and a section of $\mathbf{N}^*$.

It can be straightforwardly scrutinized that $\nabla^{\mathbf{N}}$ is a partial connection. In addition, the Bott partial connection is flat since $R(X,Y) = [\mathcal{L}_X, \mathcal{L}_Y]\nu - \mathcal{L}_{[X,Y]}\nu = 0$.

The Bott partial connection induces a partial connection $\nabla^{\mathbf{Q}}$ on the bundle $\mathbf{Q}^*$ designated by $\nabla^{\mathbf{Q}}_X\nu = \pi(\nabla^{\mathbf{N}}_X\nu)$ for X and ν are tangential vector field and a section of $\mathbf{Q}^*$ where $\pi : \mathbf{N}^* \longrightarrow \mathbf{Q}^*$ demonstrates the projection onto $\mathbf{Q}^*$ (Taking into account the decomposition $\mathbf{N}^* = \mathbf{E}^* \oplus \mathbf{Q}^*$). Then $\nabla^{\mathbf{Q}}$ is denoted by the *induced partial connection*. Suppose that $\{\mathfrak{S}^\alpha\}$ be a local frame for $\mathbf{Q}^*$. Then $\{\mathfrak{S}^\alpha, \eta\}$ is a local frame for $\mathbf{N}^*$, and the relative partial connection form $\mathfrak{B} = (\mathfrak{B}^i_j)$ of $\nabla^{\mathbf{N}}$ is identically characterized by virtue of the subsequent relation:

$$\nabla^{\mathbf{N}}\begin{pmatrix}\mathfrak{S}^\alpha\\ \\ \eta\end{pmatrix} = -\begin{pmatrix}\mathfrak{B}^\alpha_\beta & \mathfrak{B}^\alpha_{n+1}\\ \\ \mathfrak{B}^{n+1}_\beta & 0\end{pmatrix}\bigotimes\begin{pmatrix}\mathfrak{S}^\beta\\ \\ \eta\end{pmatrix}. \tag{3.11}$$

Significantly, since for $X \in \Gamma(\mathbf{L})$ as a section of $\mathbf{Q}^*$, we have: $\nabla^{\mathbf{N}}_X\eta = \mathscr{X}(X) \in \Gamma(\mathbf{Q}^*)$, it is inferred that: $\mathfrak{B}^{n+1}_{n+1} = 0$. Meanwhile, $(\mathfrak{B}^\alpha_\beta)$ is unequivocally the relative connection form for $\nabla^{\mathbf{Q}}$. Moreover, considering the point that the curvature of the Bott connection vanishes, the relative connection from $(\mathfrak{B}^i_j)$ explicitly satisfies the following identities (cf. [**18**] for further explanations):

$$\begin{cases}(\mathbf{a}) : d_{\mathcal{F}}\mathfrak{B}^\alpha_\beta + \mathfrak{B}^\alpha_\gamma \wedge \mathfrak{B}^\gamma_\beta + \mathfrak{B}^\alpha_{n+1} \wedge \mathfrak{B}^{n+1}_\beta = 0,\\ (\mathbf{b}) : d_{\mathcal{F}}\mathfrak{B}^\alpha_{n+1} + \mathfrak{B}^\alpha_\gamma \wedge \mathfrak{B}^\gamma_{n+1} = 0,\\ (\mathbf{c}) : d_{\mathcal{F}}\mathfrak{B}^{n+1}_\beta + \mathfrak{B}^{n+1}_\gamma \wedge \mathfrak{B}^\gamma_\beta = 0,\\ (\mathbf{d}) : \mathfrak{B}^{n+1}_\gamma \wedge \mathfrak{B}^\gamma_{n+1} = 0.\end{cases} \tag{3.12}$$

Ultimately, assume that V_α be the local frame for $\mathbf{L}$ which is explicitly designated via the equation $\mathscr{X}(V_\alpha) = -\mathfrak{S}^\alpha$. Thus, the independence of $\mathfrak{B}^{n+1}_\alpha$ is directly construed from the fact that $\mathfrak{B}^{n+1}_\alpha$ is a dual coframe of V_α:

$$\mathfrak{S}^\alpha = -\mathscr{X}(V_\alpha) = -i(V_\alpha)d\eta = -\nabla^{\mathbf{N}}_{V_\alpha}\eta = \mathfrak{B}^{n+1}_\beta(V_\alpha)\mathfrak{S}^\beta.$$

Conclusively, it is educed that the relative 1-forms $\mathfrak{B}^{n+1}_\alpha$ organize a local frame for $\mathbf{L}^*$.

4. Metric Legendre foliated cocycles compatible with SODE structure

Let M be an m-dimensional smooth manifold and TM be its tangent bundle. Throughout the paper, if not stated otherwise, we use indices $i, j, k, \dots \in \{1, \dots, m\}$, $a, b, c, \dots \in \{1, \dots, n\}$ and $\alpha, \beta, \dots \in \{n+1, \dots, m\}$. A coordinate system in M is defined by $\{(\mathcal{U}, \varsigma) : x^1, \dots, x^m\}$ or briefly $\{(\mathcal{U}, \varsigma) : x^i\}$, where $\mathcal{U}$ is an open subset of M, $\varsigma : \mathcal{U} \longrightarrow \mathbb{R}^m$ is a diffeomorphism of $\mathcal{U}$ onto $\varsigma(\mathcal{U})$, and $(x^1, \dots, x^m) = \varsigma(x)$ for any $x \in \mathcal{U}$. The canonical projection of TM on M and by T_xM the fibre at $x \in M$ is designated by π, i.e. $T_xM = \pi^{-1}(x)$. The coordinate system $\{(\mathcal{U}, \varsigma) : x^i\}$ in M canonically characterizes a coordinate system $\{(\mathcal{U}^*, \Sigma) : x^1, \dots, x^m, y^1, \dots, y^m\} = \{(\mathcal{U}^*, \Sigma) : x^i, y^i\}$ in TM, where $\mathcal{U}^* = \pi^{-1}(\mathcal{U})$ and $\Sigma : \mathcal{U}^* \longrightarrow \mathbb{R}^{2m}$ is a diffeomorphism of $\mathcal{U}^*$ on $\varsigma(\mathcal{U}) \times \mathbb{R}^m$, and $(x^1, \dots, x^m, y^1, \dots, y^m) = \Sigma(y_x)$, for any $x \in \mathcal{U}$ and $y_x \in T_xM$. For short we denote by (x, y) the coordinates of y_x. Subsequently, consider another coordinate system $\{(\tilde{\mathcal{U}}, \tilde{\varsigma}); x^i\}$ in M in a such a manner that $\mathcal{U} \cap \tilde{\mathcal{U}} \neq \emptyset$.

Then the local coordinates (x, y) and $(\tilde{x}, \tilde{y})$ on TM are related by:

$$\tilde{x}^i = \tilde{x}^i(x^1, x^2, \ldots, x^m), \quad \tilde{y}^i = \frac{\partial \tilde{x}^i}{\partial \tilde{x}^j} y^j, \tag{4.1}$$

Consequently, from (4.1) the local frame fields $\{\frac{\partial}{\partial x^i}, \frac{\partial}{\partial y^i}\}$ and $\{\frac{\partial}{\partial \tilde{x}^i}, \frac{\partial}{\partial \tilde{y}^i}\}$ are related as follows:

$$\frac{\partial}{\partial x^i} = \frac{\partial \tilde{x}^j}{\partial x^i}\frac{\partial}{\partial \tilde{x}^j} + \frac{\partial^2 \tilde{x}^j}{\partial x^i \partial x^k} y^k \frac{\partial}{\partial \tilde{y}^j}, \quad \frac{\partial}{\partial y^i} = \frac{\partial \tilde{x}^j}{\partial x^i}\frac{\partial}{\partial \tilde{y}^j}. \tag{4.2}$$

In an analogous manner, it is identically educed that the local coframe fields $\{dx^i, dy^i\}$ and $\{d\tilde{x}^i, d\tilde{y}^i\}$ on the cotangent bundle T^*M of M are related by

$$d\tilde{x}^i = \frac{\partial \tilde{x}^i}{\partial x^j}\, dx^j, \quad d\tilde{y}^i = \frac{\partial^2 \tilde{x}^i}{\partial x^j \partial x^k} y^j\, dx^k + \frac{\partial \tilde{x}^i}{\partial x^j}\, dy^j. \tag{4.3}$$

Assume that there exists a function $F : TM \longrightarrow [0, \infty)$ which vanishes only on the zero section of TM and is smooth on $TM^0 = TM \setminus \{0\}$. Furthermore, on the domain of any local chart on TM^0 with local coordinates (x^i, y^i), F satisfies the following conditions:

(i): It is positively homogenous of degree one with respect to (y^i), i.e. we have:

$$F(x^i, ky^i) = kF(x^i, y^i), \quad \forall\, k > 0,$$

(ii): The matrix

$$\big[g_{ij}(x, y)\big] = \left[\frac{1}{2}\frac{\partial^2 F^2}{\partial y^i \partial y^j}\right]$$

is positive definite at any point of the domain of the local chart. Then $\mathbb{F}^m = (M, F)$ is denoted by a Finsler manifold and F is the fundamental function of $\mathbb{F}^m$.

Thereupon, we denote by $\mathcal{F}_V$ the foliation on TM^0 characterized by virtue of fibers of $\pi : TM^0 \longrightarrow M$, and call it the vertical foliation on TM^0. Then the tangent distribution to $\mathcal{F}_V$ is the vertical distribution VTM^0 on TM^0. A complementary distribution HTM to VTM in TTM is specified by a nonlinear connection or a horizontal distribution on TM. This connection is of particular geometric significance and its existence yields the following decomposition:

$$TTM = HTM \oplus VTM \tag{4.4}$$

Subsequently, we take a local frame field $\{X_i, \frac{\partial}{\partial y^i}\}$ on $\acute{\mathcal{V}} \subset TM$ adapted to (4.4), i.e., $X_i \in \Gamma(HTM)$ and $\frac{\partial}{\partial y^i} \in \Gamma(VTM)$. Hence, we have

$$\frac{\partial}{\partial x^i} = E_i^j(x, y) X_j + G_i^j(x, y)\frac{\partial}{\partial y^j}, \tag{4.5}$$

where E_i^j and G_i^j are smooth functions which are locally defined on TM. Hence the transition matrix from the local frame field $\{\frac{\partial}{\partial x^i}, \frac{\partial}{\partial y^i}\}$ to $\{X_j, \frac{\partial}{\partial y^j}\}$ is explicitly describes as follows [**5**]:

$$\Delta = \begin{pmatrix} E_i^j(x, y) & 0 \\ G_i^k(x, y) & \delta_h^k \end{pmatrix}. \tag{4.6}$$

As Δ is a nonsingular matrix it follows that the $m \times m$ matrix $[E_i^j(x, y)]$ is also a nonsingular matrix. Thus the set of local vector fields $\left\{\frac{\delta}{\delta x^1}, \ldots, \frac{\delta}{\delta x^m}\right\}$ given by $\frac{\delta}{\delta x^i} = E_i^j(x,y)X_j$ is a basis in $\Gamma(HTM|_{\acute{\mathcal{U}}})$. In this way (4.5) converts to the following relation:

$$\frac{\delta}{\delta x^i} = \frac{\partial}{\partial x^i} - G_i^j(x,y)\frac{\partial}{\partial y^j}. \tag{4.7}$$

Denote by $\tilde{G}_h^k$ the functions in (4.7) given with respect to another coordinate system $\{(\tilde{\acute{\mathcal{V}}}, \tilde{\acute{\chi}}); \tilde{x}^i, \tilde{y}^i\}$ on TM such that $\acute{\mathcal{V}} \cap \tilde{\acute{\mathcal{V}}} \neq \emptyset$. Then by utilizing (4.2) and (4.7), the following predominant relations are educed [**5**]:

$$G_i^j(x,y)\frac{\partial \tilde{x}^k}{\partial x^j} = \tilde{G}_h^k(\tilde{x},\tilde{y})\frac{\partial \tilde{x}^h}{\partial x^i} + \frac{\partial^2 \tilde{x}^k}{\partial x^i \partial x^h}y^h, \qquad \frac{\delta}{\delta x^i} = \frac{\partial \tilde{x}^j}{\partial x^i}\frac{\delta}{\delta \tilde{x}^j}. \tag{4.8}$$

A nonlinear connection HTM enables us to define an almost product structure on TM as follows. Consider a vector field X on TM. Then locally we have $X = X^i\frac{\delta}{\delta x^i} + \dot{X}^i\frac{\partial}{\partial y^i}$, where X^i and $\dot{X}^i$ satisfy

$$\tilde{X}^j = \frac{\partial \tilde{x}^j}{\partial x^i}X^i \quad ; \quad \dot{\tilde{X}}^j = \frac{\partial \tilde{x}^j}{\partial x^i}\dot{X}^i, \tag{4.9}$$

with respect to (4.1). Then we define

$$\mathcal{Q}:\ \Gamma(TTM) \longrightarrow \Gamma(TTM), \qquad \mathcal{Q}X = \dot{X}^i\frac{\delta}{\delta x^i} + X^i\frac{\partial}{\partial y^i}. \tag{4.10}$$

By (4.2), (4.8) and (4.9) it follows from (4.10) that $\mathcal{Q}X$ is totally independent of the local chart on TTM. In addition, $\mathcal{Q}^2 = I$ and therefore $\mathcal{Q}$ is an almost product structure on TM. We call $\mathcal{Q}$ the *almost product structure* corresponding to the nonlinear connection HTM [**5**]. Further, denote by h and v the projection morphisms of TTM to HTM and VTM respectively. Then it is inferred that:

$$(a):\ \mathcal{Q}\circ h = v\circ\mathcal{Q} \qquad \text{and} \qquad (b):\ \mathcal{Q}\circ v = h\circ\mathcal{Q}. \tag{4.11}$$

By virtue of $\mathcal{Q}$ and v the following $\mathcal{F}(TM)$-bilinear mapping is explicitly designated:

$$\begin{cases} T_1 : \Gamma(VTM)\times\Gamma(VTM) \longrightarrow \Gamma(VTM) \\ T_1(X,Y) = -v[\mathcal{Q}X, \mathcal{Q}Y], \quad \forall\ X,Y \in \Gamma(VTM). \end{cases} \tag{4.12}$$

According to [**5**], it is observed that T_1 is a tensor field of Type $(1,2)$. By straightforward computations utilizing (4.12), (4.10) and (4.7) the subsequent predominant relation is concluded:

$$T_1\left(\frac{\partial}{\partial y^j}, \frac{\partial}{\partial y^i}\right) = \left[\frac{\delta}{\delta x^i}, \frac{\delta}{\delta x^j}\right] = R_{ij}^k\frac{\partial}{\partial y^k}, \qquad R_{ij}^k = \frac{\delta G_i^k}{\delta x^j} - \frac{\delta G_j^k}{\delta x^i}. \tag{4.13}$$

From (4.12) it follows that $T_1 = (R_{ij}^k)$ is the obstruction to the integrability of the horizontal distribution. More precisely, we have the following result:

The horizontal distribution HTM is involutive if and only if $T_1 = 0$ on TM, or equivalently, on each coordinate neighborhood $\acute{\mathcal{U}} \subset TM$ we have $R_{ij}^k = 0$. Eventually, by applying (4.7) and direct calculations it is construed that:

$$\left[\frac{\delta}{\delta x^i}, \frac{\partial}{\partial y^j}\right] = {G_i}^k{}_j\frac{\partial}{\partial y^k}, \qquad {G_i}^k{}_j = \frac{\partial {G_i}^k}{\partial y^j}. \tag{4.14}$$

Next, we consider two globally defined vector fields on TM^0 which are locally given by [**2**]:

$$(4.15) \qquad (\mathbf{a}) : \mathbb{C} = y^i \frac{\partial}{\partial y^i}, \qquad (\mathbf{b}) : \mathbb{C}^* = y^i \frac{\delta}{\delta x^i}.$$

We call $\mathbb{C}$ and $\mathbb{C}^*$ the vertical and horizontal Liouville vector fields on TM^0, respectively. Accordingly, the line distributions $\mathfrak{C} = \mathrm{span}\{\mathbb{C}\}$ and $\mathfrak{C}^* = \mathrm{span}\{\mathbb{C}^*\}$ are called the vertical and horizontal Liouville distributions on TM^0 respectively. We denote by $\mathfrak{C}'$ and $\mathfrak{C}^\perp$ the complementary orthogonal distributions to $\mathfrak{C}$ in VTM^0 and TTM^0 respectively.

A *foliation* $\mathcal{F}$ of codimension p on the $m = (n+p)$-dimensional smooth manifold M is geometrically designated via an open cover $\mathcal{V} = \{\mathcal{V}_i\}_{i\in I}$ and for each i, a local diffeomorphism $\Upsilon : \mathbb{R}^{n+p} \longrightarrow \mathcal{V}_i$ such that on each nonempty $\mathcal{V}_i \cap \mathcal{V}_j$ the following change of coordinates satisfies [**16**]:

$$\Upsilon_j^{-1} \circ \Upsilon_i : \ (x,y) \in \Upsilon_i^{-1}(\mathcal{V}_i \cap \mathcal{V}_j) \longrightarrow (\tilde{x},\tilde{y}) \in \Upsilon_j^{-1}(\mathcal{V}_i \cap \mathcal{V}_j)$$

where $\tilde{x} = \Upsilon_{ij}(x,y)$ and $\tilde{y} = \varrho_{ij}(y)$. Consequently, the manifold M is decomposed into n-dimensional connected submanifolds which are characterized by a leaf of $\mathcal{F}$. Moreover, $\mathcal{V} \subset M$ is called saturated for the foliation $\mathcal{F}$ if it can be regarded as the union of leaves. In other words, if $x \in \mathcal{V}$ then the leaf passing through x is contained in $\mathcal{V}$. Now assume that $\mathcal{F}$ is a foliation of codimension p on the manifold M and $\pi : \mathbb{R}^{n+p} = \mathbb{R}^n \times \mathbb{R}^p \longrightarrow \mathbb{R}^p$ is the second projection. Then the map $\mathcal{X}_i = \pi \circ \Upsilon_i^{-1} : \mathcal{V}_i \longrightarrow \mathbb{R}^p$ is a submersion and on $\mathcal{V}_i \cap \mathcal{V}_j \neq \emptyset$ the following identity holds: $\mathcal{X}_j = \varrho_{ij} \circ \mathcal{X}_i$. In addition, the fibres of the submersion $\mathcal{X}_i$ are scrutinized as the $\mathcal{F}$-plaques of $(\mathcal{V}_i, \mathcal{F})$ and the foliation $\mathcal{F}$ is explicitly determined by virtue of the submersions $\mathcal{X}_i$ and the local diffeomorphisms ϱ_{ij} of $\mathbb{R}^p$. Overall, a foliation $\mathcal{F}$ of codimension p on M is identically described via an open cover $\{\mathcal{V}_i\}_{i\in I}$ and submersions $\mathcal{X}_i : \mathcal{V}_i \longrightarrow \mathsf{T}$ over an p-dimensional manifold T and a diffeomorphism $\varrho_{ij} : \mathcal{X}_i(\mathcal{V}_i \cap \mathcal{V}_j) \longrightarrow \mathcal{X}_j(\mathcal{V}_i \cap \mathcal{V}_j)$ such that $\mathcal{X}_j = \varrho_{ij} \circ \mathcal{X}_i$ for $\mathcal{V}_i \cap \mathcal{V}_j \neq \emptyset$. Then $\mathscr{C} = \{\mathcal{V}_i, \mathcal{X}_i, \mathsf{T}, \varrho_{ij}\}$ is demonstrated by means of a *foliated cocycle* describing the foliation $\mathcal{F}$. According to above definition, the p-manifold $\mathsf{T}^* = \bigsqcup \mathsf{T}_i^*$, $\mathsf{T}_i^* = \mathcal{X}_i(\mathcal{V}_i)$, is specified as the *transverse manifold* concomitant to the cocycle $\mathscr{C}$ and the pseudogroup $\mathbf{H}$ of local diffeomorphisms of T^* generated by Υ_{ij} the *holonomy pseudogroup representative* on T^* which is associated to the cocycle $\mathscr{C}$. It is remarkable that T^* is a complete transverse manifold and the equivalence class of $\mathbf{H}$ is denoted by the holonomy pseudogroup of $(M, \mathcal{F})$. In this paper, in what follows, it is supposed that $\mathcal{F}$ is established by means of a cocycle $\mathscr{C}$ and the transverse manifold and holonomy pseudogroup corresponding to $\mathscr{C}$ are determined by T^* and $\mathcal{H}$, respectively. The foliation $\mathcal{F}$ is signified to be *(transversally) metric*, if it is created by means of a cocycle $\mathscr{C} = \{\mathcal{V}_i, \mathcal{X}_i, \mathsf{T}, \varrho_{ij}\}$ modeled on a manifold (T_0^*, g_0) in such a manner that the concomitant transformations ϱ_{ij} are local isometries of the metric structure g_0 [**12, 16**].

In [**19, 20**] Popesco et al. defined the notion of the Lagrangian adopted to the lifted foliation. Moreover, in [**1, 14**] a comprehensive analysis regarding the metric foliated cocycles adapted to SODE structure is presented from distinct standpoints. In this section, according to [**1**] by inducing the following three principle constraints, we will establish a powerful framework for constructing structurally transverse foliated cocycles on the tangent space TM. As it will be demonstrated the notion of formal integrability is utilized as a fundamental tool.

DEFINITION 4.1. Let $\mathcal{F} = \{\mathcal{V}_i, \mathcal{X}_i, \mathsf{T}, \varrho_{ij}\}$ be a foliated cocycle of codimension n on M and $\mathcal{F}^{\mathscr{T}} = \{\mathcal{V}_i^{\mathscr{T}}, \mathcal{X}_i^{\mathscr{T}}, \mathsf{T}^{\mathscr{T}}, \varrho_{ij}^{\mathscr{T}}\}$ be the foliated cocycle on the tangent space TM. Let S be a semispray which is locally demonstrated by $S = y^i \frac{\partial}{\partial x^i} - 2G^i(x,y)\frac{\partial}{\partial y^i}$. Then the metric g is called the **adopted transverse metric (ATM)** whenever the following three essential conditions are satisfied:

(1): The following partial differential operator

$$\mathcal{P}_{\mathrm{L}} = (d_{\mathcal{J}}, d_h, d_{\mathbf{\Phi}}, \nabla d) : \operatorname{Sec}(T_v^*) \longrightarrow \operatorname{Sec}\left(\oplus^{(4)} \Lambda^2 T_v^*\right) \tag{4.16}$$

is formally integrable.

(2): $g_{b\beta} = g(\frac{\partial}{\partial y^b}, \frac{\partial}{\partial y^\beta}) = 0$.

(3): The local functions $(g_{\alpha\beta})$ and $(g^{\alpha\beta})$ are basic functions, namely they are totally independent of the tangent variables (x^a, y^a).

Significantly, the foliated cocycle $\mathcal{F}^{\mathscr{T}} = \{\mathcal{V}_i^{\mathscr{T}}, \mathcal{X}_i^{\mathscr{T}}, \mathsf{T}^{\mathscr{T}}, \varrho_{ij}^{\mathscr{T}}\}$ equipped with g which is an adopted transverse metric (ATM) is called a **metric foliated cocycle**.

Taking into account above discussion, we can now state the first prominent theorem of this paper as follows:

THEOREM 4.2. *Let (M, η) be an $m = (2n+1)$-dimensional contact manifold which is equipped with an n-dimensional Legendre foliated cocycle $\mathcal{F} = \{\mathcal{V}_i, \mathcal{X}_i, \mathsf{T}, \varrho_{ij}\}$. Suppose that $S = y^i \frac{\partial}{\partial x^i} - 2G^i(x,y)\frac{\partial}{\partial y^i}$ is an spray on M. If the following three main conditions are satisfied:*

(A): *By virtue of the Bott partial connection and moving frames, for the relative 1-forms $\mathfrak{B}_\alpha^{n+1}$ and $\mathfrak{B}_\beta^{n+1}$, and a symmetric matrix valued function $(\mathfrak{D}^{\alpha\beta})$, the following expression of the fundamental quadratic symmetric 2-form local invariant*

$$\mathbf{\Pi} = -\mathfrak{D}^{\alpha\beta}\mathfrak{B}_\alpha^{n+1} \otimes \mathfrak{B}_\beta^{n+1} \tag{4.17}$$

is positive definite on the tangent distribution to $\mathcal{F}$ which is denoted by $T(\mathcal{F})$.

(B): *In the local coordinate system, the coefficients of the nonlinear connection concomitant to the spray S i.e. $(G_j^i) = \begin{pmatrix} G_b^a & G_b^\alpha \\ & \\ G_\beta^a & G_\beta^\alpha \end{pmatrix}$ satisfy the following relations:*

$$\begin{cases} (i) : G_b^\alpha = \frac{\partial G^\alpha}{\partial y^b} = 0 \ , & (ii) : \frac{\partial G^\alpha}{\partial x^b} = 0; \ \ or, \\ (i') : G_\alpha^b = \frac{\partial G^b}{\partial y^\alpha} = 0 \ , & (ii)' : \frac{\partial G^b}{\partial x^\alpha} = 0. \end{cases} \tag{4.18}$$

(C): *For any $X, Y \in \Gamma(TM)$, the torsion tensor field of the Vrăceanu connection:*

$$\begin{aligned} \nabla_X^{\mathbf{V}} Y = {} & \pi^{\mathfrak{p}} \nabla_{\pi^{\mathfrak{p}} X}^{\mathbf{L}} \pi^{\mathfrak{p}} Y + \pi^{\mathfrak{q}} \nabla_{\pi^{\mathfrak{q}} X}^{\mathbf{L}} \pi^{\mathfrak{q}} Y \\ & + \pi^{\mathfrak{p}}[\pi^{\mathfrak{q}} X, \pi^{\mathfrak{p}} Y] + \pi^{\mathfrak{q}}[\pi^{\mathfrak{p}} X, \pi^{\mathfrak{q}} Y]. \end{aligned} \tag{4.19}$$

is explicitly expressed as follows:

$$T^{\mathbf{V}}(X, Y) = -2h^{\mathfrak{q}}(\pi^{\mathfrak{q}} X, \pi^{\mathfrak{q}} Y), \quad \forall \ X, Y \in \Gamma(TM). \tag{4.20}$$

where $\pi^{\mathfrak{p}}$ and $\pi^{\mathfrak{q}}$ are the projection morphisms of $\Gamma(TM)$ on $\Gamma(T(\mathcal{F}))$ and $\Gamma(T(\mathcal{F})^{\perp})$ respectively, $\nabla^{\mathbf{L}}$ denotes the Levi-Civita connection on M and $h^{\mathfrak{q}}$ represents the second fundamental form of $T(\mathcal{F})^{\perp}$ the complementary orthogonal distribution to $T(\mathcal{F})$ in TM.

Then there exists a metric foliated cocycle $\mathcal{F}^{\mathscr{T}} = \{\mathcal{V}_i^{\mathscr{T}}, \mathcal{X}_i^{\mathscr{T}}, \mathsf{T}^{\mathscr{T}}, \varrho_{ij}^{\mathscr{T}}\}$ on TM whose leaves are identically the c-indicatrix bundle over M and is identically the same as the foliation defined via $\mathfrak{C}^{\perp}$ the complementary orthogonal distribution to $\mathfrak{C}$ in TTM^0. Specifically, by considering a fixed point $x_0 = (x_0^i)$ on M, the foliated cocycle $\mathcal{F}^{\mathbb{T}} = \{\widetilde{\mathcal{V}}_i, \widetilde{\mathcal{X}}_i, \widetilde{\mathsf{T}}, \widetilde{\varrho}_{ij}\}$ leads to creation of a foliated cocycle which is a totally umbilical subfoliation of the vertical foliation $\mathcal{F}_V$ and whose leaves are c-indicatrices at x_0.

PROOF. We will prove this theorem in four main steps as follows:

✠ **Step (1):** In this step, we will explore the existence of the foliated cocycle $\mathcal{F}^{\mathscr{T}} = \{\mathcal{V}_i^{\mathscr{T}}, \mathcal{X}_i^{\mathscr{T}}, \mathsf{T}^{\mathscr{T}}, \varrho_{ij}^{\mathscr{T}}\}$ on TM and condition **(1)** of the ATM as declared in definition (4.1).

First of all, we recall that for the Legendre foliated cocycle $\mathcal{F} = \{\mathcal{V}_i, \mathcal{X}_i, \mathsf{T}, \varrho_{ij}\}$ of the contact manifold (M, η) a Lagrangian foliation $\pi_E^*\mathcal{F}$ is resulted via symplectification [**18**]. Let $\mathcal{N}$ be a neighborhood of an arbitrary point in M and suppose that the leaf space $\mathcal{N}/\mathcal{F}$ of the restriction of $\mathcal{F}$ to $\mathcal{N}$ is a manifold. If we define $\mathcal{W} = \pi_E^{-1}(\mathcal{N}) \subset E^*$, then for the concomitant Lagrangian foliated cocycle $\widetilde{\mathcal{F}} = \{\widetilde{\mathcal{V}}_i, \widetilde{\mathcal{X}}_i, \widetilde{\mathsf{T}}, \widetilde{\varrho}_{ij}\}$ since $\mathcal{W}/\widetilde{\mathcal{F}} = \mathcal{N}/\mathcal{F}$ it is straightforwardly inferred that $\mathcal{W}$ is an open set in such a manner that the leaf space $\mathcal{W}/\widetilde{\mathcal{F}}$ is a manifold. Hence, according to [**18**] by applying theorem (3.2) to the Lagrangian foliation $\pi_E^*\mathcal{F}$, a local diffeomorphism $\mathscr{H} : \mathcal{W} \longrightarrow T^*(\mathcal{N}/\mathcal{F})$ is explicitly characterized satisfying the declared properties (1) and (2). If we consider η as a map from $\mathcal{N}$ to $\mathcal{W}$, then $\mathscr{F}_\eta$ is designated as the composition map:

$$\mathscr{F}_\eta : \mathcal{N} \stackrel{\eta}{\longrightarrow} \mathcal{W} \stackrel{\mathscr{H}}{\longrightarrow} T^*(\mathcal{N}/\mathcal{F}).$$

Let $F : T^*(\mathcal{N}/\mathcal{F}) \longrightarrow \mathbb{R}^+ \cup \{0\}$ and $S_F^*(\mathcal{N}/\mathcal{F}) = \{\vartheta \in T^*(\mathcal{N}/\mathcal{F}) \mid F(\vartheta) = 1\}$, then $\mathscr{F}_\eta$ maps $\mathcal{N}$ diffeomorphically into $S_F^*(\mathcal{N}/\mathcal{F})$ and due to theorem (3.2) has the following properties:

(a): For the canonical 1-form ζ on $T^*(\mathcal{W}/\widetilde{\mathcal{F}})$ we have: $\mathscr{F}_\eta^*\zeta = \eta$.

(b): $\mathscr{F}_\eta^*\mathcal{F}_F = \mathcal{F}_{|\mathcal{N}}$ where $\mathcal{F}_F$ is the Legendre foliation on $S_F^*(\mathcal{N}/\mathcal{F})$ and $\mathcal{F}_{|\mathcal{N}}$ is the restriction of the Legendre foliated cocycle $\mathcal{F} = \{\mathcal{V}_i, \mathcal{X}_i, \mathsf{T}, \varrho_{ij}\}$ to $\mathcal{N}$.

In order to characterize F, the trivialization of the line bundle $\mathcal{W} \simeq \mathcal{N} \times \mathbb{R}$ determined by the correspondence $s\eta_p \longleftrightarrow (p, s)$ is considered. Conclusively, denote $F : \mathscr{H}(\mathcal{W}) \longrightarrow \mathbb{R}$ by $F = \pi_2 \circ \mathscr{H}^{-1}$ where $\pi_2 : U \longrightarrow \mathbb{R}$ is the projection defined by $\pi_2(p, s) = s$. Taking into account the fact that $\mathscr{H}$ is a vector bundle map it is deduce that F satisfies the homogeneity condition. According to theorem (3.2), for each point $p \in \mathcal{N}$:

$$\mathscr{H}(p, s) = \mathcal{H}(\widetilde{\eta}_{(p,s)}) = \mathcal{H}(s\eta_{(p,s)}) = s\mathcal{H}(\eta_{(p,s)}) = s\mathcal{H}(\widetilde{\eta}_{(p,1)}) = s\mathscr{H}(p, 1).$$

Consequently, according to the Normal Form Theorem proposed in [**18**] it is demonstrated that the foliated cocycle $\mathcal{F}^{\mathscr{T}} = \{\mathcal{V}_i^{\mathscr{T}}, \mathcal{X}_i^{\mathscr{T}}, \mathsf{T}^{\mathscr{T}}, \varrho_{ij}^{\mathscr{T}}\}$ on TM is precisely the same as the Legendre foliation $\mathcal{F}_F$ which is intrinsically characterized by the function F on $S_F^*(\mathcal{N}/\mathcal{F})$ with contact 1-form η_F defined by the pull back of ζ to $S_F^*(\mathcal{N}/\mathcal{F})$. In addition, $\mathscr{F}_\eta$ is a diffeomorphism from $\mathcal{N}$ onto $S_F^*(\mathcal{N}/\mathcal{F})$.

Since $\mathfrak{B}_\alpha^{n+1}$ are linearly independent and taking into account relation (3.12.d), it is concluded that there exists explicitly a symmetric matrix valued function $\mathfrak{D}^{\alpha\beta}$ such that: $\mathfrak{B}_{n+1}^\alpha = \mathfrak{D}^{\alpha\beta}\mathfrak{B}_\beta^{n+1}$. Therefore according to [**18**], identity (4.17) declared in condition **(A)** is a result of the following straightforward computation: when $X = V_\alpha$, $Y = V_\beta$, $Z = V_\gamma$ for a local frame (V_ρ) of L on a neighborhood $\mathcal{V} \subset S_F^* M$ we have:

$$\begin{aligned}\mathbf{\Pi}(V_\alpha, V_\beta) &= -\Big\{\mathcal{L}_{V_\alpha}\mathcal{L}_{V_\beta}\eta\Big\}(\xi) = -\Big\{\nabla^{\mathbf{N}}_{V_\alpha}\Big(-\mathfrak{B}_\sigma^{n+1}(V_\beta)\mathfrak{S}^\sigma\Big)\Big\}(\xi)\\ &= \Big\{\nabla^{\mathbf{N}}_{V_\alpha}\mathfrak{S}^\beta\Big\}(\xi) = -\mathfrak{D}^{\alpha\beta}.\end{aligned} \tag{4.21}$$

Moreover, on the unit cotangent bundle $S_F^*(M)$ the characteristic vector field can be calculated directly by using the induced coordinates on T^*M:

$$\xi = \sum_{i=1}^{n+1}\Big\{\frac{\partial F}{\partial y^i}\frac{\partial}{\partial x^i} - \frac{\partial F}{\partial x^i}\frac{\partial}{\partial y^i}\Big\}$$

According to [**18**], by the Normal Form Theorem, we may suppose that $V_\alpha = v_\alpha^i\frac{\partial}{\partial y^i}$ where v_α^i are smooth functions defined on T^*M satisfying $dF(V_\alpha) = F_i v_\alpha^i = 0$. Furthermore, we have $\mathcal{L}_{V_\alpha}dx^i = d(V_\alpha x^i) = 0$ and $dx^i(\xi) = F_i$. Thus by direct calculations it is resulted that [**18**]:

$$\begin{aligned}\mathbf{\Pi}(V_\alpha, V_\beta) &= -\{\mathcal{L}_{V_\alpha}\mathcal{L}_{V_\beta}\eta\}(\xi) = -\Big\{\mathcal{L}_{V_\alpha}\mathcal{L}_{V_\beta}\Big(\textstyle\sum_{i=1}^{n+1} y^i dx^i\Big)\Big\}(\xi)\\ &= -\textstyle\sum_{i=1}^{n+1}\Big\{\mathcal{L}_{V_\alpha}(v_\beta^i dx^i)\Big\}(\xi) = -\textstyle\sum_{i=1}^{n+1}\Big\{(V_\alpha v_\beta^i)dx^i\Big\}(\xi)\\ &= -\textstyle\sum_{i=1}^{n+1}\Big\{(V_\alpha v_\beta^i)F_i\Big\} = \textstyle\sum_{i=1}^{n+1}\Big\{v_\beta^i(V_\alpha F_i)\Big\} = F_{ij}v_\alpha^j v_\beta^i.\end{aligned} \tag{4.22}$$

Consequently, since the foliated cocycle $\mathcal{F}^{\mathscr{T}} = \{\mathcal{V}_i^{\mathscr{T}}, \mathcal{X}_i^{\mathscr{T}}, \mathrm{T}^{\mathscr{T}}, \varrho_{ij}^{\mathscr{T}}\}$ on TM is identically the same as the Legendre foliation $\mathcal{F}_F$, the local invariant $\mathbf{\Pi}$ which is also designated as the fundamental quadratic form, can be expressed via the restriction to L of the following symmetric form:

$$\mathbf{\Pi} = F_{ij}dy^i \otimes dy^j \tag{4.23}$$

Noticeably, due to homogeneity of F we have: $F = F_i y^i$. So, by differentiating this equality again, it is inferred that $F_{ij}y^i = 0$. Thus, restricted to fibers of T^*M, $dF = F_i dy^i$. Besides, on fibers of the unit cotangent bundle $S_F^*(M)$ we have $F = 1$ and $0 = dF = F_i dy^i$. Ultimately, it is resulted that:

$$\begin{aligned}\tfrac{1}{2}(F^2)_{ij}dy^i \otimes dy^j &= \Big\{F_iF_j + FF_{ij}\Big\}dy^i \otimes dy^j\\ &= F_{ij}dy^i \otimes dy^j = \mathbf{\Pi}.\end{aligned} \tag{4.24}$$

According to condition **A**, since $\mathbf{\Pi}$ is positive definite from (4.17) it is inferred that F is a Finsler metric.

Let S_F be the geodesic spray and assume that $\Theta = d_{\mathcal{J}}F$ is the concomitant Euler-Poincaré 1-form of the Finsler function F. Then there exists a 1-homogenous function $\mathscr{P}$ on $TM \setminus \{0\}$ such that $S = S_F - 2\mathscr{P}\mathbb{C}$. Taking into account [**9**], since F is 1-homogenous it is resulted that $i_S\Theta = \mathbb{C}(F) = F > 0$. Besides, $\operatorname{rank}(d\theta) = 2n-2$ is implied because of the non-degeneracy of the Finsler function F. The following predominant points can be inherently demonstrated for the spary S:

(i): The 0-homogeneity of the Euler-Poincaré 1-form Θ implies that: $\mathcal{L}_{\mathbb{C}}\Theta = 0$.

(ii): Since $d_{\mathcal{J}}\Theta = d^2_{\mathcal{J}}F = 0$ it follows that $d_{\mathcal{J}}\Theta = 0$.

(iii): The Euler-Poincaré 2-form of F^2, $\mathfrak{W}_{F^2} = dd_{\mathcal{J}}F^2$ is a symplectic structure. Subsequently, the geodesic spray S_F is uniquely determined by the condition $i_{S_F}dd_{\mathcal{J}}F^2 = -dF^2$, from which it follows that $S_F(F^2) = 0$ and thus $S_F(F) = 0$. Since S_F also satisfies the condition $\mathcal{L}_S d_{\mathcal{J}}F^2 = dF^2$ it follows that $\mathcal{L}_{S_F}\Theta = dF$. Moreover, because $S = S_F - 2\mathscr{P}\mathbb{C}$ it is acquired that $\mathcal{L}_S\Theta - 2\mathcal{L}_{\mathscr{P}\mathbb{C}}\Theta = dF$. Considering the 0-homogeneity of the semi-basic 1-form Θ yields $\mathcal{L}_{\mathscr{P}\mathbb{C}}\Theta = \mathscr{P}\mathcal{L}_{\mathbb{C}}\Theta = 0$ and accordingly $\mathcal{L}_S\Theta = dF$. Ultimately, utilizing $d_{\mathcal{J}}$ to both sides of this last relation and applying the commutation rules $\mathcal{L}_S d_{\mathcal{J}} - d_{\mathcal{J}}\mathcal{L}_S = d_{[S,\mathcal{J}]} = -d_h + d_v$ and $dd_{\mathcal{J}} + d_{\mathcal{J}}d = 0$ leads to the following identity:

$$\begin{aligned} -d_h\Theta - d_v\Theta &= -dd_{\mathcal{J}}F = d_{\mathcal{J}}dF = d_{\mathcal{J}}\mathcal{L}_S\Theta \\ &= \mathcal{L}_S d_{\mathcal{J}}\Theta + d_h\Theta - d_v\Theta, \end{aligned} \tag{4.25}$$

from where it is deduced that $d_h\Theta = 0$.

Consequently, the projectively metrizability of the spray S yields the following predominant relations:

$$\mathcal{L}_{\mathbb{C}}\Theta = 0, \quad d_{\mathcal{J}}\Theta = 0, \quad d_h\Theta = 0. \tag{4.26}$$

It is noticeable that according to [**9**] in order to address the formal integrability of the partial differential system (4.26) the following first order partial differential operator should be reckoned:

$$\mathcal{P}_1 = (\mathcal{L}_{\mathbb{C}}, d_{\mathcal{J}}, d_h) : \operatorname{Sec}(T_v^*) \longrightarrow (T_v^* \oplus \Lambda^2 T_v^* \oplus \Lambda^2 T_v^*) \tag{4.27}$$

Furthermore, $\mathcal{P}_1$ induces the following morphism of vector bundles defined by:

$$p_0(\mathcal{P}_1) : J_1 T_v^* \longrightarrow F := T_v^* \oplus \Lambda^2 T_v^* \oplus \Lambda^2 T_v^* \tag{4.28}$$

Subsequently the lth-order jet prolongation is expressed by: $p_l(\mathcal{P}_1) : J_{l+1}T_v^* \longrightarrow J_l F$. Meanwhile, for $\mathscr{E} \in TM$, $R_{l+1,\mathscr{E}} = \operatorname{Ker}(p_l(\mathcal{P}_1)_{\mathscr{E}}) \subset J_{l+1,\mathscr{E}}T_v^*$ can be regarded as the space of solutions of order $(l+1)$ of $\mathcal{P}_1$ at $\mathscr{E}$. In addition $\mathcal{P}_1$ is formally integrable at $\mathscr{E}$ if R_l is a vector bundle and for all $l \geq 1$, $\bar{\pi}_{l,\mathscr{E}} : R_{l+1,\mathscr{E}} \longrightarrow R_{l,\mathscr{E}}$ is onto. Consequently in the analytic case the concept of formal integrability implies the existence of a locally defined section in R_1 and therefore a solution of (4.26). Overall, according to Cartan-Kählar theorem if $\bar{\pi}_1 : R_2 \longrightarrow R_1$ is onto and the symbol $\sigma_2(\mathcal{P}_1)$ is involutive then $\mathcal{P}_1$ can be regarded as a formally integrable partial differential operator. Note the following diagram [**9**]:

$$
\begin{array}{ccccccccccc}
 & & 0 & & 0 & & 0 & & & & \\
 & & \downarrow & & \downarrow & & \downarrow & \overset{\mathfrak{T}}{\dashrightarrow} & & & \\
0 & \longrightarrow & g_2(\mathcal{P}_1) & \xrightarrow{i} & \mathcal{S}^2T^*\otimes T_v^* & \xrightarrow{\sigma_2} & T^*\otimes F & \longrightarrow & \mathcal{K} & \longrightarrow & 0 \\
 & & \downarrow & & \downarrow & \overset{p_1(\mathcal{P}_1)}{\dashrightarrow} & \nabla \uparrow\downarrow & & & & \\
0 & \longrightarrow & R_2 & \xrightarrow{i} & \mathcal{J}_2T_v^* & \longrightarrow & \mathcal{J}_1F & & & & \\
 & & \downarrow \bar{\pi}_1 & \overset{i}{\dashrightarrow} & \uparrow\downarrow \pi_1 & & \downarrow \pi & & & & \\
0 & \longrightarrow & R_1 & \longrightarrow & \mathcal{J}_1T_v^* & \xrightarrow{p_0(\mathcal{P}_1)} & F & & & & \\
 & & & & \downarrow & & \downarrow & & & & \\
 & & & & 0 & & 0 & & & &
\end{array}
$$

In order to determine the surjectivity of $\bar{\pi}_1 :\ R_2 \longrightarrow R_1$ it is required to investigate the following map: $\mathfrak{T} :\ T^* \otimes \big(T_v^* \oplus \Lambda^2 T_v^* \oplus \Lambda^2 T_v^*\big) \longrightarrow \mathcal{K}$, where $\mathcal{K} = \mathrm{Coker}(\sigma_2(\mathcal{P}_1))$. Since $\dim \mathcal{K} = \frac{m^2(m-1)}{2}$, it is deduced that: $\mathcal{K} \cong \oplus^{(2)} \Lambda^2 T_v^* \oplus^{(3)} \Lambda^3 T_v^*$. Besides the five main components of the map $\mathfrak{T}$ are given by:

$$
\begin{aligned}
&\mathfrak{T}_1\big(\mathcal{A}, \mathcal{B}_1, \mathcal{B}_2\big) = \mathfrak{T}_{\mathcal{J}}\mathcal{A} - i_{\mathbb{C}}\mathcal{B}_1; \quad \mathfrak{T}_2\big(\mathcal{A}, \mathcal{B}_1, \mathcal{B}_2\big) = \mathfrak{T}_{\mathcal{J}}\mathcal{A} - i_{\mathbb{C}}\mathcal{B}_2; \\
&\mathfrak{T}_3\big(\mathcal{A}, \mathcal{B}_1, \mathcal{B}_2\big) = \mathfrak{T}_{\mathcal{J}}\mathcal{B}_1; \quad \mathfrak{T}_4\big(\mathcal{A}, \mathcal{B}_1, \mathcal{B}_2\big) = \mathfrak{T}_h\mathcal{B}_2; \\
&\mathfrak{T}_5\big(\mathcal{A}, \mathcal{B}_1, \mathcal{B}_2\big) = \mathfrak{T}_h\mathcal{B}_1 + \mathfrak{T}_{\mathcal{J}}\mathcal{B}_2.
\end{aligned} \tag{4.29}
$$

Additionally $\sigma_2(\mathcal{P}_1)$ is involutive, namely it admits a quasi-regular basis which specifically satisfies the Cartan's test if and only if all the groups of the Spencer cohomology vanish (refer to [**9**] for more complete explanations).

Consequently, a first order formal solution $\Theta \in \Lambda^1 T_v^*$ of the system (4.26) can be lifted into a second order solution (in other words $\bar{\pi} :\ R_2 \longrightarrow R_1$ is onto) if and only if $d_R\Theta = 0$ where R is the curvature tensor defined by: $R = \frac{1}{2}[h, h] = -\frac{1}{3}[\mathcal{J}, \Phi]$.

♣ **Conclusively, at this stage, we have structurally constructed the foliated cocycle $\mathcal{F}^{\mathscr{T}} = \{\mathcal{V}_i^{\mathscr{T}}, \mathcal{X}_i^{\mathscr{T}}, \mathsf{T}^{\mathscr{T}}, \varrho_{ij}^{\mathscr{T}}\}$ on TM and proved that the condition (1) of the adopted transverse metric (ATM) is satisfied.**

✠ **Step (2):** In this step, we will thoroughly explore the existence of condition **(2)** of the ATM as declared in definition (4.1).

It is worth mentioning that the spray $S = y^i \frac{\partial}{\partial x^i} - 2G^i(x, y)\frac{\partial}{\partial y^i}$, canonically characterizes a nonlinear connection $\mathscr{N}$ with concomitant local coefficients $G_j^i = \frac{\partial G^i}{\partial y^j}$. The nonlinear connection $\mathscr{N}$ has the following local components: $(G_j^i) = \begin{pmatrix} G_b^a & G_b^\alpha \\ G_\beta^a & G_\beta^\alpha \end{pmatrix}$. Moreover, each of the local portions G_b^a, G_b^α, G_β^a, G_β^α has x^a, x^α, y^b, y^β as variables.

For $a, b \in \Big\{1, \ldots, n\Big\}$ and $\alpha, \beta \in \Big\{n+1, \ldots, m\Big\}$, the nonlinear connection $\mathscr{N}$ determines locally a basis of its horizontal vector fields which are explicitly asserted as follows:

$$
\frac{\delta}{\delta x^a} = \frac{\partial}{\partial x^a} - G_a^b \frac{\partial}{\partial y^b} - G_a^\beta \frac{\partial}{\partial y^\beta} \tag{4.30}
$$

$$
\frac{\delta}{\delta x^\alpha} = \frac{\partial}{\partial x^\alpha} - G_\alpha^b \frac{\partial}{\partial y^b} - G_\alpha^\beta \frac{\partial}{\partial y^\beta} \tag{4.31}
$$

So, taking into account the relation (4.18) declared in condition **(B)** it is noticeably concluded that: $G_a^{\beta} = 0$. Subsequently, from relation (4.30), it follows that: $\frac{\delta}{\delta x^a} = \frac{\partial}{\partial x^a} - G_a^b \frac{\partial}{\partial y^b}$ which inherently signifies the fact that $H\mathcal{F}^{\mathscr{T}}$ is locally spanned by the vector fields $\{\frac{\delta}{\delta x^a}\}$, $a \in \{1, \ldots, n\}$. Besides, since $\{\frac{\partial}{\partial y^a}\}$ is a local base in $\Gamma(V\mathcal{F}^{\mathscr{T}})$ it is literally demonstrated that: $T(\mathcal{F}^{\mathscr{T}}) = V\mathcal{F}^{\mathscr{T}} \oplus H\mathcal{F}^{\mathscr{T}}$ where $H\mathcal{F}^{\mathscr{T}} = T(\mathcal{F}^{\mathscr{T}}) \cap HTM$ and $V\mathcal{F}^{\mathscr{T}} = T(\mathcal{F}^{\mathscr{T}}) \cap VTM$. Consequently, this implies that $V\mathcal{F}^{\mathscr{T}}$ and $H\mathcal{F}^{\mathscr{T}}$ determine the nonlinear connection (G_b^a) , $a, b \in \{1, \ldots, n\}$, on the leaves of $\mathcal{F}^{\mathscr{T}}$. In addition, due to (4.18) we have: $G_{\alpha}^b = 0$. Accordingly, taking into account (4.31) it follows that: $\frac{\delta}{\delta x^{\alpha}} = \frac{\partial}{\partial x^{\alpha}} - G_{\alpha}^{\beta} \frac{\partial}{\partial y^{\beta}}$. In consequence, these significant relations are derived:

$$\frac{\partial G_{\alpha}^{\beta}}{\partial x^a} = \frac{\partial^2 G^{\beta}}{\partial x^a \partial y^{\alpha}} = 0 \text{ and } \frac{\partial G_{\alpha}^{\beta}}{\partial y^a} = \frac{\partial^2 G^{\beta}}{\partial y^a \partial y^{\alpha}} = 0. \tag{4.32}$$

Hence, it is definitely explicated that $\left\{\frac{\delta}{\delta x^{\alpha}}, \alpha = n+1, \ldots, m\right\}$ are local foliated vector fields. Overall, by utilizing g a metric $\mathcal{G}^{\mathscr{T}}$ on TM can be designated as follows, where $I,\ J \in \{1, \ldots, 2m\}$ and $i,\ j \in \{1, \ldots, m\}$:

$$\mathcal{G}_{IJ}^{\mathscr{T}}(x, y) = \begin{pmatrix} g_{ij}(x, y) & 0 \\ & \\ 0 & g_{ij}(x, y) \end{pmatrix} \tag{4.33}$$

This crucially insinuates that with respect to the frame field $\left\{\frac{\delta}{\delta x^i}, \frac{\partial}{\partial y^i}\right\}$ which is locally characterized on TM, the following can be unequivocally expressed:

$$\begin{aligned} &(\mathbf{a}) : \mathcal{G}^{\mathscr{T}}\left(\frac{\delta}{\delta x^i}, \frac{\delta}{\delta x^j}\right) = \mathcal{G}^{\mathscr{T}}\left(\frac{\partial}{\partial y^i}, \frac{\partial}{\partial y^j}\right) = g_{ij}, \\ &(\mathbf{b}) : \mathcal{G}^{\mathscr{T}}\left(\frac{\delta}{\delta x^i}, \frac{\partial}{\partial y^j}\right) = 0. \end{aligned} \tag{4.34}$$

As absolutely illustrated above, $\left\{\frac{\delta}{\delta x^{\alpha}}, \frac{\partial}{\partial y^{\beta}}\right\}$, $\alpha, \beta \in \{n+1, \ldots, m\}$, is a local base of foliated vector fields for the foliated cocycle $\mathcal{F}^{\mathscr{T}} = \{\mathcal{V}_i^{\mathscr{T}}, \mathcal{X}_i^{\mathscr{T}}, \mathsf{T}^{\mathscr{T}}, \varrho_{ij}^{\mathscr{T}}\}$ on TM. Moreover, it follows that:

$$\begin{aligned} &\mathcal{G}^{\mathscr{T}}\left(\frac{\delta}{\delta x^{\alpha}}, \frac{\delta}{\delta x^{\beta}}\right) = \mathcal{G}^{\mathscr{T}}\left(\frac{\partial}{\partial y^{\alpha}}, \frac{\partial}{\partial y^{\beta}}\right) = g_{\alpha\beta}, \\ &\mathcal{G}^{\mathscr{T}}\left(\frac{\delta}{\delta x^{\alpha}}, \frac{\partial}{\partial y^{\beta}}\right) = 0. \end{aligned} \tag{4.35}$$

♣ **Conclusively, we have proved that the condition (2) of the adopted transverse metric (ATM) is satisfied.**

✠ **Step (3):** In this step, we will precisely explore the existence of the condition **(3)** of the ATM as declared in definition (4.1).

For all $X, Y \in \Gamma(TM)$, the torsion tensor field of the Vrăceanu connection (4.19) on $\left(M, \eta, g, \mathcal{F} = \{\mathcal{V}_i, \mathcal{X}_i, \mathsf{T}, \varrho_{ij}\}\right)$ is given by [**4**]:

$$T^{\mathbf{V}}(X, Y) = h^{\mathfrak{q}}(\pi^{\mathfrak{q}} Y, \pi^{\mathfrak{q}} X) - h^{\mathfrak{q}}(\pi^{\mathfrak{q}} X, \pi^{\mathfrak{q}} Y). \tag{4.36}$$

The second fundamental form $h^{\mathfrak{q}} : \Gamma(T(\mathcal{F})^{\perp}) \times \Gamma(T(\mathcal{F})^{\perp}) \longrightarrow \Gamma(T(\mathcal{F}))$ of $T(\mathcal{F})^{\perp}$ is determined by:

$$h^{\mathfrak{q}}(\pi^{\mathfrak{q}} X, \pi^{\mathfrak{q}} Y) = \pi^{\mathfrak{q}}\left(\nabla^{\mathbf{L}}_{\pi^{\mathfrak{q}} X} \pi^{\mathfrak{q}} Y\right), \quad \forall\, X, Y \in \Gamma\left(T(\mathcal{F})^{\perp}\right). \tag{4.37}$$

Meanwhile, the symmetric second fundamental form $h^{\mathfrak{q}^s}$ of $T(\mathcal{F})^{\perp}$ is defined by:

$$h^{\mathfrak{q}^s}\left(\pi^{\mathfrak{q}}X,\pi^{\mathfrak{q}}Y\right)=\frac{1}{2}\Big(h^{\mathfrak{q}}\left(\pi^{\mathfrak{q}}X,\pi^{\mathfrak{q}}Y\right)+h^{\mathfrak{q}}\left(\pi^{\mathfrak{q}}Y,\pi^{\mathfrak{q}}X\right)\Big). \tag{4.38}$$

Besides, taking into account the relation (4.20) declared in condition **(C)** and (4.36), it is noticeably inferred that $h^{\mathfrak{q}^s}$ vanishes identically on M. Accordingly, by utilizing (4.37) and (4.38) it follows that:

$$g\Big(\pi^{\mathfrak{p}}Z,\nabla^{\mathbf{L}}_{\pi^{\mathfrak{q}}X}\pi^{\mathfrak{q}}Y+\nabla^{\mathbf{L}}_{\pi^{\mathfrak{q}}Y}\pi^{\mathfrak{q}}X\Big)=0,\quad\forall\ X,Y,Z\in\Gamma(TM). \tag{4.39}$$

Congruently, according to [**2**] the last identity is equivalent to the following relation:

$$\frac{\partial g_{\alpha\beta}}{\partial x^i}=0,\qquad\forall\, i\in\Big\{1,\ldots,m\Big\},\quad\alpha,\beta\in\Big\{n+1,\ldots,m\Big\}. \tag{4.40}$$

This significantly implies that $g_{\alpha\beta}$ is a basic function.

♣ **Conclusively, we have proved that the condition (3) of the adopted transverse metric (ATM) is also satisfied.**

Ultimately, it is substantiated that all the three conditions of definition (4.1) regarding the adopted transverse metric (ATM) are unequivocally fulfilled. Hence, $\mathcal{F}^{\mathscr{T}}=\big\{\mathcal{V}_i^{\mathscr{T}},\mathcal{X}_i^{\mathscr{T}},\mathsf{T}^{\mathscr{T}},\varrho_{ij}^{\mathscr{T}},\mathcal{G}^{\mathscr{T}}\big\}$ is a metric foliated cocycle on TM.

✠ **Step (4):** In this step, we will exhaustively explore the geometric structure of the resulted metric foliated cocycle $\mathcal{F}^{\mathscr{T}}=\big\{\mathcal{V}_i^{\mathscr{T}},\mathcal{X}_i^{\mathscr{T}},\mathsf{T}^{\mathscr{T}},\varrho_{ij}^{\mathscr{T}},\mathcal{G}^{\mathscr{T}}\big\}$ on TM.

As we explicitly demonstrated above, the foliated cocycle $\mathcal{F}^{\mathscr{T}}$ on TM is identically the same as the Legendre foliation $\mathcal{F}_F$ with F a Finsler metric. Moreover, due to the decomposition (4.4) we represent each $X\in\Gamma(TTM^0)$ as follows:

$$X=HX+VX=(HX)^i\frac{\delta}{\delta x^i}+(VX)^i\frac{\partial}{\partial y^i}, \tag{4.41}$$

First of all, it is worth noticing that F is positive-valued smooth function on TM^0. In addition, it has no critical points. To explore this point, by utilizing the Euler theorem on positively homogenous functions, it follows that:

$$F^2(x,y)=g_{ij}(x,y)y^iy^j. \tag{4.42}$$

Let $IM(c)$ be a leaf of $F_{\mathcal{F}}$ expressed by the equation $F(x,y)=c$, where c is a positive constant. The normal vector field to $IM(c)$ is identically the gradient of F designated by $\operatorname{grad}F$ and defined as follows [**3**]:

$$\mathcal{G}^{\mathscr{T}}\Big(\operatorname{grad}F,X\Big)=X(F),\qquad\forall\, X\in\Gamma(TTM^0). \tag{4.43}$$

Subsequently, according to (4.43) it is concluded that X is tangent to $IM(c)$ if and only if

$$(HX)^i\frac{\delta F}{\delta x^i}+(VX)^i\frac{\partial F}{\partial y^i}=0. \tag{4.44}$$

Moreover, taking into account the vertical Liouville vector field $\mathbb{C}=y^i\frac{\partial}{\partial y^i}$ and by applying the second equality in (4.34.a) the relation (4.42) is expressed as follows:

$$F^2=\mathcal{G}^{\mathscr{T}}\left(\mathbb{C},\mathbb{C}\right) \tag{4.45}$$

Significantly, according to [**3**] by implementing $\frac{\delta}{\delta x^i}$ to this identity it is apparently derived that:

$$\frac{\delta F}{\delta x^i} = \frac{2}{F}\mathcal{G}^{\mathscr{T}}\left(\nabla^{\mathbf{L}}_{\frac{\delta}{\delta x^i}}\mathbb{C},\mathbb{C}\right) = 0, \quad \forall\, i \in \{1,\ldots,m\}. \tag{4.46}$$

Consequently, X is tangent to $IM(c)$ if and only if $(VX)^i\frac{\partial F}{\partial y^i} = 0$, which is explicitly equivalent to the following identity (refer to [**3**] for more details):

$$g_{ij}(x,y)(VX)^i y^j = 0. \tag{4.47}$$

Hence, it is deduced that the leaves of the constructed metric foliated cocycle $\mathcal{F}^{\mathscr{T}} = \{\mathcal{V}_i^{\mathscr{T}},\mathcal{X}_i^{\mathscr{T}},\mathsf{T}^{\mathscr{T}},\varrho_{ij}^{\mathscr{T}},\mathcal{G}^{\mathscr{T}}\}$ are identically $IM(c) = \bigcup_{x\in M} I_xM(c)$ the c-indicatrix bundle over M. Predominantly according to [**3**], $\mathcal{F}^{\mathscr{T}}$ is equivalent to the foliation determined by the integrable distribution $\mathfrak{C}^{\perp}$ i.e. the complementary orthogonal distribution to foliation designated by virtue of the integrable distribution $\mathfrak{C}$ in TTM^0. Specifically, whenever a fixed point $x_0 = (x_0^i)$ on M is considered, then the hypersurface $I_{x_0}M(c)$, $c > 0$, in $T_{x_0}M^0 = T_{x_0}M \setminus \{0\}$ is defined by the equation

$$F(x_0,y) = c, \quad \forall\, y \in T_{x_0}M^0. \tag{4.48}$$

By means of the identical reasoning as for $IM(c)$ it follows that the vertical Liouville vector field $\mathbb{C}$ is orthogonal to each hypersurface $I_{x_0}M(c)$ of the Riemannian manifold $\left(T_{x_0}M^0, g_{x_0}\right)$. In addition, each $I_{x_0}M(c)$ is a level hypersurface of the function $F_{x_0}(y) = F(x_0,y)$, which does not have critical points on $T_{x_0}M^0$. As a consequence, our well-established metric foliated cocycle $\mathcal{F}^{\mathscr{T}} = \{\mathcal{V}_i^{\mathscr{T}},\mathcal{X}_i^{\mathscr{T}},\mathsf{T}^{\mathscr{T}},\varrho_{ij}^{\mathscr{T}},\mathcal{G}^{\mathscr{T}}\}$ structurally characterizes a foliation whose leaves are the set of all c-indicatrices at x_0. Furthermore, according to [**3**] this foliation is identically the same as the foliation designated by means of the integrable distribution $\mathfrak{C}'$ i.e. the complementary orthogonal distribution to $\mathfrak{C}$ in VTM^0. Besides, it can be regarded as the subfoliation of the vertical foliation $\mathcal{F}_V$ whose leaves are totally umbilically immersed in $\left(T_{x_0}M^0, g_{x_0}\right)$ (refer to [**2, 3**] for more details). □

Taking into account relation (3.10.b), the other geometric representation of $\mathbf{G}$ i.e. the invariant concomitant to the Legendre foliated cocycle $\mathcal{F} = \{\mathcal{V}_i,\mathcal{X}_i,\mathsf{T},\varrho_{ij}\}$ which is intrinsically a symmetric 3-form on the tangent bundle to $\mathcal{F}$ can be derived in terms of the fundamental function F as follows. By $\mathcal{L}_{V_\alpha}\mathcal{L}_{V_\beta}\eta = \sum_{i=1}^{n+1}\left(V_\alpha v_\beta^i\right)dx^i$ and straightforward calculations it is inferred that [**18**]:

$$\begin{aligned} &l\{\mathcal{L}_{V_\alpha}\mathcal{L}_{V_\beta}\mathcal{L}_{V_\gamma}\eta\}(\xi)\\ &= \left\{\mathcal{L}_{V_\alpha}\left(\textstyle\sum_{i=1}^{n+1}\left(V_\beta v_\gamma^i\right)dx^i\right)\right\}(\xi)\\ &= \textstyle\sum_{i=1}^{n+1}\left\{\left(V_\alpha V_\beta v_\gamma^i\right)dx^i\right\}(\xi) = V_\alpha\left\{\left(V_\beta v_\gamma^i\right)F_i\right\} - \left(V_\beta v_\gamma^i\right)\left(V_\alpha F_i\right)\\ &= -V_\alpha\left\{v_\gamma^i\left(V_\beta F_i\right)\right\} - \left(V_\beta v_\gamma^i\right)\left(V_\alpha F_i\right) = -V_\alpha\left\{v_\gamma^i v_\beta^j F_{ij}\right\} - \left(V_\beta v_\gamma^i\right)v_\alpha^j F_{ij}\\ &= -V_\alpha\left\{\mathbf{\Pi}\left(V_\beta,V_\gamma\right)\right\} - \left(V_\beta v_\gamma^i\right)v_\alpha^j F_{ij}. \end{aligned} \tag{4.49}$$

Subsequently, by virtue of (4.49), the right-hand side of (3.10.b) can be precisely derived as follows [**18**]:

$$\begin{aligned}(4.50)\quad &\tfrac{1}{2}\Big\{V_\alpha\big(F_{ij}v^i_\beta v^j_\gamma\big)-\big(V_\alpha v^i_\beta\big)v^j_\gamma F_{ij}-\big(V_\alpha v^i_\gamma\big)v^j_\gamma F_{ij}\Big\}\\ &\qquad=\tfrac{1}{2}\big(V_\alpha F_{ij}\big)v^i_\beta v^j_\gamma=\tfrac{1}{2}F_{ijk}v^i_\beta v^j_\gamma v^k_\alpha=\mathbf{G}\big(V_\alpha,V_\beta,V_\gamma\big).\end{aligned}$$

Now, taking into account theorem (4.2) we can state the following theorem:

THEOREM 4.3. *Let (M,η) be an $m=(2n+1)$-dimensional contact manifold which is equipped with an n-dimensional Legendre foliated cocycle $\mathcal{F}=\{\mathcal{V}_i,\mathcal{X}_i,\mathsf{T},\varrho_{ij}\}$. Suppose that $S=y^i\frac{\partial}{\partial x^i}-2G^i(x,y)\frac{\partial}{\partial y^i}$ is an spray on M. If we add the following condition to the constraints **(A)** and **(B)** proposed in theorem (4.2):*

(Ć): *By virtue of the Bott partial connection and moving frames, for the relative 1-forms $\mathfrak{B}^{n+1}_\alpha$ and $\mathfrak{B}^{n+1}_\beta$, and a symmetric matrix valued function $(\mathfrak{D}^{\alpha\beta})$, the following identity holds:*

$$(4.51)\qquad \mathbf{G}=\frac{1}{2}\Big\{d_{\mathcal{F}}\mathfrak{D}^{\alpha\beta}+\mathfrak{D}^{\alpha\gamma}\mathfrak{B}^\beta_\gamma+\mathfrak{D}^{\beta\gamma}\mathfrak{B}^\alpha_\gamma\Big\}\otimes\mathfrak{B}^{n+1}_\alpha\otimes\mathfrak{B}^{n+1}_\beta=0$$

Then there exists a metric foliated cocycle $\mathcal{F}^{\mathscr{T}}=\{\mathcal{V}^{\mathscr{T}}_i,\mathcal{X}^{\mathscr{T}}_i,\mathsf{T}^{\mathscr{T}},\varrho^{\mathscr{T}}_{ij}\}$ on TM which is also totally geodesic. In addition, this foliated cocycle is apparently the same as the vertical foliation $\mathcal{F}_V$.

PROOF. First of all, taking into account [**18**], we analyze the relation (4.51) declared in condition **(Ć)** in terms of Lie derivatives:

$$(4.52)\qquad\begin{aligned}\Big\{\mathcal{L}_{V_\beta}\mathcal{L}_{V_\alpha}\mathcal{L}_{V_\gamma}\eta\Big\}(\xi)&=\Big\{\nabla^{\mathbf{N}}_{V_\beta}\nabla^{\mathbf{N}}_{V_\alpha}\nabla^{\mathbf{N}}_{V_\gamma}\eta\Big\}(\xi)=\Big\{\nabla^{\mathbf{B}}_{V_\beta}\nabla^{\mathbf{B}}_{V_\alpha}\mathfrak{S}^\gamma\Big\}(\xi)\\ &=\Big\{\nabla^{\mathbf{N}}_{V_\beta}\Big(\mathfrak{B}^\gamma_\sigma(V_\alpha)\mathfrak{S}^\sigma+\mathfrak{B}^\gamma_{n+1}(V_\alpha)\eta\Big)\Big\}(\xi)\\ &=-\mathfrak{B}^\gamma_\sigma(V_\alpha)\mathfrak{B}^\sigma_{n+1}(V_\beta)+\Big\{\nabla^{\mathbf{N}}_{V_\beta}\big(\mathfrak{D}^{\gamma\alpha}\eta\big)\Big\}(\xi)\\ &=-\mathfrak{D}^{\beta\sigma}\mathfrak{B}^\gamma_\sigma(V_\alpha)+V_\beta\mathfrak{D}^{\gamma\alpha}.\end{aligned}$$

Accordingly, by applying (4.52) and due to the identity $V_\alpha\Big(\mathbf{\Pi}(V_\beta,X_\gamma)\Big)=V_\alpha\mathfrak{D}^{\beta\gamma}$ it follows that [**18**]:

$$(4.53)\qquad \mathbf{G}\big(V_\alpha,V_\beta,V_\gamma\big)=\frac{1}{2}\Big\{V_\alpha\mathfrak{D}^{\beta\gamma}+\mathfrak{D}^{\gamma\sigma}\mathfrak{B}^\beta_\sigma(V_\alpha)+\mathfrak{D}^{\beta\sigma}\mathfrak{B}^\gamma_\sigma(V_\alpha)\Big\}.$$

Consequently, by virtue of identities (3.10.b), (4.52) and (4.53) it is inferred that relation (4.51) stated in condition **(Ć)** is identically equivalent to this significant formula:

$$(4.54)\qquad \mathbf{G}=\frac{1}{2}F_{ijk}dy^i\otimes dy^j\otimes dy^k=0$$

Predominantly, considering theorem (4.2) and relations (4.17) and (4.54) it is inferred that: $(F^2)_{ijk}dy^i\otimes dy^j\otimes dy^k=0$ in the direction tangent to $\mathcal{F}_F$. Hence, it is sufficient to illustrate that $(F^2)_{ijk}dy^i\otimes dy^j\otimes dy^k=0$ in the direction of the vertical Liouville vector field $\mathbb{C}=y^i\frac{\partial}{\partial y^i}$, which is explicitly transversal to the hypersurface

$F(x, y) = c$. By straightforward calculations it can be demonstrated that (refer to [**18**] for more details):

$$\begin{aligned}\tfrac{1}{2}\left(F^2\right)_{ijk}y^k &= \left\{F_{ik}F_j + F_{jk}F_i + F_{ij}F_k + FF_{ijk}\right\}y^k \\ &= F_{ij}F + FF_{ijk}y^k = F_{ij}F - F_{ij}F = 0.\end{aligned} \tag{4.55}$$

Conclusively, $\left(F^2\right)_{ijk}dy^i \otimes dy^j \otimes dy^k = 0$ and $\left(F^2\right)_{ijk} = 0$. Moreover, considering the fact that F^2 is locally a homogenous function of degree 2, it is explicitly concluded that $F^2 = g_{ij}(x)y^iy^j$ for some $\left(g_{ij}(x)\right)$ depending only upon $x = \left(x^i\right)$. Therefore, by utilizing g the metric $\mathcal{G}^{\mathscr{T}}$ on TM can be unequivocally designated as demonstrated exhaustively via relations (4.33) and (4.34).

In the next step, we consider the Levi-Civita connection $\nabla^{\mathbf{L}}$ on $(TM^0, \mathcal{G}^{\mathscr{T}})$ whose local coefficients are expressed as follows [**2, 3**]:

$$\begin{cases} (\mathbf{a}) : \nabla^{\mathbf{L}}_{\frac{\delta}{\delta x^j}} \frac{\delta}{\delta x^i} = -\left(C_i{}^k{}_j + \frac{1}{2}R^k{}_{ij}\right)\frac{\partial}{\partial y^k} + F_i{}^k{}_j\frac{\delta}{\delta x^k}, \\ (\mathbf{b}) : \nabla^{\mathbf{L}}_{\frac{\partial}{\partial y^j}} \frac{\partial}{\partial y^i} = C_i{}^k{}_j\frac{\partial}{\partial y^k} - \frac{1}{2}g_{ij|h}g^{hk}\frac{\delta}{\delta x^k}, \\ (\mathbf{c}) : \nabla^{\mathbf{L}}_{\frac{\delta}{\delta x^j}} \frac{\partial}{\partial y^i} = F_i{}^k{}_j\frac{\partial}{\partial y^k} + \left(C_i{}^k{}_j + \frac{1}{2}g_{ih}R^h{}_{lj}g^{lk}\right)\frac{\delta}{\delta x^k} \\ \qquad\qquad\quad = \nabla_{\frac{\partial}{\partial y^i}}\frac{\delta}{\delta x^j} + G_i{}^k{}_j\frac{\partial}{\partial y^k} \end{cases} \tag{4.56}$$

where $C_i{}^k{}_j$, $F_i{}^k{}_j$ and $g_{ij|h}$ are determined by these identities [**2, 3**]:

$$\begin{cases} (\mathbf{a}) : C_i{}^k{}_j = \frac{1}{2}g^{kh}\frac{\partial g_{ij}}{\partial y^j}, \\ (\mathbf{b}) : F_i{}^k{}_j = \frac{1}{2}g^{kh}\left(\frac{\delta g_{hi}}{\delta x^j} + \frac{\delta g_{hj}}{\delta x^i} - \frac{\delta g_{ij}}{\delta x^h}\right), \\ (\mathbf{c}) : g_{ij|h} = \frac{\delta g_{ij}}{\delta x^h} - g_{kj}G_i{}^k{}_h - g_{ik}G_j{}^k{}_h. \end{cases} \tag{4.57}$$

Significantly, according to (4.34) it is deduced that $\mathcal{G}^{\mathscr{T}}$ is parallel with respect to $\nabla^{\mathbf{L}}$, that is for all $X, Y, Z \in \Gamma(TTM^0)$ the following identity holds:

$$X\left(\mathcal{G}^{\mathscr{T}}(Y, Z)\right) = \mathcal{G}^{\mathscr{T}}\left(\nabla^{\mathbf{L}}_X Y, Z\right) + \mathcal{G}^{\mathscr{T}}\left(Y, \nabla^{\mathbf{L}}_X Z\right). \tag{4.58}$$

Subsequently, by applying (4.34.b) and (4.58), it is deduced that:

$$\mathcal{G}^{\mathscr{T}}\left(\nabla^{\mathbf{L}}_{\frac{\partial}{\partial y^j}}\frac{\partial}{\partial y^i}, \frac{\delta}{\delta x^k}\right) + \mathcal{G}^{\mathscr{T}}\left(\frac{\partial}{\partial y^i}, \nabla^{\mathbf{L}}_{\frac{\partial}{\partial y^j}}\frac{\delta}{\delta x^k}\right) = 0. \tag{4.59}$$

Crucially, (4.56.b) and the second equality in (4.56.c) yield this identity:

$$g_{ij|k} = 2g_{ih}\left(F_j{}^h{}_k - G_j{}^h{}_k\right). \tag{4.60}$$

Conclusively, relation (4.56.b) is re-expressed as follows [**3**]:

$$\nabla^{\mathbf{L}}_{\frac{\partial}{\partial y^j}}\frac{\partial}{\partial y^i} = C_i{}^k{}_j\frac{\partial}{\partial y^k} + g_{ih}\left(G_j{}^h{}_l - F_j{}^h{}_l\right)g^{lk}\frac{\delta}{\delta x^k}. \tag{4.61}$$

As we demonstrated above, the metric g_{ij} depend on $\left(x^k\right)$ alone, so due to (4.57.a) we have:

$$C_i{}^k{}_j = 0, \quad \forall\, i, j, k \in \{1, \ldots, m\}, \tag{4.62}$$

Then by scrutinizing the Levi-Civita connection $\nabla^{\mathbf{L}}$ on $(TM^0, \mathcal{G}^{\mathscr{T}})$ and due to (4.56.a), (4.62 and (4.34) we deduce that:

$$G^{\mathscr{T}}\Big(\nabla^{\mathbf{L}}_{\frac{\delta}{\delta x^j}}\frac{\delta}{\delta x^i} + \nabla^{\mathbf{L}}_{\frac{\delta}{\delta x^i}}\frac{\delta}{\delta x^j}, \frac{\partial}{\partial y^k}\Big) = 0, \quad \forall\, i,j,k \in \{1,\dots,m\}, \tag{4.63}$$

Furthermore, the metric g_{ij} satisfies the following relation (refer to [**3**] for more details):

$$F_j{}^h{}_k = G_j{}^h{}_k, \quad \forall\, i,j,k \in \{1,\dots,m\}, \tag{4.64}$$

In consequence, taking into account (4.64), the identity (4.61) is represented as follows:

$$\nabla^{\mathbf{L}}_{\frac{\partial}{\partial y^j}}\frac{\partial}{\partial y^i} \in \Gamma(VTM^0), \quad \forall\, i,j \in \{1,\dots,m\}. \tag{4.65}$$

Ultimately, according to (4.63) and (4.65) we have explicitly demonstrated that $\mathcal{F}^{\mathscr{T}} = \{\mathcal{V}_i^{\mathscr{T}}, \mathcal{X}_i^{\mathscr{T}}, \mathsf{T}^{\mathscr{T}}, \varrho_{ij}^{\mathscr{T}}, \mathcal{G}^{\mathscr{T}}\}$ is a metric foliated cocycle on TM whose leaves are totally geodeically immersed in $(TM^0, \mathcal{G}^{\mathscr{T}})$ and it is apparently the same as the vertical foliation $\mathcal{F}_V$. □

Remark: In the following, the spray $S = y^i\frac{\partial}{\partial x^i} - 2G^i(x,y)\frac{\partial}{\partial y^i}$ satisfying the identities (4.18) declared in condition **(B)** of theorem (4.2), is denoted by an an ***adopted foliated spray (AFS)*** on M.

THEOREM 4.4. *Let $(M, g, \varphi, \xi, \eta)$ be an $m = (2n+1)$-dimensional contact metric manifold. Suppose that $S = y^i\frac{\partial}{\partial x^i} - 2G^i(x,y)\frac{\partial}{\partial y^i}$ is an adopted foliated spray (AFS) on M. If the following identity holds:*

$$(\nabla^{\mathbf{L}}_X\varphi)Y = g(X,Y)\xi - \eta(Y)X, \quad \forall\, X,Y \in \Gamma(TM). \tag{4.66}$$

Then the characteristic vector field ξ defines canonically a metric foliated cocycle $\mathcal{F}^{\mathscr{T}} = \{\mathcal{V}_i^{\mathscr{T}}, \mathcal{X}_i^{\mathscr{T}}, \mathsf{T}^{\mathscr{T}}, \varrho_{ij}^{\mathscr{T}}, \mathcal{G}^{\mathscr{T}}\}$ on TM which is totally geodesic.

PROOF. According to [**6**], the relation (4.66) yields the following two predominant identities which are satisfied by the characteristic vector field ξ:

$$(\mathbf{a}) : \nabla^{\mathbf{L}}_\xi\xi = 0 \qquad \text{and} \qquad (\mathbf{b}) : \nabla^{\mathbf{L}}_X\xi = -\varphi X, \qquad \forall\, X \in \Gamma(TM). \tag{4.67}$$

where $\nabla^{\mathbf{L}}$ is the Levi-Civita connection on (M, g). In addition, by utilizing (4.67.b) and (3.3.h) and taking into account the structure of the metric $\mathcal{G}^{\mathscr{T}}$ proposed by (4.34), it is literally construed that:

$$\mathcal{G}^{\mathscr{T}}(X, \nabla^{\mathbf{L}}_Y\xi) + \mathcal{G}^{\mathscr{T}}(Y, \nabla^{\mathbf{L}}_X\xi) = 0, \quad \forall\, X,Y \in \Gamma(TTM). \tag{4.68}$$

this yields that ξ is a Killing vector field on the tangent bundle. Meanwhile, the contact distribution $T(\mathcal{F}_\xi)^\perp$ on $(M, g, \varphi, \xi, \eta)$ is identically the complementary orthogonal distribution to $T(\mathcal{F}_\xi) = \operatorname{span}\{\xi\}$. Consequently, from (4.68) it is straightforwardly concluded that specifically ξ is $T(\mathcal{F}_\xi)^\perp$-killing. Hence, according to Theorem 3.3 of [**2**], it is demonstrated that $\mathcal{F}^{\mathscr{T}} = \mathcal{F}_\xi = \{\mathcal{V}_i^{\mathscr{T}}, \mathcal{X}_i^{\mathscr{T}}, \mathsf{T}^{\mathscr{T}}, \varrho_{ij}^{\mathscr{T}}, \mathcal{G}^{\mathscr{T}}\}$ is a metric foliated cocycle which is according to (4.67.a) also totally geodesic. □

Remark: The contact metric manifold $(M, g, \varphi, \xi, \eta)$ satisfying the relation (4.66) is denoted by the *Sasakian manifold.*

Another viable type of Legendre foliations was comprehensively investigated by [**7, 13, 17**]. If $T(\mathcal{F})$ denotes the tangent distribution to the Legendre foliation $\mathcal{F}$, then $T(\mathcal{F})^{\perp}$ designates identically the complementary orthogonal distribution to $T(\mathcal{F})$ in TM with respect to g. Considering the fact that any integral manifold of $T(\mathcal{F})$ is anti-invariant with respect to φ, it follows that:

$$g(X, \varphi Y) = 0, \quad \forall \ X, Y \in \Gamma(TM). \tag{4.69}$$

Thus, the contact distribution $\mathcal{H}$ has the following decomposition:

$$\mathcal{H} = T(\mathcal{F}) \oplus \varphi T(\mathcal{F}), \tag{4.70}$$

where $\varphi(T(\mathcal{F})) = \Big\{\varphi X \, : \, \forall X \in \Gamma\big(T(\mathcal{F})\big)\Big\}$. So, when the distribution $\varphi T(\mathcal{F})$ is integrable, the foliation $\widetilde{\mathcal{F}}$ concomitantly characterized by virtue of $\varphi T(\mathcal{F})$ is denoted by the conjugate foliation to $\mathcal{F}$. Specifically, in the case of flat Legendre foliation on (M, η), Jane [**13**] proposed canonically a contact metric structure on M with respect to which the concomitant conjugate foliation exists and is also flat.

Now, according to [**2, 7, 13, 18**] we can state the following theorem:

THEOREM 4.5. *Let $\big(M, g, \varphi, \xi, \eta\big)$ be an $m = (2n+1)$-dimensional contact metric manifold. Suppose that $S = y^i \frac{\partial}{\partial x^i} - 2G^i(x,y)\frac{\partial}{\partial y^i}$ is an adopted foliated spray (AFS) on M. If the following two main conditions are satisfied:*

(A): *For the $(1,1)$-tensor field ψ on M designated by:*

$$\psi X = \frac{1}{2}(\mathcal{L}_\xi \varphi), \quad \forall X \in \Gamma(TM). \tag{4.71}$$

The characteristic vector field ξ belongs to the (k, μ)-nullity distribution:

$$\begin{aligned} &N(k,\mu) : x \longrightarrow N_x(k,\mu), \quad (k,\mu) \in \mathbb{R}^2 \\ &N_x(k,\mu) = \Big\{Z \in T_xM : \ R(X,Y)Z = k\Big(g(Y,Z)X \\ &\qquad - g(X,Z)Y\Big) + \mu\Big(g(Y,Z)\psi(X) - g(X,Z)\psi(Y)\Big)\Big\}. \end{aligned} \tag{4.72}$$

(B): *For any $X, Y \in \Gamma\big(T(\mathcal{F})\big)$ the following identity holds:*

$$g\big([X, \varphi Y], \xi\big) + g\big([X, \xi], \varphi Y\big) = 0. \tag{4.73}$$

Then there exists canonically two conjugate metric Legendre foliated cocycles: $\mathcal{F}^{\mathscr{T}} = \{\mathcal{V}_i^{\mathscr{T}}, \mathcal{X}_i^{\mathscr{T}}, \mathrm{T}^{\mathscr{T}}, \varrho_{ij}^{\mathscr{T}}\}$ and $\widetilde{\mathcal{F}}^{\mathscr{T}} = \{\widetilde{\mathcal{V}}_i^{\mathscr{T}}, \widetilde{\mathcal{X}}_i^{\mathscr{T}}, \widetilde{\mathrm{T}}^{\mathscr{T}}, \widetilde{\varrho}_{ij}^{\mathscr{T}}\}$ on TM. Moreover, the concomitant leaves of these two conjugate foliations are identically the c-indicatrix bundle over M: $IM(c) = \bigcup_{x \in M} I_xM(c)$.

PROOF. The $(1,1)$-tensor field ψ designated by (4.71) satisfies the following predominant identities [**2**]:

$$(\mathbf{a}) : \psi\xi = 0 \qquad \text{and} \qquad (\mathbf{b}) : \psi\varphi + \varphi\psi = 0. \tag{4.74}$$

Meanwhile, it is illustrated in [**7**] that the tensor fields φ and ψ are connected geometrically by the following relation:

$$\psi^2 = (k-1)\varphi^2 \tag{4.75}$$

Thus, taking into account that φ is an almost complex structure on $\mathcal{H}$, the identity (4.75) yields:

$$\psi^2 X = (1-k)X, \quad \forall X \in \Gamma(\mathcal{H}). \tag{4.76}$$

Accordingly, $\sqrt{1-k}$ and $-\sqrt{1-k}$ are explicitly two of the eigenvalues of ψ. In addition, due to (4.74.b) whenever X is eigenvector for $\sqrt{1-k}$, then φX is identically eigenvector for $-\sqrt{1-k}$. Consequently, two mutually orthogonal and integrable distributions $\mathscr{D}(\lambda)$ and $\mathscr{D}(-\lambda)$ leads to creation of two conjugate Legendre foliated cocycles denoted by $\mathcal{F} = \{\mathcal{V}_i, \mathcal{X}_i, \mathsf{T}, \varrho_{ij}\}$ and $\widetilde{\mathcal{F}} = \{\widetilde{\mathcal{V}}_i, \widetilde{\mathcal{X}}_i, \widetilde{\mathsf{T}}, \widetilde{\varrho}_{ij}\}$.

Moreover, according to [**2**] by straightforward computations and applying (3.3.d) and (3.7 this expression of the local invariant $\mathbf{\Pi}$ which is absolutely independent of either the metric g or the tensor field φ of any contact metric structure (g, φ, ξ, η) is determined as follows:

$$\mathbf{\Pi}(X,Y) = \eta\big(Y, [\xi, X]\big), \quad \forall\, X, Y \in \Gamma\big(T(\mathcal{F})\big) \tag{4.77}$$

Nevertheless, the relations (4.77), (3.6), (3.7), (3.3.e) and (3.3.g) yields this re-expression of the invariant $\mathbf{\Pi}$ which is also presented in [**13**]:

$$\mathbf{\Pi}(X,Y) = 2g\big([\xi, X], \varphi Y\big). \tag{4.78}$$

Significantly, from the relation (4.73) declared in condition **(B)**, it is concluded that g is a bundle-like metric for $\mathcal{F}$ (cf. [**2**], Theorem 3.3.1). Afterwards, by applying (4.78) and (3.2.b), the identity (4.73 converts to the subsequent identity:

$$\eta\big([X, \varphi Y\big) - \frac{1}{2}\mathbf{\Pi}(X,Y) = 0 \tag{4.79}$$

Ultimately, by inserting (3.6), (3.2.c) and (3.2.a) in (4.79 it follows that:

$$\mathbf{\Pi}(X,Y) = 4g(X,Y), \quad \forall X, Y \in \Gamma\big(T(\mathcal{F})\big) \tag{4.80}$$

The last equation apparently indicates that $\mathcal{F}$ is positive definite and thereupon according to the Normal Form Theorem [**18**], $\mathcal{F}$ is locally equivalent to one of the form $\mathcal{F}_F$ with F a Finsler metric. By the reason identically analogous to $\mathcal{F}$, the similar result is explicitly educed for the conjugate foliation $\widetilde{F}$. As a consequence, according to theorem (4.2) it is unequivocally illustrated that the two analyzed conjugate foliated cocycles $\mathcal{F} = \{\mathcal{V}_i, \mathcal{X}_i, \mathsf{T}, \varrho_{ij}\}$ and $\widetilde{\mathcal{F}} = \{\widetilde{\mathcal{V}}_i, \widetilde{\mathcal{X}}_i, \widetilde{\mathsf{T}}, \widetilde{\varrho}_{ij}\}$ lead to establishment of two conjugate metric Legendre foliated cocycles: $\mathcal{F}^{\mathscr{T}} = \{\mathcal{V}_i^{\mathscr{T}}, \mathcal{X}_i^{\mathscr{T}}, \mathsf{T}^{\mathscr{T}}, \varrho_{ij}^{\mathscr{T}}\}$ and $\widetilde{\mathcal{F}}^{\mathscr{T}} = \{\widetilde{\mathcal{V}}_i^{\mathscr{T}}, \widetilde{\mathcal{X}}_i^{\mathscr{T}}, \widetilde{\mathsf{T}}^{\mathscr{T}}, \widetilde{\varrho}_{ij}^{\mathscr{T}}\}$ on TM. Overall, according to our thorough discussion in the proof of theorem (4.2), it is definitely construed that the leaves of the constructed foliated cocycles are exactly the same as the c-indicatrix bundle over M i.e. $IM(c) = \bigcup_{x \in M}$. □

Conclusions

In this paper, we have comprehensively scrutinized the Legendre foliated cocycles on contact manifolds which are totally compatible with second order ordinary differential fields (SODE). For this prominent objective, first of all, we have quickly reviewed some basic features of the Frölicher-Nijenhuis formalism and utilize it to the geometric objects that live canonically on the tangent space such as vertical distribution, Liouville vector field and semi-basic forms. Afterwards, by virtue of this geometric setting, the reformulation of the projective metrizability problem in terms of the first-order partial differential operator $\mathcal{P}_1$ which acts on semi-basic 1-forms is briefly discussed. Consequently, this expression of the Helmholtz metrizability conditions, enables us to establish a transverse metric on the tangent bundle of a given contact manifold which leads to creation of the specific type of metric Legendre foliated cocycles which are entirely compatible with SODE structure. Furthermore, by means of the fundamental geometric objects concomitant to Legendre foliations, such as relative 1-forms, Bott connection and the Vrăceanu connection, we have precisely analyzed the geometric structure of the resulted Legendre foliated cocycles equipped with bundle-like metric on the tangent space. Specially, we have concentrated on the existence of the metric Legendre foliated cocycles adapted to SODE structure on two specific class of contact metric manifolds comprising the Sasakian manifolds and (k, μ) spaces. In addition, considering the circumstances under which a Legendre foliation on a contact metric manifold concomitantly possesses a second Legendre foliation denoted by the conjugate Legendre foliation, we have profoundly explored the conditions for the existence and the geometric characteristics of the conjugate Legendre foliated cocycles on the tangent space adapted to the ambient SODE structure.

References

[1] Fatemeh Ahangari, *Construction of the holonomy invariant foliated cocycles on the tangent bundle via formal integrability*, Turkish J. Math. **43** (2019), no. 1, 81–102, DOI 10.3906/mat-1705-19. MR3909277

[2] Aurel Bejancu and Hani Reda Farran, *Foliations and geometric structures*, Mathematics and Its Applications (Springer), vol. 580, Springer, Dordrecht, 2006. MR2190039

[3] Aurel Bejancu and Hani Reda Farran, *Finsler geometry and natural foliations on the tangent bundle*, Rep. Math. Phys. **58** (2006), no. 1, 131–146, DOI 10.1016/S0034-4877(06)80044-3. MR2273571

[4] Aurel Bejancu and Hani Reda Farran, *Vrănceanu connections and foliations with bundle-like metrics*, Proc. Indian Acad. Sci. Math. Sci. **118** (2008), no. 1, 99–113, DOI 10.1007/s12044-008-0006-8. MR2434196

[5] Aurel Bejancu and Hani Reda Farran, *Geometry of pseudo-Finsler submanifolds*, Kluwer Academic Publishers, Dordrecht, 2000, DOI 10.1007/978-94-015-9417-2. MR1861512

[6] David E. Blair, *Contact manifolds in Riemannian geometry*, Lecture Notes in Mathematics, Vol. 509, Springer-Verlag, Berlin-New York, 1976. MR0467588

[7] David E. Blair, Themis Koufogiorgos, and Basil J. Papantoniou, *Contact metric manifolds satisfying a nullity condition*, Israel J. Math. **91** (1995), no. 1-3, 189–214, DOI 10.1007/BF02761646. MR1348312

[8] Ioan Bucataru and Matias F. Dahl, *Semi-basic 1-forms and Helmholtz conditions for the inverse problem of the calculus of variations*, J. Geom. Mech. **1** (2009), no. 2, 159–180, DOI 10.3934/jgm.2009.1.159. MR2525756

[9] Ioan Bucataru and Zoltán Muzsnay, *Projective metrizability and formal integrability*, SIGMA Symmetry Integrability Geom. Methods Appl. **7** (2011), Paper 114, 22, DOI 10.3842/SIGMA.2011.114. MR2861227

[10] John W. Gray, *Some global properties of contact structures*, Ann. of Math. (2) **69** (1959), 421–450, DOI 10.2307/1970192. MR112161

[11] Joseph Grifone and Zoltán Muzsnay, *Variational principles for second-order differential equations*, World Scientific Publishing Co., Inc., River Edge, NJ, 2000. Application of the Spencer theory to characterize variational sprays, DOI 10.1142/9789812813596. MR1769337

[12] Detlef Gromoll and Gerard Walschap, *Metric foliations and curvature*, Progress in Mathematics, vol. 268, Birkhäuser Verlag, Basel, 2009, DOI 10.1007/978-3-7643-8715-0. MR2500106

[13] N. Jane, *Legendre Foliations on contact metric manifolds*, Ph.D. Thesis, Massey University, 1992.

[14] A. Laleh, M. M. Rezaii, and F. Ahangari, *Identification of Riemannian foliations on the tangent bundle via SODE structure*, Bull. Iranian Math. Soc. **38** (2012), no. 3, 669–688. MR3028463

[15] Paulette Libermann, *Legendre foliations on contact manifolds*, Differential Geom. Appl. **1** (1991), no. 1, 57–76, DOI 10.1016/0926-2245(91)90022-2. MR1109814

[16] I. Moerdijk and J. Mrčun, *Introduction to foliations and Lie groupoids*, Cambridge Studies in Advanced Mathematics, vol. 91, Cambridge University Press, Cambridge, 2003, DOI 10.1017/CBO9780511615450. MR2012261

[17] Beniamino Cappelletti Montano, *The foliated structure of contact metric (κ,μ)-spaces*, Illinois J. Math. **53** (2009), no. 4, 1157–1172. MR2741183

[18] Myung-Yull Pang, *The structure of Legendre foliations*, Trans. Amer. Math. Soc. **320** (1990), no. 2, 417–455, DOI 10.2307/2001683. MR1016808

[19] Marcela Popescu and Paul Popescu, *Projectable non-linear connections and foliations*, Summer School on Differential Geometry (Coimbra, 1999), Univ. Coimbra, Coimbra, 1999, pp. 159–165. MR1859952

[20] Paul Popescu and Marcela Popescu, *Lagrangians adapted to submersions and foliations*, Differential Geom. Appl. **27** (2009), no. 2, 171–178, DOI 10.1016/j.difgeo.2008.06.017. MR2501730

[21] Alan Weinstein, *Symplectic manifolds and their Lagrangian submanifolds*, Advances in Math. **6** (1971), 329–346 (1971), DOI 10.1016/0001-8708(71)90020-X. MR286137

Department of Mathematics, Faculty of Mathematical Sciences, Alzahra University, Tehran, P.O.Box 1993893973, Iran

Email address: f.ahangari@alzahra.ac.ir, fa.ahangari@gmail.com

Contemporary Mathematics
Volume **788**, 2023
https://doi.org/10.1090/conm/788/15818

Special Vinberg cones, invariant admissible cubics and special real manifolds

Dmitri V. Alekseevsky, Alessio Marrani, and Andrea Spiro

ABSTRACT. By Vinberg theory any homogeneous convex cone $\mathcal{V}$ may be realised as the cone of positive Hermitian matrices in a T-algebra of generalised matrices. The level hypersurfaces $\mathcal{V}_q \subset \mathcal{V}$ of homogeneous cubic polynomials q with positive definite Hessian (symmetric) form $g(q) := -\operatorname{Hess}(\log(q))|_{T\mathcal{V}_q}$ are the *special real manifolds*. Such manifolds occur as scalar manifolds of the vector multiplets in $N = 2$, $D = 5$ supergravity and, through the r-map, correspond to Kähler scalar manifolds in $N = 2$ $D = 4$ supergravity. We offer a simplified exposition of the Vinberg theory in terms of Nil-algebras (= the subalgebras of upper triangular matrices in Vinberg T-algebras) and we use it to describe all rational functions on a special Vinberg cone that are G_0- or G'-invariant, where G_0 is the unimodular subgroup of the solvable group G acting simply transitively on the cone, and G' is the unipotent radical of G_0. The results are used to determine G_0- and G'-invariant cubic polynomials q that are *admissible* (i.e. such that the hypersurface $\mathcal{V}_q = \{q = 1\} \cap \mathcal{V}$ has positive definite Hessian form $g(q)$) for rank 2 and rank 3 special Vinberg cones. We get in this way examples of continuous families of non-homogeneous special real manifolds of cohomogeneity less than or equal to two.

1. Introduction

In [**24**] Vinberg proved that any homogeneous convex cone may be realised as the cone $\mathcal{V}$ of positive definite matrices in an appropriate algebra $\mathcal{H} = Herm_m$ of generalised Hermitian matrices of rank m. In [**3**], Alekseevsky and Cortés described a class of rank 3 homogeneous convex cones, called *special Vinberg cones*. Such a class of cones is associated to $\mathbb{Z}_2$-graded Clifford modules with invariant metrics. If the metric is indefinite, Vinberg's construction produces homogeneous not necessarily convex cones, named *indefinite* special Vinberg cones. Alekseevsky and Cortés also proved that the associated *determinant hypersurface* $\mathcal{V}_d = \{d = 1\} \subset \mathcal{V}$, given by the so-called cubic determinant d, is a homogeneous special real manifold in the sense of de Wit and Van Proeyen. We recall that a *special real manifold* is a hypersurface $S \subset \mathbb{R}^n$, which is a subset of the level set $\{q = 1\}$ of a homogeneous cubic

2020 *Mathematics Subject Classification.* Primary 83E50, 15B48, 32M15.

Key words and phrases. Homogeneous convex cones, Vinberg T-algebras, special real manifolds, special Kähler manifolds, supergravity scalar manifolds.

The second author was supported by a "Maria Zambrano" distinguished researcher fellowship, financed by the European Union within the NextGeneration EU program.

The third author is the corresponding author.

©2023 American Mathematical Society

polynomial q, and such that the restriction to S of the Hessian (symmetric bilinear) form $g(q) := -\operatorname{Hess}\log q$, defined on a neighbourhood of S, is a Riemannian metric on S (see [**16**]). If this holds, the cubic q is called *admissible.*

The special real manifolds occur in Physics as target spaces – usually called *scalar manifolds* – for the maps that represent the scalar fields in the vector multiplets in $N = 2$, $D = 5$ supergravity. Furthermore, by dimensional reduction, any real special scalar manifold $\mathcal{V}_d$ corresponds to a $2n$-dimensional Kähler manifold $\mathcal{S}$, called *very special Kähler manifold*, which is a target space for the scalar fields in the vector multiplets in $N = 2$, $D = 4$ supergravity. This correspondence is the *supergravity r-map* defined by De Wit and Van Proeyen and is a special case of the Piatetski-Shapiro construction that associates with any real convex cone $\mathcal{V} \subset \mathbb{R}^n$ a complex Siegel domain of the first kind $\mathcal{S} = \mathbb{R}^n + i\mathcal{V} \subset \mathbb{R}^n + i\mathbb{R}^n \subset \mathbb{C}^n$. In turn, again by dimensional reduction, any (not necessarily very) special Kähler manifold corresponds to a $(4n + 4)$-dimensional quaternionic manifold, which is the target space for the scalar fields (after full dualization of the 1-forms) in $N = 4$, $D = 3$ supergravity, and also the target space for the scalar fields of hypermultiplets coupled to N-extended supergravity in $D = 6$ ($N = (1, 0)$), $5, 4$ ($N = 2$), and 3 ($N = 4$) [**9**]. This correspondence is the *supergravity c-map* [**13**]. As mentioned, the real, Kähler and quaternionic geometries related by dimensional reduction (or, equivalently, by the r-map and c-map) are named *special* geometries; see e.g. [**18, 21**] for a comprehensive treatment and a list of references.

Vinberg [**24**] developed a theory of homogeneous convex cones and gave a description of them as cones of positive definite Hermitian matrices in generalised matrix algebras, named *Vinberg T-algebras.* This class of homogeneous spaces has important applications in several areas, as for instance supergravities with eight supersymmetries in dimensions $3 \leq D \leq 6$ as well as in the description of the entropy of their extremal black hole solutions and in geometric approaches to information theory (for the latter, see e.g. [**10, 14, 22**]).

Alekseevsky [**1**] gave the classification of homogeneous quaternionic Kähler manifolds, which admit a solvable transitive Lie group of isometries, and associated an homogeneous Kähler manifold of half dimension to any such manifold, with the exception of the hyperbolic quaternionic spaces. Cecotti [**12**] remarked that such Kähler manifolds are exactly the homogeneous target spaces of scalar fields of vector multiplets coupled to $N = 2$, $D = 4$ supergravity. In [**17**] de Wit and Van Proeyen classified the admissible cubics corresponding to homogeneous special real manifolds. By the supergravity r-map, this in turn yields the classification of homogeneous special Kähler manifolds with holomorphic prepotentials given by the aforementioned admissible cubics. A few years later, Cortés [**15**] gave another proof of this classification, using the theory of homogeneous Kähler manifolds of Piatetski-Shapiro, Gindikin and Vinberg. The classification of such manifolds, determined in [**17**] and [**15**], shows that they are exactly the homogeneous special Kähler manifolds associated to the quaternionic Kähler manifolds classified in [**1**]. It is here worth remarking that Alekseevsky conjectured that any homogeneous quaternionic Kähler manifold with negative scalar curvature should be one of those classified in [**1**]. This conjecture has been recently proved to be true by Böhm and Lafuente in [**11**].

The homogeneous special real manifolds, that are determined by *irreducible* admissible cubics, are the determinant hypersurfaces $\mathcal{V}_d = \{d = 1\} \cap \mathcal{V}$ of the special Vinberg cones $\mathcal{V}$. Several properties of such determinant hypersurfaces and of their analogues in dual spaces have relevant implications in supergravity. In [**4**] the theory of special Vinberg cones and their duals has been applied to compute the Bertotti-Robinson mass, and therefore – by the Bekenstein-Hawking formula – the entropy of static extremal BPS black holes (see §4.4 for a detailed definition) in ungauged $N = 2$, $D = 4$ supergravity with an arbitrary homogeneous vector multiplets' scalar manifold. The theory of homogeneous convex cones has many other applications, for example, in information geometry, multivariate statistics, Frobenius manifolds, convex programming, etc.. For instance, for what concerns the differential geometry of the convex (not necessarily homogeneous) cones, Vinberg showed that the points of $\mathcal{V}$ define a probability measure on the adjoint cone $\mathcal{V}^*$, and determined a canonical Riemannian Hessian metric g on $\mathcal{V}$. This means that the homogeneous convex cones represent a distinguished class of statistical Hessian manifolds, which in information geometry is called *exponential family*. For such Hessian manifolds, the above canonical metrics coincide with the *Fisher metrics* in statistics [**5, 10**].

In this paper, we present a revision of Vinberg theory of homogeneous convex cones and of their adjoint cones, which is based on the notion of Nil-*algebras*, a particular class a nilpotent generalised matrix algebras introduced by Vinberg in [**24**] (there, they are named "N-algebras") and from which all Hermitian matrix algebras corresponding to the Vinberg cones can be recovered.

In order to clarify our approach, let us first briefly recall the notion of homogeneous convex cone and of its adjoint one. Let $V = \mathbb{K}^n$ be a vector space over $\mathbb{K} = \mathbb{R}$ or $\mathbb{C}$ and denote by $\mathbb{K}^*_o = \mathbb{R}^+, \mathbb{C}^*$ the connected component of the identity of the multiplicative group $\mathbb{K}^*$. A *cone* $\mathcal{V} \subset V$ is an open $\mathbb{K}^*_o$-invariant domain, which contains no one-dimensional subspace. It is called *homogeneous* if the linear automorphism group

$$\mathrm{Aut}(\mathcal{V}) = \{\ A \in \mathrm{GL}(V)\ ,\ A(\mathcal{V}) = \mathcal{V}\ \}$$

acts transitively on $\mathcal{V}$. It is called *of Vinberg type* if there exists a solvable subgroup $G \subset \mathrm{Aut}(\mathcal{V})$ which acts transitively on $\mathcal{V}$ with a finite stabiliser G_{x_o} at a point $x_o \in \mathcal{V}$. In other words, $\mathcal{V}$ is of Vinberg type if it has the form

$$\mathcal{V} = G(x_o) = G/G_{x_o} \quad \text{with } G \text{ solvable and } G_{x_o} \text{ finite}\ .$$

Vinberg proved that *any homogeneous real convex cone $\mathcal{V}$ is of this type* [**24**]. Now, given a real vector space $V = \mathbb{R}^n$, the *adjoint cone* of a cone $\mathcal{V} \subset V$ is the cone in $V^* = \mathrm{Hom}(V, \mathbb{R})$ defined by

$$\mathcal{V}^* = \{\xi \in V^*,\ \xi(x) > 0 \ \text{ for all } x \in \mathcal{V}\}\ .$$

It is known that if $\mathcal{V} \subset V$ is convex, then also the adjoint cone $\mathcal{V}^*$ is convex and that a convex cone $\mathcal{V}$ is homogeneous if and only if its adjoint cone $\mathcal{V}^*$ is homogeneous. A homogeneous real convex cone $\mathcal{V} \subset V = \mathbb{R}^n$ is called *self-adjoint* (or *self-dual*) if there exists a Euclidean inner product $\langle\cdot,\cdot\rangle$ such that $\mathcal{V}^* = \langle\mathcal{V},\cdot\rangle$.

Vinberg proved that any homogeneous convex cone is the cone of positive definite matrices in some algebra $\mathcal{H}$ of generalised Hermitian matrices. In [**24**], such algebra $\mathcal{H}$ is introduced as a subspace of a larger algebra of generalised matrices, called T-*algebra*, which is in turn completely determined by its associative subalgebra of nilpotent upper triangular matrices, called Nil-*algebra*. In this paper, we re-formulate Vinberg theory of homogeneous cones and their adjoint cones in terms of Nil-algebras. As a corollary, we get that the pairs formed by a homogeneous cone and its adjoint cone are in bijection with the pairs of a Nil-algebra N and its dual Nil-algebra $N^{t'}$, obtained from N applying the anti-transposition (= reflection with respect to its anti-diagonal) to all matrices. This implies that a homogeneous cone is self-adjoint if and only if it is determined by a Nil-algebra N that is isomorphic to its dual $N^{t'}$.

We consider the unimodular subgroup G_0 and the unipotent radical G' of the solvable Lie group G acting simply transitively on a rank 2 and special rank 3 Vinberg cone $\mathcal{V} = G(x_o)$ and determine the fields of the G_0-invariant and the G'-invariant rational functions on $\mathcal{V}$. As an application, we determine the G_0- and the G'- invariant admissible cubics. These give one-parameter families of cohomogeneity one and two special real manifolds corresponding to rank 2 and special rank 3 Vinberg cones. In this way, not only we recover the well-known examples of homogeneous scalar manifolds determined by irreducible admissible cubics, but we also obtain

- a one-parameter family of cohomogeneity one scalar manifolds in each Vinberg cone of rank 2;
- a one-parameter family of cohomogeneity two scalar manifolds in each special Vinberg cone of rank 3.

The paper is structured as follows. In §2 we give the definition of Nil-algebra and we introduce related objects that are needed for the construction of the homogeneous Vinberg cones. In §3, we describe the G_0- and G'-invariant rational functions on special Vinberg cones. In §4 we describe the Nil-algebras that correspond to self-adjoint homogeneous convex cones (which include all rank 2 homogeneous convex cones) and those associated with the special Vinberg cones of rank 3. The advertised description of the admissible cubics, corresponding to scalar manifolds of cohomogeneity one or two, is given in §5.

Notation. In the following, for any $m \times m$ matrix $X = (x_{ij})$, $1 \leq i, j \leq m$, we denote by $t(X)$ and $t'(X)$ the *transposed* and *anti-transposed* matrix of X, that is the matrix which is obtained from X by reflection with respect to the diagonal and the anti-diagonal, respectively:

$$X = (x_{ij}) \longmapsto t(X) := (x_{ji}) \qquad \text{(transposition)}$$

$$X = (x_{ij}) \longmapsto t'(X) := (x_{m+1-j\, m-1+i}) \qquad \text{(anti-transposition)}$$

2. Nil-algebras and associated Vinberg cones

2.1. Nil-algebras and associated dual Nil-algebras. A Nil-*algebra of rank* m over $\mathbb{K} = \mathbb{R}, \mathbb{C}$ is an associative algebra $\mathcal{N}$ of upper triangular $m \times m$ matrices

$$N = \begin{pmatrix} 0 & x_{12} & x_{13} & \cdots & \cdots & x_{1\,m-1} & x_{1\,m} \\ 0 & 0 & x_{23} & \cdots & & x_{2\,m-1} & x_{2\,m} \\ 0 & 0 & 0 & \cdots & & x_{2\,m-1} & x_{3\,m} \\ \vdots & \vdots & \ddots & \ddots & \ddots & \vdots & \vdots \\ 0 & 0 & 0 & \cdots & \cdots & 0 & x_{m-1\,m} \\ 0 & 0 & 0 & \cdots & \cdots & 0 & 0 \end{pmatrix}. \tag{2.1}$$

where the entries x_{ij} belong to metric vector spaces $\mathcal{N}_{ij}$, $i < j$, that is vector spaces with non-degenerate inner products $\langle \cdot, \cdot \rangle_{\mathcal{N}_{ij}}$. As a vector space, $\mathcal{N}$ is isomorphic to

$\mathcal{N} \simeq \sum_{i<j} \mathcal{N}_{ij}$ and it is equipped with the inner product $\langle \cdot, \cdot \rangle = \sum_{i<j} \langle \cdot, \cdot \rangle_{\mathcal{N}_{ij}}$. The matrix multiplication in $\mathcal{N}$ is defined by a system of *isometric maps*, i.e. bilinear maps

$$\cdot : \mathcal{N}_{ij} \times \mathcal{N}_{jk} \to \mathcal{N}_{ik} \ , \ (x_{ij}, x_{jk}) \longmapsto x_{ij} \cdot x_{jk} \ , \qquad i < j < k \ , \tag{2.2}$$

such that

$$\langle x_{ij} \cdot x_{jk}, x_{ij} \cdot x_{jk} \rangle = \langle x_{ij}, x_{ij} \rangle \langle x_{jk}, x_{jk} \rangle \ , \tag{2.3}$$

which satisfy the following additional axiom:

for $x_{ik} \in \mathcal{N}_{ik}$, $x_{jk} \in \mathcal{N}_{jk}$ *with* $i < j$

$$\langle x_{ik}, \mathcal{N} \cdot x_{jk} \rangle = \{0\} \qquad \Longrightarrow \qquad \langle \mathcal{N} \cdot x_{ik}, \mathcal{N} \cdot x_{jk} \rangle = \{0\} \ . \tag{2.4}$$

Two Nil-algebras $(\mathcal{N}, \cdot)$, $(\mathcal{N}', \cdot')$ of rank m are called *isomorphic* if there is an algebra isomorphism $\varphi : \mathcal{N} \to \mathcal{N}'$ which preserves the inner products and for which there is a permutation σ of $\{1, \ldots, m\}$ such that each subspace $\mathcal{N}_{ij}$ is mapped into the subspace $\mathcal{N}'_{\sigma(i)\ \sigma(j)} \subset \mathcal{N}'$ and with the property

$$i < j \textit{ and } \sigma(i) > \sigma(j) \qquad \Longrightarrow \qquad \mathcal{N}_{ij} = \mathcal{N}'_{\sigma(i)\sigma(j)} = \{0\} \ .$$

If $\mathbb{K} = \mathbb{R}$, a Nil-algebra $\mathcal{N}$ is called *pseudo-Euclidean of signature* (p, q) (resp. *Euclidean*) if its inner product $\langle \cdot, \cdot \rangle$ has signature (p, q) (resp. is Euclidean).

For any Nil-algebra $\mathcal{N}$ we associate an algebra with inner product, the *dual algebra* $\mathcal{N}^{t'}$, defined as follows. Let $\mathcal{N}^{t'}$ be the vector space of the matrices of the form

$$N^{t'} = \begin{pmatrix} 0 & x^{t'}_{12} & x^{t'}_{13} & \cdots & \cdots & x^{t'}_{1\,m-1} & x^{t'}_{1\,m} \\ 0 & 0 & x^{t'}_{23} & \cdots & & x^{t'}_{2\,m-1} & x^{t'}_{2\,m} \\ 0 & 0 & 0 & \cdots & & x^{t'}_{2\,m-1} & x^{t'}_{3\,m} \\ \vdots & \vdots & \ddots & \ddots & \ddots & \vdots & \vdots \\ 0 & 0 & 0 & \cdots & \cdots & 0 & x^{t'}_{m-1\,m} \\ 0 & 0 & 0 & \cdots & \cdots & 0 & 0 \end{pmatrix} . \tag{2.5}$$

with entries in the vector spaces $\mathcal{N}^{t'}_{ij} := \mathcal{N}_{m+1-j\,m+1-i}$, equipped with the inner product $\langle \cdot, \cdot \rangle = \sum_{i<j} \langle \cdot, \cdot \rangle_{\mathcal{N}^{t'}_{ij} := \mathcal{N}_{m+1-j\,m+1-i}}$. The matrix multiplication $\cdot$ between elements of $\mathcal{N}^{t'}$ is given by the family of isometric maps

$$\left(x^{t'}_{ij}, x^{t'}_{jk}\right) = \left(x_{(m+1-j)\,(m+1-i)}, x_{(m+1-k)\,(m+1-j)}\right) \longmapsto$$
$$\longmapsto x^{t'}_{ij} \cdot x^{t'}_{jk} := x_{(m+1-k)\,(m+1-j)} \cdot x_{(m+1-j)\,(m+1-i)} \qquad \text{for any } i < j < k$$

The following holds.

LEMMA 2.1.

(1) *The dual algebra* $(\mathcal{N}^{t'}, \cdot)$ *is a* Nil-*algebra;*

(2) *The anti-transposition operator*

$$
(2.6)\quad N = \begin{pmatrix}
0 & x_{12} & x_{13} & \dots & \dots & x_{1\,m-1} & x_{1\,m} \\
0 & 0 & x_{23} & \dots & & x_{2\,m-1} & x_{2\,m} \\
0 & 0 & 0 & \dots & & x_{2\,m-1} & x_{3\,m} \\
\vdots & \vdots & \ddots & \ddots & \ddots & \vdots & \vdots \\
0 & 0 & 0 & \dots & \dots & 0 & x_{m-1\,m} \\
0 & 0 & 0 & \dots & \dots & 0 & 0
\end{pmatrix} \longmapsto
$$

$$
\longmapsto N^{t'} = \begin{pmatrix}
0 & x_{m-1\,m} & \dots & & \dots & \dots & x_{2\,m} & x_{1\,m} \\
0 & 0 & x_{m-2\,m-1} & \dots & & & x_{2\,m-1} & x_{1\,m-1} \\
0 & 0 & 0 & \dots & & & x_{2\,m-2} & x_{1\,m-2} \\
\vdots & \vdots & \ddots & & \ddots & \ddots & \vdots & \vdots \\
0 & 0 & 0 & \dots & \dots & & 0 & x_{1\,2} \\
0 & 0 & 0 & \dots & \dots & & 0 & 0
\end{pmatrix}
$$

determines a linear (but, in general, not algebra) isomorphism between $\mathcal{N}$ *and* $\mathcal{N}^{t'}$.

2.2. Solv-**algebras and Vinberg groups.** Given a Nil-algebra $\mathcal{N}$ of rank m over $\mathbb{K}$, the matrix multiplication between elements in $\mathcal{N}$ naturally extends to a matrix multiplication between upper triangular matrices of the form

$$
(2.7)\quad \begin{gathered}
A = D + N = \begin{pmatrix}
a_{11} & a_{12} & a_{13} & \dots & \dots & a_{1m-1} & a_{1m} \\
0 & a_{22} & a_{23} & \dots & & a_{2m-1} & a_{2m} \\
0 & 0 & a_{33} & \dots & & a_{2m-1} & a_{3m} \\
\vdots & \vdots & \ddots & \ddots & \ddots & \vdots & \vdots \\
0 & 0 & 0 & \dots & \dots & a_{m-1m-1} & a_{m-1m} \\
0 & 0 & 0 & \dots & \dots & 0 & a_{mm}
\end{pmatrix} \\
\text{with } N \in \mathcal{N}\ ,\ D{:=}\,\mathrm{diag}(a_{11}, \dots, a_{mm}) \text{ with } a_{ii}{\in}\mathbb{K}\ ,
\end{gathered}
$$

by means of the natural maps

$$
(2.8)\qquad a_{ii}, a_{ik} \mapsto a_{ii}{\cdot}a_{ik}{:=}a_{ii}a_{ik} \ (= \text{usual product between scalar and vector}).
$$

The extended solvable algebra $\mathcal{T}{=}\mathcal{T}(\mathcal{N}){=}\mathbb{K}^m{+}\mathcal{N}$ of the matrices (2.7) is called Solv-*algebra associated with* $\mathcal{N}$.

DEFINITION 2.1. The *Vinberg group associated with* $\mathcal{N}$ is the connected component $G = G(\mathcal{N})$ of the identity $I_m = \mathrm{diag}(1, \dots, 1)$ of the solvable Lie group of the invertible elements in $\mathcal{T} = \mathcal{T}(\mathcal{N})$.

Note that a matrix $A \in \mathcal{T}$ is invertible if and only if its diagonal elements a_{ii} are non zero. Hence, if $\mathbb{K} = \mathbb{R}$ (resp. $\mathbb{K} = \mathbb{C}$), the connected Lie group $G = G(\mathcal{N})$ consists of the matrices with positive diagonal elements (resp. of all invertible matrices) in $\mathcal{T}$.

The Lie algebra $\mathfrak{g} = \mathfrak{g}(\mathcal{N})$ of $G(\mathcal{N})$ is given by the vector space $\mathcal{T} = \mathcal{T}(\mathcal{N})$ equipped with the commutator

$$
[A, A'] = A{\cdot}A' - A'{\cdot}A\ .
$$

Note that the Vinberg groups $G(\mathcal{N})$, $G(\mathcal{N}^{t'})$, associated with a Nil-algebra $\mathcal{N}$ and its dual algebra $\mathcal{N}^{t'}$, respectively, are isomorphic if and only if $\mathcal{N}, \mathcal{N}^{t'}$ are isomorphic Nil-algebras.

2.3. The T-algebra associated with a Nil-algebra. Given a Nil-algebra $\mathcal{N}$ of rank m, we denote by $\mathbb{T}(\mathcal{N})$ the vector space of all $m \times m$ matrices of the form

$$M = \begin{pmatrix} v_{1\,1} & v_{1\,2} & v_{1\,3} & \dots & \dots & v_{1\,m-1} & v_{1\,m} \\ v_{2\,1} & v_{2\,2} & v_{2\,3} & \dots & & v_{2\,m-1} & v_{2\,m} \\ v_{3\,1} & v_{3\,2} & v_{3\,3} & \dots & & v_{2\,m-1} & v_{3\,m} \\ \vdots & \vdots & \ddots & \ddots & \ddots & \vdots & \vdots \\ v_{m-1\,1} & v_{m-1\,2} & v_{m-1\,3} & \dots & \dots & v_{m-1\,m-1} & v_{m-1\,m} \\ v_{m\,1} & v_{m\,2} & v_{m\,3} & \dots & \dots & 0 & v_{m\,m} \end{pmatrix}, \tag{2.9}$$

whose entries v_{ij} are elements of the vector spaces

$$v_{ij} \in \begin{cases} \mathcal{N}_{ij} & \text{if } i < j\ , \\ \mathbb{K} & \text{if } i = j\ , \\ \mathcal{N}^*_{ji} = \mathrm{Hom}(\mathcal{N}_{ij}, \mathbb{K}) & \text{if } i > j\ . \end{cases}$$

The space $\mathbb{T}(\mathcal{N})$ has a canonical structure of *non-associative algebra over* $\mathbb{K}$ which extends the algebra structures of its subspaces $\mathcal{N}$ and $\mathcal{T}(\mathcal{N})$ [**24**]. Such a structure is given by the unique system of bilinear maps $v_{ij} \times v_{jk} \mapsto v_{ik}$ for any i, j, k, which coincide with the maps (2.2), (2.8) if $i \leq j \leq k$ and satisfies the following axioms:

$$\begin{aligned} (v_{ij}\cdot v_{jk})^\flat &= v^\flat_{jk}\cdot v^\flat_{ij}, \\ \langle v^\flat_{ij}\cdot v_{ik}, v_{jk}\rangle &= \langle v_{ik}, v_{ij}\cdot v_{jk}\rangle, \\ \langle v_{ik}\cdot v^\flat_{jk}, v_{ij}\rangle &= \langle v_{ik}, v_{ij}\cdot v_{jk}\rangle, \end{aligned} \tag{2.10}$$

where we use the notation $v^\flat_{ij} := \langle v_{ij}, \cdot\rangle$. The algebra $(\mathbb{T}(\mathcal{N}), \cdot)$ is called the *Vinberg T-algebra* determined by $\mathcal{N}$. It is in general *non-associative* but it contains the *associative* subalgebras $\mathcal{N}$ and $\mathcal{T}(\mathcal{N})$.

By the results in [**24**], a Nil-algebra $\mathcal{N}$ is isomorphic to its dual algebra $\mathcal{N}^{t'}$ if and only if the associated Vinberg T-algebras $\mathbb{T}(\mathcal{N})$ and $\mathbb{T}(\mathcal{N}^{t'})$ are isomorphic.

2.4. Action of a Vinberg group on the space $\mathcal{H}$ of Hermitian matrices. In the following, for any $X = (x_{ij}) \in \mathbb{T}(\mathcal{N})$, we denote by X^* the transposed matrix

$$X^* := t(X^\flat) \in \mathbb{T}(\mathcal{N})$$

where $X^\flat = (x^\flat_{ij})$ is the matrix with co-vector entries $x^\flat_{ij}$, defined by

$$x^\flat_{ij} = \begin{cases} \langle x_{ij}, \cdot\rangle & \text{if } i \neq j \\ x_{ii} & \text{if } i = j \end{cases}.$$

Using this notation, for any upper triangular matrix

$$A = D + N = \begin{pmatrix} a_{11} & a_{12} & a_{13} & \dots & \dots & a_{1m-1} & a_{1m} \\ 0 & a_{22} & a_{23} & \dots & & a_{2m-1} & a_{2m} \\ 0 & 0 & a_{33} & \dots & & a_{2m-1} & a_{3m} \\ \vdots & \vdots & \ddots & \ddots & \ddots & \vdots & \vdots \\ 0 & 0 & 0 & \dots & \dots & a_{m-1m-1} & a_{m-1m} \\ 0 & 0 & 0 & \dots & \dots & 0 & a_{mm} \end{pmatrix} \in \mathcal{T}(\mathcal{N}) \subset \mathbb{T}(\mathcal{N})$$

we may consider the *associated Hermitian matrix*

$$H_A := N^* + D + N = \begin{pmatrix} a_{11} & a_{12} & a_{13} & \dots & \dots & a_{1m-1} & a_{1m} \\ a^\flat_{12} & a_{22} & a_{23} & \dots & & a_{2m-1} & a_{2m} \\ a^\flat_{13} & a^\flat_{23} & a_{33} & \dots & & a_{2m-1} & a_{3m} \\ \vdots & \vdots & \ddots & \ddots & \ddots & \vdots & \vdots \\ a^\flat_{1m-1} & a^\flat_{2m-1} & a^\flat_{3m-1} & \dots & \dots & a_{m-1\,m-1} & a_{m-1\,m} \\ a^\flat_{1m} & a^\flat_{2m} & a^\flat_{3m} & \dots & \dots & a^\flat_{m-1\,m} & a_{m\,m} \end{pmatrix} \in \mathbb{T}(\mathcal{N})\ .$$

By construction, $H_A^* = H_A$. The vector space of such matrices is called *space* $\mathcal{H} = \mathit{Herm}(\mathcal{N})$ *of Hermitian matrices* associated with $\mathcal{N}$. Notice that the map $A \longmapsto H_A$ is a linear isomorphism between $\mathcal{T}(\mathcal{N})$ and $\mathit{Herm}(\mathcal{N})$.

The next theorem collects some of the crucial results by Vinberg [**24**] on homogeneous cones in spaces of Hermitian matrices.

THEOREM 2.2.

(1) *The map*

$$L : \mathfrak{g} \times \mathit{Herm}(\mathcal{N}) \longrightarrow \mathit{Herm}(\mathcal{N}) , \qquad L_A(X) := A{\cdot}X + X{\cdot}A^*$$

is a faithful linear representation of the Lie algebra $\mathfrak{g} = \mathfrak{g}(\mathcal{N}) = \mathcal{T}(\mathcal{N})$ *of the Vinberg group* $G = G(\mathcal{N})$ *on* $\mathit{Herm}(\mathcal{N})$.

(2) *The Lie algebra representation in (1) integrates to a Lie group representation* $\rho : G \to \mathrm{GL}(\mathcal{H})$ *of the Vinberg group* G *on* $\mathit{Herm}(\mathcal{N})$.[1]

(3) *The stabiliser* G_I *of the* G*-action at the identity* I *is a finite group (more precisely, it is* $\{I\}$ *if* $\mathbb{K} = \mathbb{R}$ *and it is* $\{\mathrm{diag}(\pm 1, \ldots, \pm 1)\} \simeq \mathbb{Z}_2^m$ *if* $\mathbb{K} = \mathbb{C}$*) and the orbit* $G(I) \simeq G/G_I$ *is equal to the open cone in the vector space* $\mathcal{H} = \mathit{Herm}(\mathcal{N})$ *given by*

$$\mathcal{V} = \mathcal{V}(\mathcal{N}) := \{A{\cdot}A^* \ , \ A \in G(\mathcal{N})\} \subset \mathcal{H} \ . \tag{2.11}$$

(4) *If* $\mathbb{K} = \mathbb{R}$ *and* $\mathcal{N}$ *is Euclidean, the cone* $\mathcal{V} = \mathcal{V}(\mathcal{N})$ *defined in* (2.11) *is homogeneous and convex. Conversely, any homogeneous convex cone in a vector space has the form* $\mathcal{V} = \mathcal{V}(\mathcal{N}) \subset \mathit{Herm}(\mathcal{N})$ *for some Euclidean* Nil*-algebra* $\mathcal{N}$.

(5) *Two convex cones* $\mathcal{V}_1 = \mathcal{V}(\mathcal{N}_1)$ *and* $\mathcal{V}_2 = \mathcal{V}(\mathcal{N}_2)$*, associated with Euclidean* Nil*-algebras* $\mathcal{N}_1$*,* $\mathcal{N}_2$*, are linearly isomorphic if and only if* $\mathcal{N}_1$*,* $\mathcal{N}_2$ *are isomorphic* Nil*-algebras.*

The cone $\mathcal{V} = \mathcal{V}(\mathcal{N}) \subset \mathcal{H}$ defined in (2.11) is called *Vinberg cone* determined by $\mathcal{N}$.

3. Invariant rational functions of a Vinberg cone

3.1. Matrix and group coordinates of a Vinberg cone. Let $\mathcal{N}$ be a Nil-algebra of rank m, $G = G(\mathcal{N}) \subset \mathcal{T}(\mathcal{N})$ the associated Vinberg group and $\mathcal{V} = G(I) \subset \mathit{Herm}(\mathcal{N})$ the corresponding Vinberg cone. There are two natural multiplicative characters on the group G.

- One of them is the G*-determinant*, that is the character that associates with any $A \in G$ the product of its diagonal entries

$$\det{}_G A = a_{11} \cdot \ldots \cdot a_{mm} \ .$$

- Another is the $\mathcal{H}$*-determinant*, that is the determinant of the representation ρ of G on $\mathcal{H} = \mathit{Herm}(\mathcal{N})$

$$\det{}_{\mathcal{H}} : G \longrightarrow \mathbb{K} \ , \qquad \det{}_{\mathcal{H}}(A) := \det(\rho_A) \ .$$

The map

$$\chi : \mathcal{V} = G(I) \longrightarrow \mathbb{K} \ , \qquad X_A = A{\cdot}A^* \longmapsto \chi(X_A) := \frac{1}{\det_{\mathcal{H}}(A)}$$

[1]Such a G-action is determined by power series and in general cannot be written in the standard form $A(X) = A{\cdot}X{\cdot}A^*$

is called the *Vinberg-Koszul characteristic function* of the cone $\mathcal{V} = \mathcal{V}(\mathcal{N})$. Note that it is the density of a G-invariant measure on $\mathcal{V}$.

Let us denote by G' and G_0 the normal subgroups of G defined by

$$G' = \{ A \in G , a_{ii} = 1\} , \qquad G_0 = \{ A \in G , \det_G A = 1\} .$$

We remark that:

(a) the (unimodular) subgroup G_0 is solvable and its Lie algebra is $\mathfrak{g}_0 = \mathcal{T}(\mathcal{N})$;
(b) G' is the unipotent radical of G and has Lie algebra $\mathfrak{g}' = \mathcal{N}$.

The entries x_{ij}, $i \leq j$, of the elements $X = (x_{ij}) \in \mathcal{T}(\mathcal{N})$ naturally determine coordinates on $\mathcal{H} \simeq \mathcal{T}(\mathcal{N})$ and on the cone $\mathcal{V} \subset \mathcal{H}$, which we call *matrix coordinates*. Besides this, on the cone $\mathcal{V} \subset \mathcal{H}$ there is another natural and very convenient set of coordinates, which we now introduce.

Recall that the matrix group G is an open subset of the vector space $\mathcal{T} = \mathcal{T}(\mathcal{N})$ and that the (vector) entries $a_{ij}, i \leq j$, of the matrices in $\mathcal{T}$ can be used to identify $\mathcal{T}$ with the vector space $\mathbb{K}^{\dim \mathcal{T}}$ and, consequently, $G \subset \mathcal{T}$ with an open subset of $\mathbb{K}^{\dim \mathcal{T}}$. Since the Vinberg cone $\mathcal{V} = G(I)$ is in bijection with G/G_I and being $G_I = \mathbb{Z}_2^m$ when $\mathbb{K} = \mathbb{C}$ and $G_I = \{I\}$ when $\mathbb{K} = \mathbb{R}$, the map

$$\psi_G : G \subset \mathbb{K}^{\dim \mathcal{T}} \longrightarrow \mathcal{V} , \qquad A \in G \subset \mathbb{K}^{\dim \mathcal{T}} \overset{\psi_G}{\longmapsto} X_A = A{\cdot}A^*,$$

is a local diffeomorphism if $\mathbb{K} = \mathbb{C}$ and it is a global diffeomorphism if $\mathbb{K} = \mathbb{R}$. The composition

$$X \longmapsto A = \xi_G(X) \longmapsto (a_{11}, a_{12}, \ldots, a_{ij}, \ldots)$$

of the (local) inverse $\xi_G = \psi_G^{-1}$ with the coordinates $a_{ij} = a_{ij}(A)$ of the matrices $A \in G \subset \mathbb{K}^{\dim \mathcal{T}}$ can be considered as a set of coordinates $a_{ij} = a_{ij}(X)$ on $\mathcal{V}$. We call them the *group coordinates* of the Vinberg cone.

The diagonal elements $a_{ii}(X)$, $1 \leq i \leq m$, constitute a subset of the group coordinates, which we call *diagonal group coordinates.* Note that if $\mathbb{K} = \mathbb{C}$, the group coordinates of $\mathcal{V}$ are *only locally* defined on $\mathcal{V}$ and, in particular, each diagonal group coordinate $a_{ii}(X)$, $1 \leq i \leq m$, is defined only up to a sign.

3.2. G'- and G_0- invariant rational functions of $\mathcal{V}$. A *G'- (resp. G_0-) invariant rational function (resp. polynomial) of* $\mathcal{V} = G(I)$ is a rational function (polynomial) $q : \mathcal{V} \to \mathbb{K}$ of the matrix coordinates, which is invariant under any transformation $X \mapsto \rho_A(X)$, $A \in G'$ (resp. G_0).

Using group coordinates, one can directly check that a rational function $\varphi : \mathcal{V} \to \mathbb{K}$ is G'-invariant if and only if it is a function only of the diagonal group coordinates $a_{ii} = a_{ii}(X)$. From this it follows that a function $\varphi : \mathcal{V} \to \mathbb{K}$ is G_0-invariant if and only if it has the form $\varphi(X) = \psi(\pi(X))$ where π is the (locally defined) $\mathbb{K}$-valued map given by

$$\pi(X) := \prod_{i=1}^{m} a_{ii}(X)$$

and ψ is a rational function of one variable on $\mathbb{K}$. We recall that, in case $\mathbb{K} = \mathbb{R}$, the diagonal group coordinates $a_{ii}(X)$ (and hence also the function $\pi(X)$) are globally defined on $\mathcal{V}$. On the other hand, if $\mathbb{K} = \mathbb{C}$, the maps $a_{ii}(X)$ (and hence $\pi(X)$ too) are locally defined with values determined only up to a sign. This implies that the maps $a_{ii}^2(X)$ and $\pi(X)^2$ are globally defined in all cases.

Since for any $X_A = AA^* \in \mathcal{V}$, the value $\pi^2(X_A)$ is the square of the G-determinant of A, with a small abuse of language, we call $\pi^2 : \mathcal{V} \to \mathbb{K}$ the *squared G-determinant.*

The next lemma is a crucial tool to determine the invariant rational functions and polynomials. It is a direct generalisation of the results of Vinberg in [**24**].

LEMMA 3.1. *There exist m homogeneous polynomials $p_i : \mathcal{V} \to \mathbb{K}$, $1 \le i \le m$, of degree 2^{m-i} in the matrix coordinates of $\mathcal{V} \subset Herm(\mathcal{N})$ which allow to write the (*globally defined*) G'-invariant functions $a^2_{ii}(X)$, $1 \le i \le m$, in terms of the matrix coordinates as*

$$a^2_{ii}(X) = \frac{p_i(X)}{\prod_{s>i} p_s(X)} \tag{3.1}$$

In particular, in matrix coordinates, each function $a^2_{ii}(X)$ is a homogeneous rational function of degree 1. Moreover:

(i) *The* squared G-determinant $\pi^2(X) = \prod_{i=1}^m a^2_{ii}(X)$ *is a G_0-invariant homogeneous rational function of degree m in the matrix coordinates and hence it extends as a rational function over to the whole $\mathcal{H}$ by the formula*

$$\pi^2(X) = \prod_{i=1}^m p_i^{-i+2}(X) = \frac{p_1(X)}{\prod_{\ell \ge 3} p_\ell^{\ell-2}(X)} \ ; \tag{3.2}$$

(ii) *The* Vinberg-Koszul characteristic function $\chi : \mathcal{V} \to \mathbb{K}$ *is given in matrix coordinates by* ([**24**, Ch. III, formula (33)])

$$\chi(X) = \Pi_{i=1}^m p_i(X)^{n_i - n_{i-1} - \cdots - n_1} \ , \qquad \text{where} \qquad n_i := 1 + \frac{1}{2}\sum_{s \ne i} \dim \mathcal{N}_{is} \ . \tag{3.3}$$

Explicit expressions for the polynomials $p_i(X)$ can be found in [**24**, §III.3]. From such expressions, one can see that each $p_i(X)$ is a polynomial of first order in the matrix coordinate x_{ii} and is independent of the matrix coordinates x_{jj}, $j < i$. This implies that the m-tuple of polynomials $\big(p_1(X), p_2(X), \ldots, p_m(X)\big)$ is functionally independent at all points. From formula (3.2), we see that also the m-tuple $\big(\pi^2(X), p_2(X), \ldots, p_m(X)\big)$ is functionally independent at any point where π is defined. From Lemma 3.1 and these remarks we get:

PROPOSITION 3.2. *Let $p_i(X)$ be the polynomials of Lemma 3.1.Then:*

(a) *Any G'-invariant rational function $\varphi(X)$ on $\mathcal{V}$ can be locally presented in the form*

$$\begin{aligned}(3.4)\quad \varphi(X) &= \psi\left(a_{11}(X), \ldots, a_{mm}(X)\right) = \\ &= \psi\left(\sqrt{\frac{p_1(X)}{\prod_{s>1} p_s(X)}}, \ldots, \sqrt{\frac{p_{m-1}(X)}{\prod_{s>m-1} p_s(X)}}, \sqrt{p_m(X)}\right)\end{aligned}$$

for appropriate locally defined square roots and a locally defined real analytic function ψ; in case $\mathbb{K} = \mathbb{R}$, the maps $X \mapsto \sqrt{\frac{p_j(X)}{\prod_{s>j} p_s(X)}}$, $1 \le j \le m$, are globally defined on $\mathcal{V}$ and ψ is globally defined on the image of the map

$$X \longmapsto \left(\sqrt{\frac{p_1(X)}{\prod_{s>1} p_s(X)}}, \ldots, \sqrt{\frac{p_{m-2}(X)}{p_{m-1}(X) p_m(X)}}, \sqrt{\frac{p_{m-1}(X)}{p_m(X)}}, \sqrt{p_m(X)}\right) \ ;$$

In particular, a polynomial $q(X)$ is G'-invariant if and only if there is a rational function $Q(y_1, \ldots, y_m)$ that allows to write $q(X)$ as

$$(3.5)\quad q(X) = Q\left(a_{11}^2(X), \ldots, a_{mm}^2(X)\right) = \\ = Q\left(\frac{p_1(X)}{\prod_{s>1} p_s(X)}, \ldots, \frac{p_{m-2}(X)}{p_{m-1}(X)p_m(X)}, \frac{p_{m-1}(X)}{p_m(X)}, p_m(X)\right).$$

(b) *The field of rational functions $\mathbb{K}(\alpha_1, \ldots, \alpha_m)$, generated by the (*globally defined*) rational function $\alpha_i(X) := a_{ii}^2(X)$, is equal to*

$$\mathbb{K}(\alpha_1, \alpha_2 \ldots, \alpha_m) = \mathbb{K}(p_1, p_2, \ldots, p_m) = \mathbb{K}(\pi^2, p_2, \ldots, p_m)$$

and it is therefore a subfield of the field of the G'-invariant rational functions on $\mathcal{V}$. In particular, a homogeneous polynomial is G'-invariant if and only if it is a polynomial in $\mathbb{K}(\pi^2, p_2, \ldots, p_m)$.

(c) *The field $\mathbb{K}(\pi^2)$, generated by the squared G-determinant $\pi^2(X)$, is the field of the G_0-invariant homogeneous rational functions on $\mathcal{H} = Herm(\mathcal{N})$. In particular, there exists a G_0-invariant homogeneous polynomial $q(X)$ of degree m if and only if $\pi^2(X)$ is a polynomial and $\deg \pi^2(X)$ divides m.*

3.3. Theory of duality. Let $\mathbb{K} = \mathbb{R}$ and $\mathcal{N}$ be a Euclidean Nil-algebra and denote by $G = G(\mathcal{N})$ and $\mathcal{V} = \mathcal{V}(\mathcal{N}) = G(I)$ the corresponding Vinberg group and homogeneous Vinberg cone in $\mathcal{H} = Herm(\mathcal{N})$, respectively. We denote by $G^* = \{A^*, A \in G\} \subset \mathbb{T}(\mathcal{N})$ the group of lower triangular matrices $A^* = t(A^\flat)$, $A \in G$. According to [**4, 24**], the G^*-orbit of the identity is the homogenous convex cone

$$\mathcal{V}' := G^*(I) = \{BB^*\ ,\ B \in G^*\} = \{\ A^*A\ ,\ A \in G\}\ ,$$

which we call *dual cone*. The following proposition shows that the dual cone is metrically equivalent to the metric dual cone $\mathcal{V}^* := g_{\mathcal{H}}(\mathcal{V}')$ and gives the Nil-algebra of which it is the associated Vinberg cone.

PROPOSITION 3.3. *Let $g_{\mathcal{H}} : \mathcal{H} \to \mathcal{H}^*$ denote the linear isomorphism determined by the Euclidean inner product $g_{\mathcal{H}} = \langle \cdot, \cdot \rangle$ of $\mathcal{H}$.*

(i) *The dual cone $\mathcal{V}' = G^*(I)$ is metrically equivalent to $\mathcal{V}^*$ (that is, $g_{\mathcal{H}}(\mathcal{V}') = \mathcal{V}^*$).*
(ii) *The anti-transposition $t' : \mathcal{H} \to \mathcal{H}^{t'}$ maps isomorphically the dual cone $\mathcal{V}' \subset \mathcal{H}$ into the homogeneous cone $\mathcal{V}(\mathcal{N}^{t'}) \subset \mathcal{H}^{t'}$ associated with $\mathcal{N}^{t'}$.*
(iii) *The cone $\mathcal{V}$ is self-adjoint if and only if the* Nil-*algebras $\mathcal{N}$ and $\mathcal{N}^{t'}$ are isomorphic (*[**24**]*).*

4. Examples of Nil-algebras and associated Vinberg cones

4.1. Nil-**algebras of rank** 2 **and associated homogeneous cones.** The matrices of a Nil-algebra $\mathcal{N}$ of rank 2 have only one non-trivial entry, the entry x_{12}, which belongs to the vector space $\mathcal{N}_{12} = W = \mathbb{K}^n$, equipped with a non-degenerate inner product $\langle \cdot, \cdot \rangle$:

$$(4.1)\qquad \mathcal{N} = \mathcal{N}_2(W) := \left\{ \begin{pmatrix} 0 & x_{12} \\ 0 & 0 \end{pmatrix}\ ,\ x_{12} \in W \right\}.$$

The corresponding Vinberg group $G = G(\mathcal{N}_2(W))$ consists of the group

$$(4.2)\qquad G = \left\{ \begin{pmatrix} a_{11} & a_{12} \\ 0 & a_{22} \end{pmatrix} \text{ where } \begin{matrix} a_{11}, a_{22} \neq 0 & \text{if } \mathbb{K} = \mathbb{C} \\ a_{11}, a_{22} > 0 & \text{if } \mathbb{K} = \mathbb{R} \end{matrix} \quad \text{and} \quad a_{12} \in W \right\}$$

and acts on the vector space $\mathcal{H}_2(W) = Herm(\mathcal{N}_2(W))$ of Hermitian matrices

$$\mathcal{H}_2(W) = \left\{ X = \begin{pmatrix} x_{11} & x_{12} \\ x_{12}^\flat & x_{22} \end{pmatrix} \text{ where } x_{ii} \in \mathbb{K},\ x_{12} \in W \right\} . \tag{4.3}$$

This vector space has a natural structure of *Jordan algebra*, given by the Jordan multiplication

$$X \circ Y := \frac{1}{2}\left(X{\cdot}Y + Y{\cdot}X\right) .$$

If W is a Euclidean vector space, $\mathcal{H}_2(W)$ is a Minkowski space and the Jordan algebra $(\mathcal{H}_2(W), \circ)$ is called *spin factor* [**7**].

The action of G on $\mathcal{H}_2(W)$ is given by the standard formula

$$\rho_A(X) = A{\cdot}X{\cdot}A^*$$

and the homogeneous cone $\mathcal{V} = G(I) \subset \mathcal{H}_2(W)$ is

$$\begin{aligned} \mathcal{V} &= \left\{ A{\cdot}A^* = \begin{pmatrix} a_{11}^2 + \langle a_{12}, a_{12}\rangle & a_{22}a_{12} \\ a_{22}a_{12}^* & a_{22}^2 \end{pmatrix} ,\ a_{ii} \in \mathbb{K} ,\ a_{12} \in W \right\} = \\ &= \left\{ X \in W \text{ with } \begin{array}{ll} x_{11}x_{22} - \langle x_{12}, x_{12}\rangle \neq 0 ,\ x_{22} \neq 0 & \text{if } \mathbb{K} = \mathbb{C} \\ x_{11}x_{22} - \langle x_{12}, x_{12}\rangle > 0 ,\ x_{22} > 0 & \text{if } \mathbb{K} = \mathbb{R} \end{array} \right\} . \end{aligned} \tag{4.4}$$

According to the explicit expressions in [**24**], the homogeneous polynomials $p_1(X)$ and $p_2(X)$ of Lemma 3.1 are

$$\begin{aligned} p_2(X) &= a_{22}^2 = x_{22} , \\ p_1(X) &= a_{11}^2 a_{22}^2 = x_{11}x_{22} - |x_{12}|^2 \end{aligned} \tag{4.5}$$

and the squared G-determinant is the quadratic polynomial

$$\pi^2(X) = p_1(X) = x_{11}x_{22} - |x_{12}|^2 . \tag{4.6}$$

If $\mathbb{K} = \mathbb{R}$ and $(W = \mathbb{R}^n, \langle\cdot,\cdot\rangle)$ is Euclidean, then $\mathcal{V} = \mathcal{V}(\mathcal{N}_2(W))$ is the cone of the future directed time-like vectors in the Minkowski space $\mathcal{H}_2(W) \simeq \mathbb{R}^{1,n+1}$ with Lorentz metric given by

$$\langle X, X\rangle = -\pi^2(X) = -x_{11}x_{22} + |x_{12}|^2 .$$

In [**8**], Baez and Huerta gave a nice quantum mechanical interpretation of the homogeneous cone $\mathcal{V} = \mathcal{V}(\mathcal{N}_2(W))$ in case the vector space W is one of the division algebras $\mathbb{R}, \mathbb{C}, \mathbb{H}$ or $\mathbb{O}$.

4.2. Self-adjoint homogeneous cones. Let $\mathbb{D} = \mathbb{R}, \mathbb{C}, \mathbb{H}$ or $\mathbb{O}$ be a division algebra. In the following, if $\mathbb{D} \neq \mathbb{O}$, we denote by $\mathcal{N}_m(\mathbb{D})$ the Nil-algebra of rank m of the upper triangular $m \times m$ matrices with entries x_{ij}, $i < j$, in the algebra $\mathbb{D}$. If $\mathbb{D} = \mathbb{O}$, the Nil-algebra $\mathcal{N}_m(\mathbb{O})$ is defined only for $m = 3$. One can check that:

- the space of Hermitian matrices $\mathcal{H}_m(\mathbb{D}) := Herm(\mathcal{N}_m(\mathbb{D}))$ has a natural structure of Euclidean Jordan algebra given by the Jordan multiplication

$$X \circ Y := \frac{1}{2}\left(X{\cdot}Y + Y{\cdot}X\right)$$

- the associated homogeneous cone $\mathcal{V}_m(\mathbb{D}) = \mathcal{V}(\mathcal{N}_m(\mathbb{D}))$ is the self-adjoint cone of positive defined Hermitian $m \times m$ matrices over $\mathbb{D}$.

M. Köcher and E.B. Vinberg, independently, proved the so-called *Köcher-Vinberg Theorem* (see [**20, 23**]) which establishes a bijection between the self-adjoint homogeneous convex cones and the Euclidean Jordan algebras. This result together with the classification of Euclidean Jordan algebras by Jordan, von Neumann and Wigner [**19**] gives the following:

THEOREM 4.1. *There is a one-to-one correspondence between the self-adjoint homogeneous convex cones and the Euclidean Jordan algebras, and any Euclidean Jordan algebra* $(J, \circ)$ *is a direct sum of Jordan algebras of the form* $(\mathcal{H}_2(W), \circ)$ *or* $(\mathcal{H}_m(\mathbb{D}), \circ)$.

In particular, any indecomposable self-adjoint homogeneous convex cone $\mathcal{V}$ *is a symmetric space and it is either a cone* $\mathcal{V} = \mathcal{V}(W) \subset \mathcal{H}_2(W)$ *associated with a* Nil-*algebra of the form* $\mathcal{N}_2(W)$ *or a cone* $\mathcal{V} = \mathcal{V}_m(\mathbb{D}) \subset \mathcal{H}_m(\mathbb{D})$ *associated with a* Nil-*algebra of the form* $\mathcal{N}_m(\mathbb{D})$.

4.3. Special Vinberg cones of rank 3. The matrices of a Nil-algebra $\mathcal{N}$ of rank 3 have only *three* non-trivial entries, namely $x_{12} \in \mathcal{N}_{12}$, $x_{13} \in \mathcal{N}_{13}$ and $x_{23} \in \mathcal{N}_{23}$. The matrix multiplication between the matrices in $\mathcal{N}$ is determined by the isometric map

$$\bullet : \mathcal{N}_{12} \times \mathcal{N}_{23} \longrightarrow \mathcal{N}_{13} , \qquad (x_{12}, x_{23}) \longmapsto x_{12}\bullet x_{23} .$$

To simplify the notation, we denote the spaces $\mathcal{N}_{ij}$, $i < j$, by

$$V = \mathcal{N}_{23} , \qquad U = \mathcal{N}_{12} , \qquad W = \mathcal{N}_{13} .$$

and the non-degenerate (= pseudo-Euclidean if $\mathbb{K} = \mathbb{R}$) inner products of V, U and W by $g_V = \langle \cdot, \cdot \rangle_{\mathcal{N}_{12}}$, $g_U = \langle \cdot, \cdot \rangle_{\mathcal{N}_{23}}$ and $g_W = \langle \cdot, \cdot \rangle_{\mathcal{N}_{13}}$, respectively. We remark that:

- If $\dim U = \dim W$, then the isometric map $\bullet : V \times U \to W$ (i.e. bilinear map with

$$g_W(u\bullet v, u\bullet v) = g_V(v, v) \cdot g_U(u, u) \tag{4.7}$$

defines a structure of $\mathbb{Z}_2$ -graded Clifford module with respect to $\mathcal{C}\ell(V, g_V)$ on the direct sum

$$S = S_0 + S_1 := U + W ;$$

- Conversely, if $S = S_0 + S_1$ is a $\mathbb{Z}_2$ -graded metric Clifford module over $\mathcal{C}\ell(V, g)$ and g_{S_0} and g_{S_1} are non-degenerate inner products on the spaces S_0, S_1 such that (4.7) holds, then the vector spaces V, $U := S_0$ and $W := S_1$ together with their inner products and the Clifford multiplication $\mu : V \times S_0 \to S_1$ determine a Nil-algebra of rank 3 with product $u\bullet v \stackrel{\text{def}}{=} \mu(v, u)$.

Due to this, the description of all Nil-algebras of rank 3 with a prescribed metric vector space $(V, g_V) = (\mathcal{N}_{23}, \langle \cdot, \cdot \rangle_{\mathcal{N}_{23}})$ and with $\dim \mathcal{N}_{12} = \dim \mathcal{N}_{13}$ reduces to:

(a) the classification of $\mathbb{Z}_2$-graded Clifford $\mathcal{C}\ell(V, g_V)$-modules
(b) the classification of admissible metrics in S such that the Clifford multiplication is skew-symmetric.

The classifications (a) and (b) were obtained by Atiyah, Bott and Shapiro in [**6**] and by the first author and Cortés in [**2**], respectively.

These observations lead to the classification of all homogeneous Vinberg cones $\mathcal{V}(\mathcal{N}) \subset \mathit{Herm}(\mathcal{N})$ associated with Nil-algebras $\mathcal{N}$ of rank 3 with

$$\dim \mathcal{N}_{12} = \dim \mathcal{N}_{13} .$$

In case $\mathbb{K} = \mathbb{R}$, such Nil-algebras and their associated cones are called *special Nil-algebras* and *special Vinberg cones (of rank 3)*, respectively ([**3**]; see also [**4**]).[2]

In detail, given a metric vector space (V, g) and an associates metric $\mathbb{Z}_2$-graded Clifford $C\ell(V, g)$-module $(S = S_0 + S_1, g_S)$ with skew-symmetric Clifford multiplication, we uniquely associate the rank 3 associative Nil-algebra $\mathcal{N} = \mathcal{N}_3(V+S, g_V+g_S)$ given by the upper triangular matrices

$$N = \begin{pmatrix} 0 & s_0 & s_1 \\ 0 & 0 & v \\ 0 & 0 & 0 \end{pmatrix} , \qquad v \in V,\ s_0 \in S_0,\ s_1 \in S_1 , \tag{4.8}$$

with matrix multiplication given by $v \cdot s_0 := \mu(v, s_0)$, where μ stands for the Clifford multiplication. The corresponding Vinberg group $G = G(V + S, g_V + g_S)$ consists of the group

$$G = \left\{ \begin{pmatrix} \alpha_1 & t_0 & t_1 \\ 0 & \alpha_2 & w \\ 0 & 0 & \alpha_3 \end{pmatrix} \text{ where } \begin{matrix} \alpha_i \neq 0 & \text{if } \mathbb{K} = \mathbb{C} \\ \alpha_i > 0 & \text{if } \mathbb{K} = \mathbb{R} \end{matrix} \quad \text{and} \quad w \in V, t_i \in S_i \right\} \tag{4.9}$$

and acts on the vector space

$$\mathcal{H}_3(V+S) = Herm(\mathcal{N}_3(V+S, g_V+g_S))$$

of the Hermitian matrices

$$\mathcal{H}_3(V + S) = \left\{ X = \begin{pmatrix} x_1 & s_0 & s_1 \\ s_0^\flat & x_2 & v \\ s_1^\flat & v^\flat & x_3 \end{pmatrix} \text{ with } x_i \in \mathbb{K},\ v \in V, s_i \in S_i \right\} . \tag{4.10}$$

According to the explicit expressions in [**3, 24**], the homogeneous polynomials $p_i(X)$ of Lemma 3.1 are given by

$$\begin{aligned} p_3(X) &= a_{33}^2 = x_3 , \\ p_2(X) &= a_{33}^2 a_{22}^2 = x_3 x_2 - |v|^2 , \\ p_1(X) &= a_{33}^2 a_{22}^2 a_{11}^4 = x_3 d(X) , \end{aligned}$$

where $d(X) = a_{33}^2 a_{22}^2 a_{11}^2 = \pi^2(X)$ is the cubic polynomial

$$d(X) = x_1 x_2 x_3 - x_1 |s_0|^2 - x_2 |s_1|^2 - x_3 |v|^2 + 2\langle \mu_v(s_0), s_1 \rangle . \tag{4.11}$$

The homogeneous Vinberg cone $\mathcal{V} \subset \mathcal{H}_3(V+S)$ is given by Hermitian matrices, whose matrix coordinates satisfy the inequalities:

– for $\mathbb{K} = \mathbb{R}$

$$\begin{aligned} & p_3(X) = x_3 > 0 , \qquad p_2(X) = x_3 x_2 - |v|^2 > 0 , \\ & d(X) = x_1 x_2 x_3 - x_1 |s_0|^2 - x_2 |s_1|^2 - x_3 |v|^2 + 2\langle \mu_v(s_0), s_1 \rangle > 0 ; \end{aligned} \tag{4.12}$$

– for $\mathbb{K} = \mathbb{C}$

$$\begin{aligned} & x_3 \neq 0 , \qquad x_3 x_2 - |v|^2 \neq 0 , \\ & x_1 x_2 x_3 - x_1 |s_0|^2 - x_2 |s_1|^2 - x_3 |v|^2 + 2\langle \mu_v(s_0), s_1 \rangle \neq 0 . \end{aligned} \tag{4.13}$$

[2] In our definition of a special Vinberg cone $\mathcal{V}(\mathcal{N})$, the Nil-algebra $\mathcal{N}$ is different from the one in [**3, 4**]. We consider matrices $N = (N_{ij})$ in which N_{12} is in S_0 (an even spinor) and N_{23} is in V (a vector). In [**3, 4**], the opposite assumption is taken: $N_{12} \in V$ and $N_{23} \in S_0$. Only our choice implies that the squared G-determinant $d(X) = \pi^2(X)$ is a cubic polynomial, while in the opposite case it is a rational function. Due to this, some of the results of [**3, 4**] should be amended. A Corrigendum on this regard for [**3**] and [**4**] will soon be published.

We recall that if g_V and $g_S = g_{S_0} + g_{S_1}$ are Euclidean, then the Nil-algebra $\mathcal{N} = \mathcal{N}_3(V+S, g_V+g_S)$ is Euclidean and the homogeneous Vinberg cone $\mathcal{V} \subset \mathcal{H}_3(V+S)$ is convex.

4.4. The special *dual* Vinberg cones of rank 3. By the results in §3.3, the *special* dual Nil-*algebras*, associated with the special Nil-algebras over $\mathbb{K} = \mathbb{R}$, determine the dual cones of the special Vinberg cones. In detail, the dual special Nil-algebra $\mathcal{N}^{t'}$, associated with the Nil-algebra determined by (V, g_V) and $(S = S_0 + S_1, g_S)$, is the matrix algebra of upper triangular matrices

$$N^{t'} = \begin{pmatrix} 0 & v & s_1 \\ 0 & 0 & s_0 \\ 0 & 0 & 0 \end{pmatrix} , \qquad v \in V,\ s_0 \in S_0,\ s_1 \in S_1 \ , \tag{4.14}$$

with matrix multiplication given by $v \cdot' s_0 \stackrel{def}{=} \mu(v, s_0)$. The corresponding vector space of Hermitian matrices is

$$\mathcal{H}_3^{t'}(V+S) = \left\{ X = \begin{pmatrix} y_1 & w & t_1 \\ w^\flat & y_2 & t_0 \\ t_1^\flat & t_0^\flat & y_3 \end{pmatrix} \text{ with } y_i \in \mathbb{K},\ w \in V, t_i \in S_i \right\} \tag{4.15}$$

and the associated homogeneous Vinberg cone $\mathcal{V}^{t'} \subset \mathcal{H}_3^{t'}(V+S)$ is the open set characterised by the inequalities:

$$\begin{aligned} & y_3 > 0 \ , \qquad y_3 y_2 - |t_0|^2 > 0 \ , \\ & d^{t'}(Y) = y_1 y_2 y_3 - y_3 |t_0|^2 - y_2 |t_1|^2 - y_1 |w|^2 + 2\langle \mu(w, t_0), t_1 \rangle \\ & \qquad + \frac{|t_0|^2 |t_1|^2 - \langle t_1 \cdot t_0^\flat, t_1 \cdot t_0^\flat \rangle}{y_1} > 0. \end{aligned} \tag{4.16}$$

Moreover, the homogeneous polynomials of Lemma 3.1 are in this case

$$\begin{aligned} p_3^{t'}(X) &= a_{33}^2 = y_3 \ , \\ p_2^{t'}(X) &= a_{33}^2 a_{22}^2 = y_3 y_2 - |t_0|^2 \ , \\ p_1^{t'}(X) &= a_{33}^2 a_{22}^2 a_{11}^4 = y_3 d^{t'}(X) \ , \end{aligned}$$

where $d^{t'}(X) = \pi^2(X)$ is the squared G-determinant of $\mathcal{V}^{t'}$ given in (4.16).

Under the anti-transposition map $t' : \mathcal{H}^{t'} \to \mathcal{H}$, the Vinberg cone $\mathcal{V}^{t'} \subset \mathcal{H}_3^{t'}(V+S)$ is mapped onto the dual cone $\mathcal{V}'(\simeq \mathcal{V}^*) \subset \mathcal{H}_3(V+S)$ of $\mathcal{V}$, characterised by the inequalities

$$\begin{aligned} & x_1 > 0 \ , \qquad x_1 x_2 - |s_0|^2 > 0 \ , \\ & d'(Y) := x_1 x_2 x_3 - x_1 |s_0|^2 - x_2 |s_1|^2 - x_3 |v|^2 + 2\langle \mu(v, s_0), s_1 \rangle \\ & \qquad + \frac{|s_0|^2 |s_1|^2 - \langle s_0 \cdot s_1^\flat, s_0 \cdot s_1^\flat \rangle}{x_1} > 0. \end{aligned} \tag{4.17}$$

Note that the homogeneous rational function $d'(X)$ of degree 3 (associated with the dual cone $\mathcal{V}' = G^*(I) = (\simeq \mathcal{V}^*) \subset \mathcal{H}_3(V+S)$) differs from the cubic polynomial $d'(X)$ of the Vinberg cone $\mathcal{V} = G(I)$ by the term $\frac{|s_0|^2|s_1|^2 - \langle s_0 \cdot s_1^\flat, s_0 \cdot s_1^\flat \rangle}{x_1}$. This shows that for *any* special Vinberg cone, the corresponding cubic polynomial d and the homogeneous rational funcion d' of degree 3 are invariant under the actions on $\mathcal{H}$ of the groups G_0 and its dual group G_0^*, respectively. This is a crucial property which, in [**4**], allowed to compute the Bertotti-Robinson mass – and hence also the entropy – of the BPS static, spherically symmetric and asymptotically flat BPS extremal black holes in supergravity theory with homogeneous non-symmetric scalar manifolds.

From the above remarks and from Proposition 3.3, Theorem 4.1 and Hurwitz' classification of division algebras we get

COROLLARY 4.2.

(1) *For any special Vinberg cone* $\mathcal{V}$ *(not necessarily self-adjoint), the cubic polynomial* $d(X)$ *and the homogeneous rational function* $d'(X)$ *of degree* 3 *are* G_0*- and* G_0^**-invariant, respectively.*
(2) *A special Vinberg cone* $\mathcal{V} \subset \mathcal{H}_3(V+S)$ *is self-adjoint if and only if the spaces* V*,* S_0 *and* S_1 *are isomorphic Euclidean vector spaces. In this case* $d = d'$*.*

5. Construction of special real manifolds with large isometry groups

5.1. Admissible cubics. Let $\mathcal{V} = G(I)$ be a homogeneous Vinberg cone in the space of Hermitian matrices $\mathcal{H} = Herm(\mathcal{N})$ determined by a *real* Nil-algebra $\mathcal{N}$ of rank m. Given a homogeneous cubic polynomial $q : \mathcal{H} \to \mathbb{R}$, we use the following notation:

– $\mathcal{V}_q$ is the hypersurface in $\mathcal{V}$ contained in the level set $\{q = 1\}$:

$$\mathcal{V}_q = \{X \in \mathcal{V} \ : \ q(X) = 1\} \ ;$$

– $\mathrm{Aut}(q)$ is the group of linear transformations of $\mathcal{H}$ which preserve $\mathcal{V}$ and q:

$$\mathrm{Aut}(q) = \{ \ L \in \mathrm{GL}(\mathcal{H}) \ : \ L(\mathcal{V}) = \mathcal{V} \ , \ q(L(X)) = q(X) \text{ for any } X \in \mathcal{H} \ \} \ ;$$

– $g = g(q)$ is the Hessian form given by q, i.e. the field of symmetric bilinear forms on the hypersurface $\mathcal{V}_q$ defined by

$$g_X(V, W) := -\operatorname{Hess}(\log(q))|_X(V, W) \ , \qquad X \in \mathcal{V}_q \ , \ V, W \in T_X\mathcal{V}_q \ .$$

The submanifold $\mathcal{V}_q$ is called the *level hypersurface* of q and $g = g(q)$ is called the *Hessian form* (or *degenerate metric*) of $\mathcal{V}_q$.

DEFINITION 5.1. A homogenous cubic polynomial $q : \mathcal{H} \to \mathbb{R}$ is called *admissible* (resp. *locally admissible*) if the Hessian form $g = g(q)$ is a Riemannian metric on $\mathcal{V}_q$ (resp. on an open subset $\mathcal{U}$ of $\mathcal{V}_q$). In this case $(\mathcal{V}_q, g)$ is called *special real manifold* [**17**].

Remark. One can easily see that the action of $\mathrm{Aut}(q)$ preserves the degenerate metric $g(q)$ on $\mathcal{V}_q$.

LEMMA 5.1. *Let* q *be a homogeneous cubic polynomial* $q : \mathcal{H} \to \mathbb{R}$ *and* $H \subset \mathrm{Aut}(q)$ *a subgroup of the automorphism group of* q*. Then* q *is (locally) admissible if and only if the associated Hessian form* $g = g(q)$ *is positive definite at all points of (an open subset of) a submanifold, which is transversal to the* H*-orbits in* $\mathcal{V}_q$*. In particular:*

(1) *If* $G_0 \subset \mathrm{Aut}(q)$*, then* G_0 *acts transitively on* $\mathcal{V}_q$ *and* q *is admissible if and only if* $g = g(q)$ *is positive definite at the identity matrix* I*.*
(2) *If* $G' \subset \mathrm{Aut}(q)$*, then* G' *acts on* $\mathcal{V}_q$ *of cohomogeneity* $m-1$ *and* q *is (locally) admissible if and only if* g_X *is positive definite at all points* X *of (an open subset of) the* $(m-1)$*-dimensional submanifold of* $\mathcal{V}_q$

$$\mathcal{D} = \{ \ X \in \mathcal{H} \ : X = \mathrm{diag}(\lambda_1, \ldots, \lambda_m) \ \} \cap \mathcal{V}_q \ .$$

5.2. Admissible cubics on Vinberg cones of rank 2. Consider the real Nil-algebra of rank 2

$$\mathcal{N} = \mathcal{N}_2(W) := \left\{ \begin{pmatrix} 0 & w \\ 0 & 0 \end{pmatrix} \ , \quad w \in W \right\} \tag{5.1}$$

and let $\mathcal{V} = G(I)$ be the homogeneous Vinberg cone in the space of Hermitian matrices

$$\mathcal{H}_2(W) = \left\{ X = \begin{pmatrix} x_1 & w \\ w^\flat & x_2 \end{pmatrix} \text{ where } x_i \in \mathbb{R},\ w \in W \right\}$$

associated with $\mathcal{N}$.

THEOREM 5.2.

(i) *There is no G_0-invariant cubic polynomial on $\mathcal{V}$ and, in particular, no admissible cubic $q : \mathcal{V} \to \mathbb{R}$ which determines a G_0-homogeneous special real manifold.*

(ii) *Up to a positive scaling constant, the G'-invariant admissible cubics $q : \mathcal{V} \to \mathbb{R}$ are those having the form*

$$q(X) = (x_1x_2 - |w|^2)x_2 + \varepsilon x_2^3 \ , \qquad \varepsilon \in \mathbb{R} \ . \tag{5.2}$$

PROOF. (i) Since $\pi^2(X)$ is the irreducible polynomial of degree 2

$$\pi^2(X) = p_1(X) = x_1x_2 - |w|^2 \ , \tag{5.3}$$

the claim follows immediately from Proposition 3.2 (c).

(ii) By Proposition 3.2 (a), any G'-invariant polynomial $q(X)$ on $\mathcal{V}$ has the form

$$q(X) = Q\left(x_2, \frac{x_1x_2 - |w|^2}{x_2} \right) \tag{5.4}$$

for some rational function $Q(y_1, y_2)$. A function as in (5.4) is a homogeneous polynomial of degree 3 only if the non-trivial monomials of the Taylor expansion of Q at 0 are only the terms of degree 3, and thus only if Q itself is a homogeneous polynomial of degree 3. This implies that $q(X)$ is a G'-invariant homogeneous cubic if and only if it has the form

$$q(X) = ax_2^3 + bx_2\left(x_1x_2 - |w|^2\right) = x_2\left(ax_2^2 + b\left(x_1x_2 - |w|^2\right)\right) \tag{5.5}$$

for some real coefficients a, b. Since for $b = 0$ this cubic is associated with a symmetric bilinear form $g = g(q)$ on the level hypersurface $\mathcal{V}_q = \{x_2^3 = 1\}$ which is degenerate, the cubic (5.5) is admissible only if $b \neq 0$. Hence, up to a scaling, $q(X)$ has the form (5.2). By Lemma 5.1, a cubic $q(X)$ is admissible if and only if, for any diagonal matrix $X = \mathrm{diag}(x_1, x_2) \in \mathcal{V}_q$, the Hessian form $g(q) = -\mathrm{Hess}(\log(q))|_{T\mathcal{V}_q}$ is well defined and positive definite at X. A direct computation shows that for any $X = \mathrm{diag}(x_1, x_2) \in \mathcal{V}_q$, the Hessian $n \times n$ matrix $H_X = -\mathrm{Hess}(\log(q))_X$, $n = \dim \mathcal{H}$, has the form

$$-\mathrm{Hess}(\log(q))_X = \begin{pmatrix} \frac{1}{(x_1+\varepsilon x_2)^2} & \frac{\varepsilon}{(x_1+\varepsilon x_2)^2} & 0 \\ \frac{\varepsilon}{(x_1+\varepsilon x_2)^2} & \frac{1}{(x_1+\varepsilon x_2)^2}\left(2\left(\frac{x_1}{x_2}\right)^2 + 4\varepsilon\frac{x_1}{x_2} + 3\varepsilon^2\right) & \\ & 0 & \frac{2I_{n-2}}{x_1x_2+\varepsilon x_2^2} \end{pmatrix} . \tag{5.6}$$

Therefore claim (ii) follows by observing that for any $\varepsilon \in \mathbb{R}$

(a) $-\mathrm{Hess}(\log(q))_X$ (and its restriction to $g(q)$) is well defined at all points of

$$\{\mathrm{diag}(x_1, x_2)\} \cap \mathcal{V}_q = \{\mathrm{diag}(x_1, x_2) \ , \ x_2 > 0 \ , \ x_2(x_1 + \varepsilon x_2) = 1\}$$

(b) $-\operatorname{Hess}(\log(q))_X$ is positive definite at all points of $\{\operatorname{diag}(x_1, x_2)\} \cap \mathcal{V}_q$.

Indeed, by Sylvester criterion, (b) can be checked by just computing the principal second order minor

$$\frac{1}{(x_1+\varepsilon x_2)^4} \det \begin{pmatrix} 1 & \varepsilon \\ \varepsilon & \left(2\left(\frac{x_1}{x_2}\right)^2 + 4\varepsilon\frac{x_1}{x_2} + 3\varepsilon^2\right) \end{pmatrix} =$$

$$= \frac{2}{(x_1+\varepsilon x_2)^4}\left(\varepsilon + \frac{x_1}{x_2}\right)^2 = \frac{2}{x_2^2(x_1+\varepsilon x_2)^2} ,$$

which is manifestly positive on $\{\operatorname{diag}(x_1, x_2), x_1, x_2 > 0, x_2(x_1 + \varepsilon x_2) = 1\}$. □

5.3. Admissible cubics on special Vinberg cones of rank 3. Let $\mathcal{V} = G(I)$ be a special Vinberg cone in the space of Hermitian matrices

$$\mathcal{H}_3(V+S) = \left\{ X = \begin{pmatrix} x_1 & s_0 & s_1 \\ s_0^\flat & x_2 & v \\ s_1^\flat & v^\flat & x_3 \end{pmatrix} \text{ where } x_i \in \mathbb{R},\ v \in V, s_i \in S_i \right\} . \tag{5.7}$$

determined a special Nil-algebra $\mathcal{N} = \mathcal{N}_3(V + S, g_V + g_S)$.

THEOREM 5.3.

(i) *Up to a scaling, there is only one G_0-invariant cubic polynomial on $\mathcal{V}$, namely the squared G-determinant* (4.11). *This cubic polynomial is admissible if and only if the* Nil-*algebra* $\mathcal{N}_3(V + S, g_V + g_S)$ *is Euclidean.*

(ii) *The G'-invariant polynomial algebra is the polynomial algebra $\mathbb{K}[d, p_2, p_3]$. In particular any G'-invariant cubic is a linear combination of the cubics*

$$d(X) , \qquad p_2(X)p_3(X) = x_3\left(x_2x_3 - |v|^2\right) , \qquad p_3^3(X) = x_3^3 \tag{5.8}$$

and any linear combinations of these cubics is an admissible cubic only if, up to a positive scaling, it has the form

$$q(X) = d(X) + \varepsilon_1 x_3\left(x_2x_3 - |v|^2\right) + \varepsilon_2 x_3^3 \quad \textit{with } \varepsilon_1 \in \mathbb{R} \textit{ and } \varepsilon_2 \le 0 . \tag{5.9}$$

Moreover:

- *For any pair $(\varepsilon_1, \varepsilon_2 = 0)$ the corresponding cubic is admissible;*
- *For the pairs $(\varepsilon_1, \varepsilon_2)$ with $\varepsilon_2 < 0$, $\varepsilon_1 = \alpha|\varepsilon_2|$, with $\alpha > 0$, the corresponding cubics are locally admissible.*

PROOF. (i) We recall that the squared G-determinant $\pi^2(X)$ is the irreducible cubic polynomial (4.11). By Proposition 3.2 (c), up to a scaling, $d(X) = \pi^2(X)$ is the unique G_0-invariant homogeneous polynomial of degree 3. The second claim can be checked by explicitly determine the expression of the symmetric bilinear form $g|_I = -\operatorname{Hess}(\log(d))|_I$ in matrix coordinates.

(ii) The same arguments for the claim (ii) in Theorem 5.2 show that the G'-invariant cubic polynomials $q(X)$ are exactly those which are linear combinations of the three cubics (5.8), namely

$$q(X) = a\,d(X) + b\,p_2(X)p_3(X) + c\,p_3^3(X) , \qquad a, b, c \in \mathbb{R} . \tag{5.10}$$

One can directly check that for $a = 0$, the corresponding cubic is associated with a degenerate metric $g = g(q)$. Hence a cubic $q(X)$ is admissible only if, up to a scaling, it has the form

$$q(X) = d(X) + \varepsilon_1 x_3\left(x_2x_3 - |v|^2\right) + \varepsilon_2 x_3^3 , \ \varepsilon_1, \varepsilon_2 \in \mathbb{R} . \tag{5.11}$$

By Lemma 5.1, this cubic is admissible if and only if, for any matrix X in the set

$$\begin{aligned}(5.12)\quad \{\mathrm{diag}(x_1,x_2,x_3)\}\cap \mathcal{V}_q &= \\ &= \{\mathrm{diag}(x_1,x_2,x_3)\ ,\ x_i>0\ ,\ x_3\left(x_1x_2+\varepsilon_1x_2x_3+\varepsilon_2x_3^2\right)=1\}\end{aligned}$$

the symmetric form $H_X := -\,\mathrm{Hess}(\log(q))_X$ is well defined and its restriction to $T_X\mathcal{V}$ is positive definite. A direct computation shows that for any X in (5.12), $H_X = -\,\mathrm{Hess}(\log(q))_X$ reads

$$(5.13)\quad -\,\mathrm{Hess}(\log(q))_X = q^{-2}(X)\cdot$$
$$\cdot\left(\begin{array}{ccc|ccc} x_2^2x_3^2 & -\varepsilon_2x_3^4 & \mathcal{A} & & & \\ -\varepsilon_2x_3^4 & \mathcal{B} & \mathcal{C} & & 0 & \\ \mathcal{A} & \mathcal{C} & \mathcal{D} & & & \\ \hline & & & 2q(X)x^3I_{\dim V} & 0 & 0 \\ & 0 & & 0 & 2q(X)(x_1+\varepsilon_1x_3)I_{\dim S/2} & 0 \\ & & & 0 & 0 & 2q(X)x_2I_{\dim S/2}\end{array}\right)$$

where

$$\begin{aligned}&\mathcal{A} := x_2x_3^2\left(\varepsilon_1x_2+2\varepsilon_2x_3\right)\ , \qquad \mathcal{B} := x_3^2\left(x_1+\varepsilon_1x_3\right)^2\ ,\\ (5.14)\quad &\mathcal{C} := \varepsilon_2x_3^3\left(2x_1+\varepsilon_1x_3\right)\ ,\\ &\mathcal{D} := x_2^2\left(x_1+\varepsilon_1x_3\right)^2+x_3^2\left(\varepsilon_1x_2+\varepsilon_2x_3\right)^2+2\varepsilon_2x_3^3\left(\varepsilon_1x_2+\varepsilon_2x_3\right)\end{aligned}$$

From this expression it is manifest that $g(q)_X = H_X|_{T_X\mathcal{V}_q}$ is positive definite only if the following inequality holds:

$$x_1+\varepsilon_1x_3>0\ .$$

Since we are assuming that X is in (5.12), this condition is equivalent to

$$(5.15)\qquad 1-\varepsilon_2x_3^3>0\ .$$

Now, for any fixed value of $\varepsilon_2>0$, there is at least one point X_o in (5.12) with sufficiently large coordinate x_3 so that (5.15) is not satisfied. We may therefore conclude that if $\varepsilon_2>0$ the corresponding cubic is not admissible. Moreover, in the case $\varepsilon_2=0$ the Hessian $H_X = -\,\mathrm{Hess}(\log(q))_X$ has the form

$$(5.16)\qquad H_X = \left(\begin{array}{ccc|c} x_2^2x_3^2 & 0 & \varepsilon_1x_2^2x_3^2 & \\ 0 & x_3^2(x_1+\varepsilon_1x_3)^2 & 0 & 0 \\ \varepsilon_1x_2^2x_3^2 & 0 & x_2^2(x_1+\varepsilon_1x_3)^2+\varepsilon_1^2x_2^2x_3^2 & \\ \hline & 0 & & *\end{array}\right).$$

A direct computation based on Sylvester criterion shows that it is positive definite at each point of (5.12) for any choice of ε_1.

Finally, in the cases when $\varepsilon_2<0$ and $\varepsilon_1=\alpha|\varepsilon_2|$, with $\alpha>0$, a tedious but straightforward computation shows the existence of a point X_o in the set (5.12) at which $H|_{X_o}$ is positive definite. This implies that there exists a neighbourhood of X_o on which H is positive definite, meaning that the corresponding cubic is locally admissible. □

Acknowledgments

The authors are grateful to the referee for her/his useful comments and for proposing the interesting papers [**10, 22**].

References

[1] D. V. Alekseevskiĭ, *Classification of quaternionic spaces with transitive solvable group of motions* (Russian), Izv. Akad. Nauk SSSR Ser. Mat. **39** (1975), no. 2, 315–362, 472. MR0402649

[2] D. V. Alekseevsky and V. Cortés, *Classification of N-(super)-extended Poincaré algebras and bilinear invariants of the spinor representation of* $\mathrm{Spin}(p,q)$, Comm. Math. Phys. **183** (1997), no. 3, 477–510, DOI 10.1007/s002200050039. MR1462223

[3] D. V. Alekseevsky and V. Cortés, *Special Vinberg cones*, Transform. Groups **26** (2021), no. 2, 377–402, DOI 10.1007/s00031-021-09649-w. MR4283321

[4] Dmitri V. Alekseevsky, Alessio Marrani, and Andrea Spiro, *Special Vinberg cones and the entropy of BPS extremal black holes*, J. High Energy Phys. **11** (2021), Paper No. 100, 34, DOI 10.1007/jhep11(2021)100. MR4368757

[5] Shun-ichi Amari, *Information geometry and its applications*, Applied Mathematical Sciences, vol. 194, Springer, [Tokyo], 2016, DOI 10.1007/978-4-431-55978-8. MR3495836

[6] M. F. Atiyah, R. Bott, and A. Shapiro, *Clifford modules*, Topology **3** (1964), no. suppl, suppl. 1, 3–38, DOI 10.1016/0040-9383(64)90003-5. MR167985

[7] John C. Baez, *Division algebras and quantum theory*, Found. Phys. **42** (2012), no. 7, 819–855, DOI 10.1007/s10701-011-9566-z. MR2928833

[8] John C. Baez and John Huerta, *Division algebras and supersymmetry II*, Adv. Theor. Math. Phys. **15** (2011), no. 5, 1373–1410. MR2989835

[9] Jonathan Bagger and Edward Witten, *Matter couplings in $N = 2$ supergravity*, Nuclear Phys. B **222** (1983), no. 1, 1–10, DOI 10.1016/0550-3213(83)90605-3. MR708440

[10] Frédéric Barbaresco, *Higher Order Geometric Theory of Information and Heat Based on Poly-Symplectic Geometry of Souriau Lie Groups Thermodynamics and Their Contextures: The Bedrock for Lie Group Machine Learning*, Entropy **20** (2018), 840.

[11] C. Böhm and R. A. Lafuente, *Non-compact Einstein manifolds with symmetry*, `arXiv:2107.04210` [math.DG].

[12] S. Cecotti, *Homogeneous Kähler manifolds and T-algebras in $N = 2$ supergravity and superstrings*, Comm. Math. Phys. **124** (1989), no. 1, 23–55. MR1012857

[13] S. Cecotti, S. Ferrara, and L. Girardello, *Geometry of type II superstrings and the moduli of superconformal field theories*, Internat. J. Modern Phys. A **4** (1989), no. 10, 2475–2529, DOI 10.1142/S0217751X89000972. MR1017548

[14] Noémie Combe and Yuri I. Manin, *F-manifolds and geometry of information*, Bull. Lond. Math. Soc. **52** (2020), no. 5, 777–792, DOI 10.1112/blms.12411. MR4171402

[15] V. Cortés, *Homogeneous special geometry*, Transform. Groups **1** (1996), no. 4, 337–373, DOI 10.1007/BF02549212. MR1424448

[16] V. Cortés, M. Dyckmanns, M. Jüngling, and D. Lindemann, *A class of cubic hypersurfaces and quaternionic Kähler manifolds of co-homogeneity one*, Asian J. Math. **25** (2021), no. 1, 1–30, DOI 10.4310/AJM.2021.v25.n1.a1. MR4323301

[17] B. de Wit and A. Van Proeyen, *Special geometry, cubic polynomials and homogeneous quaternionic spaces*, Comm. Math. Phys. **149** (1992), no. 2, 307–333. MR1186031

[18] B. de Wit, F. Vanderseypen, and A. Van Proeyen, *Symmetry structure of special geometries*, Nuclear Phys. B **400** (1993), no. 1-3, 463–521, DOI 10.1016/0550-3213(93)90413-J. MR1227264

[19] P. Jordan, J. von Neumann, and E. Wigner, *On an algebraic generalization of the quantum mechanical formalism*, Ann. of Math. (2) **35** (1934), no. 1, 29–64, DOI 10.2307/1968117. MR1503141

[20] Max Koecher, *The Minnesota notes on Jordan algebras and their applications*, Lecture Notes in Mathematics, vol. 1710, Springer-Verlag, Berlin, 1999. Edited, annotated and with a preface by Aloys Krieg and Sebastian Walcher, DOI 10.1007/BFb0096285. MR1718170

[21] Edoardo Lauria and Antoine Van Proeyen, *$\mathcal{N} = 2$ supergravity in $D = 4, 5, 6$ dimensions*, Lecture Notes in Physics, vol. 966, Springer, Cham, [2020] ©2020, DOI 10.1007/978-3-030-33757-5. MR4181220

[22] C. M. Marle, *From tools in symplectic and Poisson geometry to Souriau's theories of statistical mechanics and thermodynamics*, Entropy **18** (2016), 370.

[23] È. B. Vinberg, *Homogeneous cones*, Soviet Math. Dokl. **1** (1960), 787–790. MR0141680

[24] È. B. Vinberg, *The theory of homogeneous convex cones* (Russian), Trudy Moskov. Mat. Obšč. **12** (1963), 303–358. MR0158414

[25] È. B. Vinberg, *Structure of the group of automorphisms of a homogeneous convex cone* (Russian), Trudy Moskov. Mat. Obšč. **13** (1965), 56–83. MR0201575

Institute for Information Transmission Problems, B. Karetnuj per. 19 Moscow 127051, Russia & University of Hradec Králové, Faculty of Science, Rokitanského 62 500 03 Hradec Králové, Czech Republic
Email address: dalekseevsky@iitp.ru

Instituto de Física Teorica, Universidad de Murcia, Campus de Espinardo, E-30100, Murcia, Spain
Email address: alessio.marrani@um.es

Scuola di Scienze e Tecnologie, Università di Camerino, Via Madonna delle Carceri, I-62032 Camerino (Macerata), Italy
Email address: andrea.spiro@unicam.it

Contemporary Mathematics
Volume **788**, 2023
https://doi.org/10.1090/conm/788/15819

Bundles of Weyl structures and invariant calculus for parabolic geometries

Andreas Čap and Jan Slovák

Dedicated to the memory of Alexandre Michailovich Vinogradov

Abstract. For more than hundred years, various concepts were developed to understand the fields of geometric objects and invariant differential operators between them for conformal Riemannian and projective geometries. More recently, several general tools were presented for the entire class of parabolic geometries, i.e., the Cartan geometries modelled on homogeneous spaces G/P with P a parabolic subgroup in a semi-simple Lie group G. Similarly to conformal Riemannian and projective structures, all these geometries determine a class of distinguished affine connections, which carry an affine structure modelled on differential 1-forms Υ. They correspond to reductions of P to its reductive Levi factor, and they are called the Weyl structures similarly to the conformal case. The standard definition of differential invariants in this setting is as affine invariants of these connections, which do not depend on the choice within the class. In this article, we describe a universal calculus which provides an important first step to determine such invariants. We present a natural procedure how to construct all affine invariants of Weyl connections, which depend only tensorially on the deformations Υ.

Differential invariants of various geometric structures are the core ingredients for numerous applications both in geometric analysis and mathematical physics. In particular, the invariants of conformal Riemannian manifolds attracted a lot of attention in the course of the last 100 years.

For smooth manifolds with an affine connection, the so called 'first invariant theorem' says that all the invariants are expressions built of the covariant derivatives of sections of natural bundles, the curvature and the torsion of the connection by means of algebraic tensorial invariants, cf. **[20]** for a modern treatment. Let us call them the *affine differential invariants* on smooth manifolds. The analogous 'first invariant theorem' for Riemannian geometries says that all differential invariants are built from affine invariants of the canonical Levi-Civita connection via (algebraic) invariants of the orthogonal group.

A conformal Riemannian geometry is defined as a class of conformally equivalent Riemannian metrics and so the above Riemannian first invariant theorem

2020 *Mathematics Subject Classification.* Primary 53A55, 58A20, 53C15, 53A40, 53C05, 58J60.

The second author is the corresponding author.

The first author was supported by the Austrian Science Fund (FWF): P33559-N, and the second author was supported by the grant GX19-28628X of GAČR..

©2023 American Mathematical Society

can be used to define conformal invariants. Thus, a conformal invariant is usually understood as a Riemannian invariant in terms of any metric from the conformal class, such that the change of the metric does not change the invariant. As proved by the extraordinary effort to understand such invariants for many decades, already the first invariant theorem is not easy in this case.

An equivalent definition of conformal structures treats them in terms of classical G-structures as reductions of the linear frame bundle to the structure group $G_0 = CO(n)$, the group of all conformally Euclidean linear transformations in the given dimension. Such a structure admits compatible torsion-free connections, which are classically called *Weyl connections*. This broader class of conformal connections was exploited by H. Weyl which motivates their name. It turns out that the Weyl connections form an affine space modelled on one-forms. Of course, the Levi-Civita connections of metrics in the conformal class are Weyl connections, they form an affine subspace modelled on exact one-forms. The study of conformal Riemannian invariants goes back to É. Cartan, T. Thomas, J.A. Schouten, and others (e.g., [**13, 25, 26**]). A lot of spectacular tricks to build invariant expressions have been developed, and some of them were turned into a quite effective calculus for conformal invariants by V. Wünsch, see [**27**].

Motivated by the rich geometry of conformal Riemannian manifolds, the Weyl structures and the preferred connections were introduced in the general framework of parabolic geometries in [**9**] (generalizing the approach from [**17**]). In particular, the notions of scales, closed and exact Weyl structures, and (Schouten's) Rho-tensors were extended, and natural generalizations of classical normal coordinates in affine geometries were discussed.

Moreover, filtered analogues of classical G-structures which are equivalent to parabolic geometries and the general Cartan-Tanaka theory for all parabolic geometries are explained in great detail in [**12**], see also [**5**]. In this setting, the Weyl connections on a parabolic geometry of type G/P correspond to reductions of the parabolic structure group P of the canonical Cartan connection ω on the principal bundle $\mathcal{G} \to M$ to its reductive Levi factor $G_0 \subset P$.

More recently, the geometry of the bundle of Weyl structures $\pi : A = \mathcal{G}/G_0 \to M$ was studied carefully in the joint work [**7**] of the first author and T. Mettler. Here the canonical Cartan connection ω induces a canonical affine connection on the manifold A as well as a canonical splitting $TA = L^- \oplus L^+$. There also is a nice relation between natural bundles over M associated to P-representations and natural bundles over A. Using this, we will identify a natural class of differential invariants of the canonical connection on A which induce affine differential invariants of the Weyl connections. The invariants obtained in this way transform tensorially in the one-forms Υ that parameterize the Weyl connections. We shall also prove a converse to this statement, which is much more subtle: If an affine invariant of Weyl connections, its curvature and its torsion transforms tensorially, then it comes from a natural affine differential invariant on A as described above.

The paper is organized as follows: After a brief review of the main tools and concepts following [**12**], [**7**], we show in Section 2 how affine invariants of the canonical connection D on A can be used to construct affine invariants of the Weyl connections. This is based on an invariant concept of jets and by construction, the resulting invariants always change tensorially under a change of Weyl structure. We introduce the terminology "nearly invariant operators" for this behavior. In the last section, we prove that all nearly invariant operators are obtained in this way. This can be viewed as a week version of the 'first invariant theorem' (providing only an ansatz of possible expressions for invariant operators) and also a universal procedure realizing a generalization of the so-called Wünsch calculus in conformal geometry, which provides universal formulae for large classes of invariant operators, cf. [**10**].

1. Parabolic geometries and the bundle of Weyl structures

We shall follow the terminology and notation of [**12**] and [**7**]. The complete procedure deriving the canonical (i.e., properly normalized) Cartan connection from more elementary data on filtered manifolds corresponding to parabolic geometries was first worked out long ago by Tanaka in [**24**], more straightforward, simpler and more general versions appeared in [**8**], [**12**], [**5**]. We do not go into details on the normalization here, but just use its implications for the curvature and torsion of Weyl connections. The calculus we develop actually does also work for non-normal Cartan connections. However in the non-normal case, in some of the results the notion of invariants have to be adapted.

1.1. The Cartan connections. A Cartan geometry of type G/P is an absolute parallelism on a principal fiber bundle $\mathcal{G} \to M$ with structure group P encoded as a one-form $\omega \in \Omega^1(\mathcal{G}, \mathfrak{g})$. One requires suitable equivariancy properties with respect to the principal action of P, similar to those of the Maurer-Cartan form ω_G on $G \to G/P$. Thus, we may view Cartan geometries as curved deformations of the homogeneous spaces. More explicitly, the *Cartan connection* $\omega \in \Omega^1(\mathcal{G}, \mathfrak{g})$ is required to obey the following properties

(1) $\omega(\zeta_X)(u) = X$ for all $X \in \mathfrak{p}$, $u \in \mathcal{G}$ (the connection reproduces the generators of fundamental vertical fields)
(2) $(r^g)^*\omega = \mathrm{Ad}(g^{-1}) \circ \omega$ for any $g \in P$ (the connection form is equivariant with respect to the principal action)
(3) $\omega_{|T_u\mathcal{G}} : T_u\mathcal{G} \to \mathfrak{g}$ is a linear isomorphism for all $u \in \mathcal{G}$ (the absolute parallelism condition).

A morphisms of Cartan geometries is a principal fiber bundle morphism $\varphi : \mathcal{G} \to \mathcal{G}'$ with the property $\varphi^*(\omega') = \omega$. *Parabolic geometries* are Cartan geometries of type (G, P), where P is a parabolic subgroup in a semisimple real Lie group G.

In the sequel, we shall consider a fixed parabolic subgroup $P \subset G$. It is well known, that P uniquely determines a grading

$$\mathfrak{g} = \mathfrak{g}_- \oplus \mathfrak{g}_0 \oplus \mathfrak{p}_+ = \mathfrak{g}_{-k} \oplus \dots \mathfrak{g}_{-1} \oplus \mathfrak{g}_0 \oplus \mathfrak{g}_1 \oplus \dots \oplus \mathfrak{g}_k \tag{1}$$

on the Lie algebra $\mathfrak{g}$ of the group G, such that $\mathfrak{p} = \mathfrak{g}_0 \oplus \mathfrak{p}_+ = \mathfrak{g}_0 \oplus \mathfrak{g}_1 \oplus \dots \oplus \mathfrak{g}_k$ is the Lie algebra of P. This comes with a so called grading element $E \in \mathfrak{p}$ with the property $\mathrm{ad}\, E_{|\mathfrak{g}_i} = i \cdot \mathrm{id}_{\mathfrak{g}_i}$ for all $i = -k, \dots, k$.

While the grading (1) is not P-invariant, there is an associated filtration

$$\mathfrak{g} = \mathfrak{g}^{-k} \supset \mathfrak{g}^{-k+1} \supset \dots \supset \mathfrak{g}^0 = \mathfrak{p} \supset \mathfrak{g}^1 = \mathfrak{p}_+ \supset \mathfrak{g}^2 \dots \supset \mathfrak{g}^k = \mathfrak{g}_k \tag{2}$$

defined by $\mathfrak{g}^j := \oplus_{\ell \geq j} \mathfrak{g}_\ell$ which is invariant under the adjoint action of P. The subgroup $G_0 \subset P$ of all elements whose adjoint action does preserve the grading has Lie algebra $\mathfrak{g}_0$. This is the reductive Levi factor of P. We also write $P_+ = \exp \mathfrak{p}_+$

for the nilradical of P, and similarly $G_- = \exp \mathfrak{g}_-$. In particular, for each $g \in P$, there are unique elements $g_0 \in G_0$, $\Upsilon \in \mathfrak{p}_+$, and $\Upsilon_i \in \mathfrak{g}_i$, $i = 1, \dots, k$, such that

$$g = g_0 \cdot \exp \Upsilon = g_0 \cdot \exp \Upsilon_1 \cdot \dots \cdot \exp \Upsilon_k. \tag{3}$$

This decomposition reflects the fixed splitting of the filtration of $\mathfrak{p}_+$ by P-submodules, i.e. our fixed isomorphism $\operatorname{gr} \mathfrak{p}_+ \to \mathfrak{p}_+$.

The Cartan connection ω provides the *constant vector fields* $\omega^{-1}(X) \in \mathfrak{X}(\mathcal{G})$ defined for all $u \in \mathcal{G}$ and $X \in \mathfrak{g}$ by

$$\omega(\omega^{-1}(X)(u)) = X.$$

These generalize the left invariant vector fields on the homogeneous model $G \to G/P$ and enjoy the same equivariancy property, i.e.

$$Tr^g \cdot \omega^{-1}(X)(u) = \omega^{-1}(\mathrm{Ad}_{g^{-1}} \cdot X)(u \cdot g),$$

where r^g is the principal right action by the element $g \in P$. In particular, due to our fixed splitting of $\mathfrak{g}$, there are the *horizontal vector fields* $\omega^{-1}(X)$ with $X \in \mathfrak{g}_-$.

On the homogeneous model, the Maurer-Cartan equation reads as $d\omega + \frac{1}{2}[\omega, \omega] = 0$, while on a general geometry the same expression provides the two-form

$$K = d\omega + \frac{1}{2}[\omega, \omega] \tag{4}$$

called the *curvature*. The equivariance properties of the Cartan connection imply that K is always a horizontal two-form and so the curvature is completely determined by the *curvature function* $\tilde{\kappa} \in C^\infty(\mathcal{G}, \Lambda^2 \mathfrak{g}_-^* \otimes \mathfrak{g})$,

$$\tilde{\kappa}(u)(X, Y) = K(\omega^{-1}(X)(u), \omega^{-1}(Y)(u)) = [X, Y] - \omega(u)([\omega^{-1}(X), \omega^{-1}(Y)]).$$

Of course, the values of both the connection and the curvature function split according to the corresponding splitting of $\mathfrak{g}$ into

$$\omega = \omega_- + \omega_0 + \omega_+, \quad \tilde{\kappa} = \tilde{\kappa}_- + \tilde{\kappa}_0 + \tilde{\kappa}_+, \tag{5}$$

but the individual components are not P-equivariant. For instance, κ_- is a well defined object only if its values are considered as elements of the quotient $\mathfrak{g}/\mathfrak{p}$. The latter component of the curvature is called the *torsion* of the Cartan connection.

Clearly, the curvature is the obstruction to the integrability of the horizontal distribution $\omega^{-1}(\mathfrak{g}_-)$ in $T\mathcal{G}$ and the Cartan connection is locally isomorphic to its homogeneous model if and only if the curvature vanishes, see e.g., [**12**, Section 1.5].

1.2. Natural bundles. Every P-representation on a vector space $\mathbb{V}$ provides the homogeneous vector bundle $G \times_P \mathbb{V} \to G/P$ and, more generally, the associated vector bundles

$$\mathcal{V}M = \mathcal{G} \times_P \mathbb{V} \to M$$

with standard fiber $\mathbb{V}$ over all manifolds with a parabolic geometry of the type G/P. Shortly, we shall talk about P-modules $\mathbb{V}$ and the induced *natural bundles.* In the sequel, we shall restrict ourselves to P-modules with a diagonalizable action of the center of G_0.

For instance, the Cartan connection ω on $p : \mathcal{G} \to M$ identifies the tangent bundle $TM = \mathcal{G} \times_P \mathfrak{g}/\mathfrak{p} = \mathcal{G} \times_P \mathfrak{g}_-$, $(u, X) \mapsto Tp(\omega^{-1}(X)(u))$. Similarly, $T^*M = \mathcal{G} \times_P \mathfrak{p}_+$ and the duality is expressed by the Cartan-Killing form on $\mathfrak{g}$. Observe that for $i < 0$, we have the P-invariant subspace $\mathfrak{g}^i/\mathfrak{p} \subset \mathfrak{g}/\mathfrak{p}$ induced by the filtration of $\mathfrak{g}$ from (2). This determines a smooth subbundle $T^iM \subset TM$, so we get a filtration of TM. Likewise, the subspaces $\mathfrak{g}^i \subset \mathfrak{p}_+$ for $i > 0$ give rise to a filtration of the cotangent bundle T^*M by smooth subbundles.

A special class of natural bundles is induced by the G-modules viewed as P-modules. They are called the *tractor bundles*, see [**1**], [**12**] for detailed description in many specific geometries and historical links. A very special role is reserved for the *adjoint tractor bundles* $\mathcal{A}$ coming from the adjoint representation of G on $\mathfrak{g}$. Of course, the Lie bracket itself is Ad-equivariant and thus there is the algebraic bracket $\{\ ,\ \}$ on the adjoint tractors.

It is well known that sections σ of the induced bundles $\mathcal{V}M$ are in bijective correspondence with smooth functions $\tilde{\sigma} : \mathcal{G} \to \mathbb{V}$ which are P-equivariant in the sense that

$$\tilde{\sigma}(u \cdot g) = g^{-1} \bullet \tilde{\sigma}(u).$$

Here $u \cdot g$ denotes the principal right action of g on u, while the bullet denotes the P-action on $\mathbb{V}$. We have already seen the curvature function $\tilde{\kappa} : \mathcal{G} \to \Lambda^2(\mathfrak{g}/\mathfrak{p})^* \otimes \mathfrak{g} = \Lambda^2\mathfrak{p}_+ \otimes \mathfrak{g}$, representing a section of the adjoint tractor valued form $\kappa \in \Omega^2(M; \mathcal{A})$, cf. (4), which is the curvature of ω viewed as a two-form on M. Of course, the adjoint tractor bundles inherit all the P-invariant objects from $\mathfrak{g}$, including the metric defined by the Cartan-Killing form. Further, the 1-forms on M live in the invariant subbundle $\mathcal{A}^1$ corresponding to $\mathfrak{p}_+$, while the vector fields on M can be viewed (with the help of the Cartan connection on $\mathcal{G}$ as noticed above) as sections of the quotient $\mathcal{A}/\mathcal{A}^0$, where $\mathcal{A}^0 = \mathcal{G} \times_P \mathfrak{p}$. In particular, the torsion κ_- of the Cartan connection is a vector valued two-form on M, exactly as we are used to see the torsions of affine connections on manifolds.

1.3. The bundle of Weyl structures. We briefly remind the impacts of reductions of the parabolic structure groups to their reductive parts. The reader can find more details in [**12**] and [**7**].

Given the Cartan geometry $(\mathcal{G} \to M, \omega)$, we can form the quotient $\mathcal{G}_0 := \mathcal{G}/P_+$ which is a principal bundle over M with structure group $P/P_+ \cong G_0$. Each reduction of $\mathcal{G}$ to the structure group $G_0 \subset P$ can then be seen as a G_0 equivariant smooth section $\bar{s} : \mathcal{G}_0 \to \mathcal{G}$ of the quotient projection. It is well known that such reductions are in bijective correspondence with the smooth sections s of the bundle $\pi : \mathcal{G}/G_0 \to M$, which in our case is equal to $\pi : A = \mathcal{G} \times_P P/G_0 \to M$. We call $\pi : A \to M$ the *bundle of Weyl structures*, see [**18**] and [**7**].

In particular, we see that Weyl structures always exist globally, and for two such sections $\hat{s}$ and s, there always exist unique equivariant functions $\Upsilon : \mathcal{G}_0 \to \mathfrak{p}_+$ and $\Upsilon_i : \mathcal{G}_0 \to \mathfrak{g}_i$ for $i = 1, \ldots, k$, such that for all $v \in \mathcal{G}_0$,

$$\bar{\hat{s}}(v) = \bar{s}(v) \cdot \exp \Upsilon(v) = \bar{s}(v) \cdot \exp \Upsilon_1(v) \cdots \exp \Upsilon_k(v).$$

Now we come to several crucial observations.

First, $\mathcal{G} \to A$ is a principal fibre bundle with the structure group G_0 and the tangent bundle is the associated bundle $TA = \mathcal{G} \times_{G_0} (\mathfrak{g}_- \oplus \mathfrak{p}_+)$ via the adjoint action of G_0 on $\mathfrak{g}$. In particular, TA naturally splits into two components $TA = L^- \oplus L^+$ corresponding to the G_0-invariant components $\mathfrak{g}_-$ and $\mathfrak{p}_+$ in $\mathfrak{g}$. Moreover, L^+ is the vertical bundle of $\pi : A \to M$, while $L^- = \pi^* TM$.

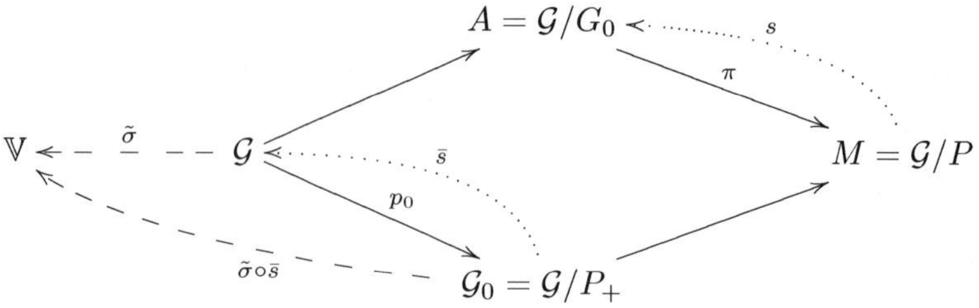

Second, the Cartan connection $\omega = \omega_- + \omega_0 + \omega_+$ can be viewed as an affine connection D on A, with soldering form $\omega_- + \omega_+$ and principal connection form ω_0. The natural splitting $TA = L^- \oplus L^+$ leads to two partial affine connections D^- and D^+ by restriction. We also observe that, up to universal algebraic terms, the curvature of ω on A encodes the torsion and curvature of D by means of the G_0-equivariant functions $\tilde\kappa_- + \tilde\kappa_+$ and $\tilde\kappa_0$ from (5), respectively. Moreover, since κ is horizontal with respect to $\mathcal{G} \to M$, the torsion and curvature of D are given by universal algebraic terms if evaluated on one argument from L^+. Evaluated on two arguments from L^-, they provide, after algebraic corrections, the components $T \in \Omega^2(A, L_-)$, $Y \in \Omega^2(A, L^+)$, and $W \in \Omega^2(A, \mathcal{G} \times_P \mathfrak{g}_0)$. We call these components the *universal torsion* T, *universal Cotton York tensor* Y, and *universal Weyl curvature* W, respectively, see **[7]** for details.

For each Weyl structure s, we can also consider the pullback of the Cartan connection $\bar{s}^*\omega \in \Omega^1(\mathcal{G}_0, \mathfrak{g})$ which now splits into the three components naturally. Let us write

$$\begin{aligned}
\theta^s &= \bar{s}^*\omega_- = \theta^s_{-k} + \cdots + \theta^s_{-1} : T\mathcal{G}_0 \to \mathfrak{g}_{-k} \oplus \cdots \oplus \mathfrak{g}_{-1} \\
\gamma^s &= \bar{s}^*\omega_0 : T\mathcal{G}_0 \to \mathfrak{g}_0 \\
\mathsf{P}^s &= \bar{s}^*\omega_+ = \mathsf{P}^s_1 + \cdots + \mathsf{P}^s_k : T\mathcal{G}_0 \to \mathfrak{p}_+ = \mathfrak{g}_1 \oplus \cdots \oplus \mathfrak{g}_k.
\end{aligned}$$

The filtration on TM introduced in Section 1.2 gives rise to the associated graded bundle $\operatorname{gr}(TM) = \oplus_{i=-k}^{-1} T^i M / T^{i+1} M$. Now the form θ^s provides an identification of $\mathcal{G}_0$ with a frame bundle for $\operatorname{gr}(TM)$ with structure group G_0 and should be understood as an analog of the *soldering form.* In particular, this defines the isomorphism $TM \simeq \operatorname{gr}(TM)$, i.e. a splitting of the filtration

$$TM = T^{-k}M \supset \cdots \supset T^{-1}M$$

determined by the choice of s.

The next component γ^s is the connection form of a principal connection on $\mathcal{G}_0$. Thus, the first two components provide a reductive Cartan connection $(\theta^s + \gamma^s)$ of type $(G_- \cdot G_0)/G_0$ on M. We call it the *Weyl connection* corresponding to the choice of the Weyl structure s. We shall write ∇^s for the covariant derivatives with respect to this connection on all vector bundles associated to $\mathcal{G}_0$.

The last component P^s measures the difference between the Cartan connection $(\theta^s + \gamma^s)$ on $\mathcal{G}_0$ and the original Cartan connection ω on $\mathcal{G}$, along the image of $\bar{s}$. We view it as a one-form on M valued in $\operatorname{gr} T^*M$ and call it the *Rho-tensor.*

Third, let us take a representation $\mathbb{V}$ of P and consider the induced bundle $\mathcal{G} \times_P \mathbb{V} = \mathcal{V}M \to M$. Sections of this bundle correspond to P-equivariant maps $\tilde\sigma : \mathcal{P} \to \mathbb{V}$, so they obviously also represent sections of the natural bundle $\mathcal{V}A =$

$\mathcal{G} \times_{G_0} \mathbb{V} \to A$. By the very construction, the natural bundles $\mathcal{V}A \to A$ are naturally identified with $\pi^*\mathcal{V}M$.

Next, the action of the grading element $E \in \mathfrak{g}_0$ decomposes $\mathbb{V}$ into G_0-invariant components

$$\mathbb{V} = \mathbb{V}_0 \oplus \cdots \oplus \mathbb{V}_\ell$$

with the property that $X \bullet \mathbb{V}_j \subset \mathbb{V}_{i+j}$, for all $X \in \mathfrak{g}_i$ and $i = 1, \ldots, k$. For instance, the Lie algebra $\mathfrak{g}_-$, viewed as $\mathrm{gr}(\mathfrak{g}/\mathfrak{p})$, or the entire $\mathfrak{g}$ are special cases illustrating the difference between the P-invariant filtrations and G_0-invariant induced gradings.

Thus while the natural bundles $\mathcal{V}M \to M$ come equipped with the natural filtrations only, the bundles $\mathcal{V}A \to A$ come with natural gradings. Once we choose a Weyl structure s, the bundles $\mathcal{V}M$ get graded, too (via the structure group reduction). This grading coincides with the pullback $s^*(\mathcal{V}A) \to M$, [**7**, Theorem 2.4].

Of course, the canonical affine connection D induces linear connections on all natural bundles $\mathcal{V}A$. Now, the required P-equivariance of the maps $\tilde{\sigma}$ representing sections of $\mathcal{V}M$ directly implies that the sections of $\mathcal{V}M$ are exactly those sections τ of $\mathcal{V}A$, whose covariant derivatives satisfy $D_\xi\sigma = -\xi \bullet \tau$ for all sections $\xi \in \Gamma(L^+)$, [**7**, Theorem 2.4].

For a section $\sigma \in \Gamma(\mathcal{V}M)$, the covariant derivative $D\sigma$ with respect to the canonical connection D defines a map $D\sigma : \mathcal{G} \to (\mathfrak{g}_- \oplus \mathfrak{p}_+)^* \otimes \mathbb{V}$. Restricting to entries from $\mathfrak{g}_-$, one obtains an operation ∇^ω which is called the *invariant differential* and which played a crucial role in [**10**] and [**11**]. However, this was viewed as an operation mapping sections of $\mathcal{V}M$ to equivariant functions on the Cartan bundle, the relation to the bundle of Weyl structures was not known at that time.

1.4. The curvatures. As mentioned in Section 1.1, the fundamental invariant of a Cartan geometry $(\mathcal{G}, \omega)$ is the Cartan curvature κ. Via the curvature function discussed there, this admits direct interpretations both on M and on A. P-equivariancy of the curvature function implies that κ can be viewed as a two-form on M with values in the adjoint tractor bundle $\mathcal{A} = \mathcal{G} \times_P \mathfrak{g}$. A choice of Weyl structure determines an isomorphism between the bundle $\Lambda^2T^*M \otimes \mathcal{A}$ with its associated graded. The latter bundle decomposes into components according to the decomposition of $\Lambda^2(\mathfrak{g}_-)^* \otimes \mathfrak{g}$ into G_0-irreducible representations.

On the level of A, the decomposition into G_0-components is available canonically. In particular, the G_0-invariant decomposition $\mathfrak{g} = \mathfrak{g}_- \oplus \mathfrak{g}_+ \oplus \mathfrak{g}_0$ decomposes $\mathcal{G} \times_{G_0} \mathfrak{g}$ with the first two summands corresponding to $L^- \oplus L^+ = TA$. Via the restriction of the adjoint action, $\mathfrak{g}_0$ canonically injects into endomorphisms of $\mathfrak{g}_- \oplus \mathfrak{g}_+$, so we can view $\mathcal{G} \times_{G_0} \mathfrak{g}_0$ naturally as a subbundle $\mathrm{End}_0(TA) \subset T^*A \otimes TA$. So on this level, we can interpret κ as being given by a section of the bundle $\Lambda^2(L^-)^* \otimes TA$ which we write as $T + Y$ according to $TA = L^- \oplus L^+$ and a section $W \in \Gamma(\Lambda^2(L^-)^* \otimes \mathrm{End}_0(TA))$. These look already like a torsion and a curvature, and indeed by [**7**, Theorem 2.12] these equivalently encode the torsion and curvature of D. More precisely, one has to add universal bundle maps induced by (G_0-equivariant) components of the Lie bracket on $\mathfrak{g}$ to $T + Y$ respectively to W in order to obtain the torsion and the curvature of D.

Next, we want to relate the Cartan curvature to the curvature and torsion of the Weyl connections associated to a Weyl structure. Consider a Weyl structure s, and the corresponding affine connection $\theta^s + \gamma^s$ on $\mathcal{G}_0$. The full Cartan curvature (i.e., torsion and curvature) of this connection is also given by the structure equations

$$T^s(\xi,\eta) = d\theta^s(\xi,\eta) + [\gamma^s(\xi),\theta^s(\eta)] + [\theta^s(\xi),\gamma^s(\eta)] + [\theta^s(\xi),\theta^s(\eta)]$$
$$R^s(\xi,\eta) = d\gamma^s(\xi,\eta) + [\gamma^s(\xi),\gamma^s(\eta)],$$

where ξ,η are vector fields on $\mathcal{G}_0$. Clearly the torsion T^s and the curvature R^s are horizontal two-forms which descend to well defined forms on the underlying manifold. Let us point out, that T^s is not the usual torsion of a linear connection but there is a correction involving natural bundle maps induced by the (G_0-equivariant) Lie bracket on $\mathfrak{g}_-$. Since the forms θ^s and γ^s are pullbacks of ω_- and ω_0, the above structure equations are easily compared with those of ω.

The latter curvature forms are clearly related to the pullback $\bar{s}^*\kappa$ of the curvature of the Cartan connection ω on $\mathcal{G}$ (i.e. also the torsion and curvature of D). The missing components of curvature are related to the P-tensor and they can be understood easily, too:

$$\bar{s}^*\kappa = T^s + R^s + Y^s + \partial \mathsf{P}^s \tag{6}$$

where we define the *Cotton-York* tensor of the Weyl structure on M as

$$Y^s(\xi,\eta) = d^{\nabla^s}\mathsf{P}^s(\xi,\eta) + \mathsf{P}^s(\{\xi,\eta\}) + \{\mathsf{P}^s(\xi),\mathsf{P}^s(\eta)\},$$

with d^{∇^s} denoting the covariant exterior differential, $\{\ ,\ \}$ is the natural bracket on the adjoint tractors, and

$$\partial\mathsf{P}^s(\xi,\eta) = \{\xi,\mathsf{P}^s(\eta)\} - \{\eta,\mathsf{P}^s(\xi)\} - \mathsf{P}^s(\{\xi,\eta\})$$

is the (G_0-equivariant) Lie algebra cohomology differential. One then defines the *Weyl curvature* of the Weyl connection determined by s as $W^s := R^s + (\partial\mathsf{P}^s)_{\mathfrak{g}_0}$.

As suggested by the notation, the quantities T^s, W^s and Y^s are directly related to the quantities on A discussed above. Indeed by [**7**, Proposition 2.10] and [**12**, Section 5.2.9] one has $s^*T = T^s + (\partial\mathsf{P}^s)_{\mathfrak{g}_-}$, $s^*Y = Y^s + (\partial\mathsf{P})_{\mathfrak{p}_+}$, and $s^*W = W^s$. There is also a nice way to interpret the Rho-tensor via A. Viewing the projection onto the second factor in $TA = L^- \oplus L^+$ as an element $\mathsf{P} \in \Omega^1(A, L^+)$, one can form $s^*\mathsf{P} \in \Omega^1(TM, \mathrm{gr}(TM))$ and by [**7**, Proposition 2.8] this coincides with P^s.

It will be very important for our results that the decomposition of the curvature and torsion of a Weyl connection into the pullback of the Cartan curvature and a part obtained from the Rho-tensor can be made explicit without reference to the Cartan bundle. This only works for *normal parabolic geometries* which anyway is the class usually studied. They are characterized by a normalization condition on their curvature κ which is usually phrased as $\partial^*(\kappa) = 0$. Here ∂^* is induced by a P-equivariant linear map $\Lambda^2(\mathfrak{g}/\mathfrak{p})^* \otimes \mathfrak{g} \to (\mathfrak{g}/\mathfrak{p})^* \otimes \mathfrak{g}$, the so-called *Kostant codifferential.* This normalization condition then implies that the parabolic geometry is determined by some underlying geometric structure known as an *infinitesimal flag structure*, for example a conformal structure.

The detailed form of this is not important for our purposes. What it tells us, however, is that the right hand side of (6) lies in the kernel of ∂^*. This allows us to express, $\partial^*\partial(\mathsf{P}^s)$ in terms of the curvature and torsion of ∇^s and of Y^s as defined above. Now on the one hand it turns out that $\partial^*\partial$ is invertible. On the other hand, the definition of Y^s is chosen in such a way our equation can be solved iteratively homogeneity by homogeneity (with respect to the grading element), see Section 5.1 of [**12**] for details. Hence P^s can be computed from curvature quantities, which in turn allows us compute the pullback of the Cartan curvature. In particular, in the case of conformal geometry this leads to the description of the Rho tensor as a trace-modification of the Ricci curvature as introduced by J.A. Schouten.

1.5. Normal Weyl structures. As we have seen, the Rho-tensor measures the deviation of the Cartan connection $\theta^s+\gamma^s$ on $\mathcal{G}_0$ determined by a Weyl structure s, from the given Cartan connection ω. Thus minimizing the values of P^s and its derivatives in a point looks like a good idea. We can easily follow the way how the normal coordinates for affine connections are built. This leads to an analog for all parabolic geometries of the so-called Graham normal coordinates (see [**16**]) from conformal geometry and their analog in CR geometry that was recently introduced by K. Hirachi in [**19**].

For each fixed frame $u \in \mathcal{G}$ and $X \in \mathfrak{g}_-$, there is the flow line $\mathrm{Fl}_t^{\omega^{-1}(X)}(u)$ of the horizontal vector field $\omega^{-1}(X)$, and for some neighborhood of the origin in $\mathfrak{g}_-$ these flow lines exist at least up to the time $t = 1$. In this way we obtain a horizontal embedding $X \mapsto \varphi_u(X) = \mathrm{Fl}_1^{\omega^{-1}(X)}(u)$ and also a local section $p(\varphi_u(X)) \mapsto \varphi_u(X)$ of the Cartan bundle $p : \mathcal{G} \to M$ through u. Consequently there is a unique local Weyl structure $\bar{s}_u$ through the frame u defined by $\bar{s}_u(p_0(\varphi_u(X))) = \varphi_u(X)$. (Here $p_0 : \mathcal{G} \to \mathcal{G}_0$ is the natural projection and the G_0-orbit of this section defines the reduction.) We call them the *normal Weyl structures*. The images $c^{u,X}$ of the defining flow lines in M, i.e. $c^{u,X}(t) = p(\mathrm{Fl}_t^{\omega^{-1}(X)})$, are called the *generalized geodesics* of the Cartan connection ω. Thus, each choice of a frame $u \in \mathcal{G}$ over the point $p(u) = x \in M$ determines a uniquely defined "geodetical parametrization" from a neighborhood of 0 in T_xM to a neighborhood of x in M.

This construction can be also nicely seen on A. We are restricting the standard normal coordinates of the affine connection D to geodesics emanating in L^--directions, which directly creates a (local) section $s : M \to A$, similarly as above. These sections evidently have the property, that the universal Rho tensor vanishes along those geodesics. Indeed, this leads to a complete characterization: The Weyl structures s that are normal at x are characterized by the fact that $\mathsf{P}^s(c(t))(c'(t)) = 0$ for all generalized geodesics through $x = p(u) \in M$. In particular, this implies that for every integer $\ell \geq 0$ of covariant derivatives, and any tangent vectors $\xi_0, \dots, \xi_\ell$, the full symmetrization of the expression

$$(\nabla^s_{\xi_\ell} \cdots \nabla^s_{\xi_1} \mathsf{P}(x))(\xi_0) \in \Omega^1(M) \tag{7}$$

over the ξ's vanishes, see [**9**] for details.

2. The invariant calculus

Whatever concept of invariance we adopt, all objects built naturally in the category of Cartan connections of given type should be invariant. Thus, we can consider any (possibly non-linear) differential operator $\Phi : \Gamma(\mathcal{V}A) \to \Gamma(\mathcal{W}A)$ between two natural vector bundles over A coming from P-modules $\mathbb{V}$ and $\mathbb{W}$, expressed in covariant derivatives with respect to D and derivatives of the torsion and curvature of D. As noted in Section 1.3, we can view $\Gamma(\mathcal{V}M)$ naturally as a subspace of $\Gamma(\mathcal{V}A)$ (corresponding to P-equivariant functions among G_0-equivariant functions). Then it may happen that the restriction of Φ to this subspace has values in $\Gamma(\mathcal{W}M) \subset \Gamma(\mathcal{W}A)$. If this is the case, the restriction has to define an invariant operator.

However, choosing a Weyl structure s, there is a possibility to "descend" any operator $\Phi : \Gamma(\mathcal{V}A) \to \Gamma(\mathcal{W}A)$ as above to an operator $\Phi^s : \Gamma(\mathcal{V}M) \to \Gamma(\mathcal{W}M)$. Namely, for a section $\sigma \in \Gamma(\mathcal{V}M)$ we can view σ as an element of $\Gamma(\mathcal{V}A)$ and form $\Phi(\sigma) \in \Gamma(\mathcal{W}A)$. Then we can take the pullback $s^*\Phi(\sigma)$ and observe that there is a unique section $\Phi^s(\sigma) \in \Gamma(\mathcal{W}M) \subset \Gamma(\mathcal{W}A)$ such that $s^*\Phi^s(\sigma) = s^*\Phi(\sigma)$. Equivalently, this can be phrased as $\Phi(\sigma)|_{s(M)} = \Phi^s(\sigma)|_{s(M)}$. In the language of equivariant functions, this simply means that we take the G_0-equivariant function $\Phi(\sigma) : \mathcal{G} \to \mathbb{W}$, restrict it to $\tilde{s}(\mathcal{G}_0) \subset \mathcal{G}$ and then define $\Phi^s(\sigma) : \mathcal{G} \to \mathbb{W}$ to be the P-equivariant extension of this G_0-equivariant function.

Observe that this construction immediately implies a property that will be of crucial importance in what follows, namely that for any $\sigma \in \Gamma(\mathcal{V}M)$ and any point $x \in M$, the value $\Phi^s(\sigma)(x)$ depends only on $s(x)$, so the dependence of Φ^s on s is of tensorial character.

An obvious first step in trying to understand this construction is to analyze the operator $\Phi : \Gamma(\mathcal{V}A) \to \Gamma((L^-)^* \otimes \mathcal{V}A)$ defined by $\Phi(\sigma) := D^-\sigma$ on some natural bundle. Choosing a Weyl structure s, this descends to an operator $(D^-)^s : \Gamma(\mathcal{V}M) \to \Gamma(T^*M \otimes \mathcal{V}M)$ which has the chance to define a covariant derivative on $\mathcal{V}M$. This case is sorted out by a result which basically is proved as a part of Theorem 2.6 in [**7**]:

2.1. Proposition. *Let $\mathbb{V}$ be any representation of P and consider the corresponding natural bundles $\mathcal{V}M \to M$ and $\mathcal{V}A \to A$, and the operator $\Phi : \Gamma(\mathcal{V}A) \to \Gamma((L^-)^* \otimes \mathcal{V}A)$ defined by $\Phi(\sigma) := D^-\sigma$. Then for any Weyl structure $s : M \to A$, the operator $\Phi^s : \Gamma(\mathcal{V}M) \to \Gamma(T^*M \otimes \mathcal{V}M)$ coincides with the Rho-corrected derivative ∇^{P^s} corresponding to s as introduced in Section 5.1.9 of* [**12**].

Proof. Consider $\bar{s}(\mathcal{G}_0) \subset \mathcal{G}$ and restrict the component $\omega_{\mathfrak{p}}$ to this subset. Then this can be uniquely extended to a principal connection form γ^s on $\mathcal{G}$ and by definition, the Rho-corrected derivative is the covariant derivative induced by this principal connection. Viewing $\sigma \in \Gamma(\mathcal{V}M)$ as a P-equivariant function $\mathcal{G} \to \mathbb{V}$, the function $\mathcal{G} \to L(\mathfrak{g}_-, \mathbb{V})$ corresponding to $\nabla^{\mathsf{P}^s}\sigma$ thus is given by taking the derivative of σ with respect to horizontal vector fields along $\bar{s}(\mathcal{G}_0)$ and then equivariantly extending to $\mathcal{G}$. But this exactly means that, along $\bar{s}(\mathcal{G}_0)$, this coincides with the G_0-equivariant function associated to $D^-\sigma$, see Section 1.3, which proves the claim. □

In view of our observation above, this gives an alternative argument for the fact that the transformation rule for Rho-corrected derivatives is tensorial in the one-form Υ describing the change between two Weyl-structures, see part (2) of [**12**, Proposition 5.1.9].

The name Rho-corrected derivative is motivated by the relation of this operation (determined by some Weyl structure s) to the Weyl connection ∇^s. This can be written as $\nabla^{\mathsf{P}^s}_\xi \sigma = \nabla^s_\xi \sigma + \mathsf{P}^s(\xi) \bullet \sigma$ for any vector field $\xi \in \mathfrak{X}(M)$, since the difference of the horizontal lifts of $\xi \in T_xM$ for γ^s and γ on the image $\bar{s}(\mathcal{G}_0)$ is the fundamental vector field corresponding to $-\omega_{\mathfrak{p}_+}(\xi)$, and σ is P-equivariant. Here the bullet comes from the bundle map $T^*M \times \mathcal{V}M \to \mathcal{V}M$ induced by the infinitesimal representation $\mathfrak{p}_+ \times \mathbb{V} \to \mathbb{V}$. A bit of care is needed in interpreting this formula, however, since both the Weyl connection ∇^s and the $\mathsf{P}^s(\xi)$ are actually carried over from the associated graded bundles $\operatorname{gr}(\mathcal{V}M)$ and $\operatorname{gr}(T^*M)$ via the isomorphism induced by s. This also explains how the transformation law under a change of Weyl structure can be tensorial in the one-form Υ in spite of the explicit occurrence of a Rho-tensor (whose transformation is not tensorial, see Section 3.1 below). Hence under the isomorphism $\Gamma(\mathcal{V}M) \cong \Gamma(\mathrm{gr}(\mathcal{V}M))$ induced by a Weyl structure s, the operator $(D^-)^s$ corresponds to a universal expression in terms of ∇^s and the Rho-tensor P^s of s.

We next want to obtain similar descriptions for more general operators built up from D^- and the torsion and curvature of D. The first step is to understand the iterated derivatives $(D^-)^k$ with respect to D^-. This is tricky, because the argument in the proof of Proposition 2.1 was based on the fact that we start with $\sigma \in \Gamma(\mathcal{V}M)$, i.e. with a P-equivariant function. However, the function corresponding to $D^-\sigma$ is not P-equivariant any more, so we cannot simply iterate. A neat way around this is via an invariant notion of jets that was crucial for the developments in [**10**] and in [**11**].

2.2. First order jets. In geometry, a differential operator $\Gamma(\mathcal{V}M) \to \Gamma(\mathcal{W}M)$ of order at most k can be equivalently viewed as a vector bundle homomorphism $J^k\mathcal{V}M \to \mathcal{W}M$, where $J^k\mathcal{V}M$ is the kth jet prolongation of $\mathcal{V}M$. Thus, in order to understand the operators on $\mathcal{V}A$ and $\mathcal{V}M$, let us look at jets.

Let us first consider Klein geometries, i.e., the flat model Cartan geometries $G \to G/P$ with the Maurer-Cartan form ω. There, natural vector bundles are exactly the homogenous bundles $\mathcal{V} = G \times_P \mathbb{V}$ determined by P-modules $\mathbb{V}$. Any jet prolongation of a homogeneous bundle is homogeneous, too, so there is a P-module $J^k\mathbb{V}$ such that $J^k\mathcal{V} = G \times_P J^k\mathbb{V}$. The P-module $J^k\mathbb{V}$ can be obtained as the fiber of $J^k\mathcal{V}$ over the origin $o \in G/P$. Thus, the invariant operators are in bijective correspondence with the intertwining maps $J^k\mathbb{V} \to \mathbb{W}$ between the P-modules. In a dual picture, this provides an identification of linear invariant operators with the singular vectors in generalized Verma modules for all parabolic models, see e.g. [**14**].

Let us now consider a parabolic geometry $\mathcal{G} \to M$ with Cartan connection ω. In 1.3, we identified the covariant derivative of the canonical affine connection D restricted to L^- with the so called invariant differential ∇^ω acting on $\mathbb{V}$-valued functions on $\mathcal{G}$ via the constant vector fields $\omega^{-1}(X)$, $X \in \mathfrak{g}_-$. This is just the usual definition of the covariant derivative of affine connections in its frame form.

Although ∇^ω does not map P-equivariant functions σ into P-equivariant results, its extension

$$D^\omega : C^\infty(\mathcal{G}, \mathbb{V}) \to C^\infty(\mathcal{G}, \mathfrak{g}^* \otimes \mathbb{V}), \quad D^\omega\sigma(u)(X) = \omega^{-1}(X)(u) \cdot \sigma \in \mathbb{V}$$

does. This operator is called the *fundamental derivative.* It is very well understood in terms of the adjoint tractor bundles and leads to the so called tractor calculus, see [**12**, Sections 1.5.7-8].

The operator $D^\omega : \Gamma(\mathcal{V}M) \to \Gamma(\mathcal{W}M)$ can be understood as the universal differential operator and the standard fiber $\mathbb{W}$ of the target is $\mathfrak{g}^* \otimes \mathbb{V}$. We may also consider the analogue of the first jet prolongation by mapping σ to $(\sigma, D^\omega\sigma)$. The natural action of $g \in P$ on this target module $\mathbb{V} \oplus (\mathfrak{g}^* \otimes \mathbb{V})$ is

$$g \cdot (v, \varphi) = (g \cdot v, B \mapsto g \cdot \varphi(\mathrm{Ad}_{g^{-1}} B)).$$

The definition of D^ω together with equivariancy of the function σ that is differentiated readily implies that for $X \in \mathfrak{p} \subset \mathfrak{g}$, we get $D^\omega\sigma(u)(X) = -X \bullet \sigma(u)$, where in the right hand side, we use the infinitesimal action of $\mathfrak{p}$ on $\mathbb{V}$. Correspondingly, there is a P-submodule $J^1\mathbb{V} \subset \mathbb{V} \oplus \mathfrak{g}^* \otimes \mathbb{V}$, and $\sigma \mapsto (\sigma, D^\omega\sigma)$ induces a functorial isomorphism $J^1\mathcal{V}M \to \mathcal{G}\times_P J^1\mathbb{V}$. This construction works for all Cartan geometries of type G/P. See [**12**, Section 1.5.10] for details.

Now as a G_0-module, $\mathfrak{g} = \mathfrak{g}_- \oplus \mathfrak{p}$ which leads to an isomorphism $J^1\mathbb{V} \cong_{G_0} \mathbb{V} \oplus (\mathfrak{g}_-)^* \otimes \mathbb{V}$ via restriction in the second component. Consequently, we can naturally identify $\mathcal{G} \times_{G_0} J^1\mathbb{V} \to A$ with $\mathcal{V}A \oplus (L^-)^* \otimes \mathcal{V}A$. Now for a section $\sigma \in \Gamma(\mathcal{V}M)$ the value of the one-jet operator $j^1\sigma \in \Gamma(J^1\mathcal{V}M)$ defines a section of $\mathcal{G} \times_{G_0} J^1\mathbb{V} \to A$, which under this identification corresponds to $(\sigma, D^-\sigma)$.

2.3. Higher order jets. The iteration of the first jet prolongation leads to the non-holonomic jet prolongations and most of the redundancies can be removed by considering semi-holonomic jets. It turns out that the target space is a natural bundle with the fiber called *semiholonomic jet module* $\bar{J}^k\mathbb{V}$ induced by $\mathbb{V}$. Hence there is a kth order *invariant semiholonomic jet* operator, $k = 1, 2, \dots$ with values in $\mathcal{G}\times_P \bar{J}^k\mathbb{V}$, see [**10**] for details. Now similarly as in order 1, we get an isomorphism of G_0-modules

$$\bar{J}^k\mathbb{V} \cong_{G_0} \mathbb{V} \oplus (\mathfrak{g}_-)^* \otimes \mathbb{V} \oplus \cdots \oplus \otimes^k(\mathfrak{g}_-)^* \otimes \mathbb{V} \tag{8}$$

which is induced by restriction of multilinear maps. In particular, since any bundle map induced by a G_0-homomorphism is parallel for D and hence for D^-, for a section $\sigma \in \Gamma(\mathcal{V}M)$, the semi-holonomic k-jet operator $\bar{j}^k\sigma$ corresponds to $(\sigma, D^-\sigma, \dots, (D^-)^k\sigma)$ under this identification. We will implement this description by viewing $\bar{J}^k\mathbb{V}$ as the G_0–module $\mathbb{V}\oplus(\mathfrak{g}_-)^*\otimes\mathbb{V}\oplus\cdots\oplus\otimes^k(\mathfrak{g}_-)^*\otimes\mathbb{V}$ endowed with a (very complicated) extension of the action to P. Still, the infinitesimal action of $\mathfrak{p}_+$ is given by a G_0–equivariant map $\mathfrak{p}_+ \otimes \bar{J}^k\mathbb{V} \to \bar{J}^k\mathbb{V}$ in this picture.

From the construction it follows easily that for $\mathcal{V}M = \mathcal{G} \times_P \mathbb{V}$ the invariant semi-holonomic jet operator defines an injective bundle map $J^k\mathcal{V}M \hookrightarrow \mathcal{G} \times_P \bar{J}^k\mathbb{V}$. However, the image does not correspond to a P-invariant subspace in $\bar{J}^k\mathbb{V}$ and there seems to be no way to realize higher order jet prolongations as associated bundles to the Cartan bundle on Cartan geometries with non-vanishing curvature. (The identification of a jet prolongation with an induced bundle on the homogeneous model G/P discussed in Section 2.2 extends to geometries with vanishing curvature, which are locally isomorphic to G/P but in general not further.)

This makes the construction of invariant differential operators via algebraic techniques subtle. On the one hand, any P-equivariant map $J^k\mathbb{V} \to \mathbb{W}$ induces a kth order invariant differential operator on locally flat geometries, but not on general curved geometries. On the other hand, any P-equivariant map $\bar{J}^k\mathbb{V} \to \mathbb{W}$ gives rise to an invariant differential operator on all Cartan geometries (via composition with the invariant semi-holonomic jet operator) but, even in the linear case, not all invariant differential operators are obtained in this way. One obtains many operators in this way, however, in particular all operators that arise in BGG-sequences are of this character, see e.g. [**14**] and [**11**]. We will use the semi-holonomic jet prolongations in a different way here.

2.4. Expansion of $(D^-)^k$. Now we are ready to prove our first main result, which is a higher order analog of Proposition 2.1, so we want to study the operators $((D^-)^k)^s$ for a Weyl structure s and any $k \geq 1$. We will show that, under the isomorphism to the associated graded, this operator can be written via a "universal formula" in the following sense. Fix an initial representation $\mathbb{V}$ of P and an order k. Then we need finitely many G_0-equivariant linear maps A_i with values in $\otimes^k\mathfrak{g}_-\otimes\mathbb{V}$, such that for each Weyl structure s and any section $\sigma \in \Gamma(\mathcal{V}M)$, we can write the image of $s^*((D^-)^k(\sigma))$ in the associated graded (under the isomorphism determined by s) as $(\nabla^s)^k s^*\sigma + \sum_i \tilde{A}_i(T_i)$. Here ∇^s denotes the Weyl connection, $\tilde{A}_i$ is the bundle map induced by A_i and T_i is an iterated tensor product of factors of the form $(\nabla^s)^\ell s^*\sigma$ and $(\nabla^s)^\ell \mathsf{P}^s$ with $0 \le \ell < k$ that is independent of s. (The form of such a tensor product also determines the domain of the map A_i.)

2.5. THEOREM. *For any representation $\mathbb{V}$ and any $k \ge 1$, there is a universal expression for the image of $s^*((D^-)^k(\sigma))$ in the associated graded under the isomorphism determined by s as $(\nabla^s)^k s^*\sigma$ plus a sum of terms containing iterated Weyl derivatives of σ and P^s in the sense described above. Moreover, in any tensor product in this expansion, the total number of covariant derivatives is $< k$.*

PROOF. We proceed by induction on k. For $k = 1$, Proposition 2.1 provides an expansion in the claimed form with $i = 1$, $A_1 : (\mathfrak{g}_-)^* \otimes \mathfrak{p}_+ \otimes \mathbb{V} \to (\mathfrak{g}_-)^* \otimes \mathbb{V}$, the tensor product of the identity with the infinitesimal representation, and $T_1 = \mathsf{P}^s \otimes \sigma$.

Assuming that the result has been proved for all $\ell \le k$, we consider the invariant semi-holonomic k-jet operator $\bar{j}^k$, which via our description of $\bar{J}^k\mathbb{V}$ is given by $\bar{j}^k\sigma = (\sigma, D^-\sigma, \dots, (D^-)^k\sigma)$. Of course, the components of the pullback $s^*(\bar{j}^k\sigma)$ are just $s^*(D^-)^\ell\sigma$ for $0 \le \ell \le k$. Hence by induction hypothesis, we can express the image of $s^*\bar{j}^k\sigma$ in the associated graded under the isomorphism induced by s as $(\nabla^s)^k s^*\sigma$ plus a sum $\sum_i \tilde{A}_i(T_i)$ of universal terms as described above which involve covariant derivatives of $s^*\sigma$ and of P^s. Moreover, the total number of covariant derivatives occurring in each T_i is $< k$.

Since $\bar{j}^k\sigma$ lies in the subspace $\Gamma(\mathcal{G} \times_P \bar{J}^k\mathbb{V})$, we can apply Proposition 2.1 to conclude that $s^*(D^-\bar{j}^k\sigma) = \nabla^s(s^*\bar{j}^k\sigma) + \mathsf{P}^s \bullet s^*\bar{j}^k\sigma$. But naturality of D^- implies that viewing $(\mathfrak{g}_-)^* \otimes \bar{J}^k\mathbb{V}$ as $\oplus_{r=1}^{k+1} \otimes^r (\mathfrak{g}_-)^* \otimes \mathbb{V}$, the components of $D^-\bar{j}^k\sigma$ are just $(D^-)^r\sigma$ for $r = 1, \dots, k+1$. In particular, we can extract an expression for $s^*(D^-)^{k+1}\sigma$ from $s^*(D^-\bar{j}^k\sigma)$ and the known expressions for the iterated derivatives of lower order.

Expanding $s^*\bar{j}^k\sigma = (\nabla^s)^k s^*\sigma + \sum_i \tilde{A}_i(T_i)$ and applying ∇^s, we obtain $(\nabla^s)^{k+1}s^*\sigma$ from the first summand. On the other hand $\mathsf{P}^s \bullet (\nabla^s)^k s^*\sigma$ is of the form that we allow for the additional terms in the expansion of $s^*((D^-)^{k+1}\sigma)$. Thus we can conclude the proof by showing that the expressions $\nabla^s(\tilde{A}_i(T_i))$ and $\mathsf{P}^s \bullet \tilde{A}_i(T_i)$ all lead to expressions of the form we allow. For the first case, we can use that the bundle map $\tilde{A}_i$ is induced by a G_0-equivariant linear map and hence it is parallel for ∇^s. Thus we obtain $(\mathrm{id} \otimes \tilde{A}_i)(\nabla^s T_i)$ and of course $\nabla^s T_i$ can be expanded as a tensor product of factors $(\nabla^s)^\ell\sigma$ and $(\nabla^s)^\ell\mathsf{P}^s$ with a total of at most k covariant derivatives by our assumptions on T_i. For the other term, we can write $\mathsf{P}^s \bullet \tilde{A}_i(T_i)$ as $\tilde{B}_i(\mathsf{P}^s \otimes T_i)$. Here $B_i = (\mathrm{id} \otimes \tau) \circ (\mathrm{id} \otimes \mathrm{id} \otimes A_i)$, where τ denotes the infinitesimal representation, so this is again of the required form. □

2.6. Involving curvature quantities. As discussed in Section 1.4, the torsion and the curvature of the canonical connection D on A can be equivalently encoded via the curvature κ of the canonical Cartan connection ω. Now the Cartan curvature κ is a section of $\mathcal{G} \times_P \mathbb{V} \to M$, where $\mathbb{V} := (\Lambda^2(\mathfrak{g}/\mathfrak{p})^* \otimes \mathfrak{g})$, so Theorem 2.5 applies to κ. Hence for any $k \in \mathbb{N}$, we can describe the pullback $s^*(D^-)^k\kappa$ in terms of $(\nabla^s)^k s^*\kappa$ and universal terms depending on iterated Weyl derivatives of order less than k of $s^*\kappa$ and of the Rho-tensor P^s. We have seen, that on A, κ decomposes as $T + W + Y$ according to the values in $\mathfrak{g} = \mathfrak{g}_- \oplus \mathfrak{g}_0 \oplus \mathfrak{p}_+$. Of course, these and further decompositions coming from G_0-invariant operations are, on the level of A, preserved by the operator D^-. They correspond to decompositions of $s^*\kappa$ on M (depending on s) which in turn are preserved by the Weyl connection ∇^s. We also discussed in Section 1.4 how to decompose $s^*\kappa$ into its components $s^*T = T^s + (\partial\mathsf{P}^s)_{\mathfrak{g}_-}$, $s^*W = R^s + (\partial\mathsf{P}^s)_{\mathfrak{g}_0}$, and $s^*Y = Y^s + (\partial\mathsf{P})_{\mathfrak{p}_+}$, which are related to the torsion and curvature of ∇^s.

Otherwise put, if we consider a component K of the curvature and torsion of D, and form $(D^-)^kK$, we can recover this from $(D^-)^k\kappa$ by G_0-equivariant operations. Applying the same operations to $s^*(D^-)^k\kappa$ we conclude that we can express $s^*(D^-)^kK$ as $(\nabla^s)^ks^*K$ plus a universal expression in the sense discussed in Section 2.4 involving iterated derivatives of order $< k$ of components of $s^*\kappa$ and of P^s.

But this is sufficient to pull back polynomial invariant operators (in a sense that will be made precise) constructed from D^- along Weyl structures and obtain a special class of affine invariants of Weyl structures. So we start with arbitrary representations $\mathbb{V}$ and $\mathbb{W}$ of P and we are looking for an operator $\Phi : \Gamma(\mathcal{V}A) \to \Gamma(\mathcal{W}A)$ such that we can write $\Phi(\sigma) = \sum_i A_i(T_i)$, where A_i is a natural bundle map induced by a G_0-equivariant map with values in $\mathbb{W}$ and T_i is an iterated tensor product of factors of the form $(D^-)^\ell(\sigma)$ and $(D^-)^\ell K$ for a component K of κ. The form of this tensor product is required to be independent of σ and it determines the representation on which the map inducing A_i is defined.

2.7. Theorem. *Let $\mathbb{V}$ and $\mathbb{W}$ be representations of P and let $\Phi : \Gamma(\mathcal{V}A) \to \Gamma(\mathcal{W}A)$ be a polynomial invariant operator constructed from D^- in the sense introduced above. Then for any Weyl structure s, the operator $\Phi^s : \Gamma(\mathcal{V}M) \to \Gamma(\mathcal{W}M)$ is induced, under the isomorphism to the associated graded bundles determined by s, by a polynomial invariant operator constructed from ∇^s, its curvature and its torsion in the same sense. Moreover, this correspondence has the property that for two Weyl structures s and $\tilde{s}$ and a point $x \in M$ such that $s(x) = \tilde{s}(x)$ one obtains $\Phi^s(\sigma)(x) = \Phi^{\tilde{s}}(\sigma)(x)$ for any $\sigma \in \Gamma(\mathcal{V}M)$.*

Proof. Theorem 2.5 provides us with an expansion for $(D^-)^\ell(\sigma)$ for any ℓ involving G_0-equivariant maps acting on tensors. The discussion in Section 2.6 gives us analogous expressions for $(D^-)^\ell(K)$ for any component K of κ. Since tensor products of G_0-equivariant maps are G_0-equivariant, we also get expressions for iterated tensor products of such terms. Applying G_0-equivariant maps to such tensor products just gives rise to compositions of G_0-equivariant maps, which are equivariant, too. The last statement has already been observed for any operator of the form Φ^s in the beginning of Section 2. □

3. The nearly invariant operators

3.1. The transformation rules. The natural approach to invariants of parabolic geometries is via Weyl-structures, i.e. to consider differential operators and differential invariants defined using the Weyl connections ∇^s, their torsion and their curvature, and request that they are independent of the choice of s. Thus, we should understand how the gradings, the covariant derivatives and Rho tensors change under the change of the Weyl structures.

Considering two Weyl structures as reductions $\bar{s}_1, \bar{s}_2 : \mathcal{G}_0 \to \mathcal{G}$, clearly there must be a function $\Upsilon : \mathcal{G}_0 \to \mathfrak{p}_+$ such that for all $u \in \mathcal{G}_0$, $g_0 \in G_0$,

$$\bar{s}_2(u) = \bar{s}_1(u) \cdot \exp \Upsilon(u),$$
$$\bar{s}_2(u \cdot g_0) = \bar{s}_1(u) \cdot \exp \Upsilon(u) \cdot g_0 = \bar{s}_1(u) \cdot g_0 \cdot \exp(\mathrm{Ad}_{g_0^{-1}} \Upsilon(u)).$$

Thus the function Υ represents a one-form on M and, fixing s_1, this is a bijective correspondence between $\Upsilon \in \Omega^1(M)$ and all Weyl structures. This reflects the affine bundle structure of A. Of course, these one-forms are also represented as functions $\Upsilon : \mathcal{G} \to \mathfrak{p}_+$ with the right equivariancy property, and then the functions $f = \exp \Upsilon : \mathcal{G} \to P/G_0 \simeq P_+$ with the equivariance property $f(u \cdot (g_0 g_+)) = g_+^{-1} g_0^{-1} f(u) g_0$ can be directly seen as the corresponding sections $s : M \to A = \mathcal{G} \times_P P/P_0$. See [**7**, Proposition 2.2] for a detailed exposition.

The formulae for the transformations in terms of the functions Υ are explained in detail in [**12**, Sections 5.1.6-9]. They look pretty complicated and we shall not need them explicitly here. Just in the simplest case of trivial filtrations (i.e., $|1|$-graded $\mathfrak{g}$) we obtain for vector fields ξ on M and sections σ of irreducible natural bundles (the hat indicates the transformed objects)

$$\hat{\nabla}_\xi \sigma = \nabla_\xi \sigma - \{\Upsilon, \xi\} \bullet \sigma \tag{9}$$

$$\hat{\mathsf{P}}(\xi) = \mathsf{P}(\xi) + \nabla_\xi \Upsilon + \frac{1}{2} \operatorname{ad}(\Upsilon)^2(\xi). \tag{10}$$

Differentiating (9) again, derivatives of Υ appear, and the relatively nice transformation rule (10) indicates a chance to balance the formulae by adding 'correction terms' including P in order to keep the transformation rules algebraic in the parameters Υ. This is exactly what happens with the Rho-corrected derivatives appearing in Proposition 2.1.

About a hundred years back this was the motivation for considering Schouten's Rho tensor in conformal Riemannian geometry.

In the sequel we shall consider *polynomial invariant operators* $\Psi_s : \Gamma(\mathcal{V}M) \to \Gamma(\mathcal{W}M)$, constructed by a fixed universal expression Ψ from from ∇^s, its curvature and its torsion in the way discussed in Section 2.7. We say that Ψ is a *nearly invariant operator* if the transformation formula for Ψ_s under the change of the Weyl structure s is tensorial in Υ. Otherwise put, the operator Ψ is nearly invariant if and only if for Weyl structures s_1, s_2 and a point $x \in M$ such that $s_1(x) = s_2(x)$ we get $\Psi_{s_1}\sigma(x) = \Psi_{s_2}\sigma(x)$ for every $\sigma \in \Gamma(\mathcal{V}M)$.

Theorem 2.7 shows, that via $\Phi \mapsto \Phi^s$, any polynomial affine invariant of the canonical connection D^- on A gives rise to a nearly invariant operator on M.

3.2. Derivatives of the Rho-tensor. The final key step towards understanding nearly invariant operators is analogous to the well known fact that in affine differential invariants one may use symmetrized iterated covariant derivatives rather then iterated covariant derivatives. We need an analogous symmetrization argument for iterated covariant derivatives of the Rho tensor associated to a Weyl structure. This will allow us to effectively use normal Weyl structures in the study of nearly invariant operators.

Recall from Section 1.5 that a characteristic of normal Weyl structures is the vanishing of certain symmetrizations of iterated covariant derivatives of the Rho-tensor. We will refer to these as *form-symmetrized* iterated derivatives. Fixing a

Weyl-structure s with Rho-tensor P and Weyl connections ∇, we view P as a one-form on M with values in $\mathrm{gr}(T^*M)$. The kth covariant derivative $(\nabla)^k\mathsf{P}$ is then a section of $\otimes^{k+1}T^*M \otimes T^*M$ (or the associated graded of this bundle). Now we define the kth *form symmetrized Weyl-derivative* $\mathcal{F}^k(\mathsf{P}) \in \Gamma(\mathrm{gr}(S^{k+1}T^*M \otimes T^*M))$ as the symmetrization of $(\nabla^s)^k\mathsf{P}$ over the first $k+1$ indices. For convenience, we put $\mathcal{F}^0(\mathsf{P}) = \mathsf{P}$.

3.3. THEOREM. *For any k, there is a universal expression for $(\nabla)^k\mathsf{P} - \mathcal{F}^k(\mathsf{P})$ obtained via bundle maps induced by G_0-equivariant maps from tensor products whose factors are of the form $\mathcal{F}^\ell(\mathsf{P})$ and $(\nabla^s)^\ell K$ with $\ell < k$, where K is a component of the pullback of the Cartan curvature along s.*

PROOF. Throughout this proof, we write ∇ for ∇^s. We start by computing $(\nabla\mathsf{P} - \mathcal{F}^1(\mathsf{P}))(\xi, \eta)$ for vector fields $\xi, \eta \in \mathfrak{X}(M)$ via

$$\nabla_\xi\mathsf{P}(\eta) - \nabla_\eta\mathsf{P}(\xi) = \nabla_\xi(\mathsf{P}(\eta)) - \mathsf{P}(\nabla_\xi\eta) - \nabla_\eta(\mathsf{P}(\xi)) + \mathsf{P}(\nabla_\eta\xi).$$

The definition of the torsion of a Weyl connection reads as

$$T(\xi, \eta) = \nabla_\xi\eta - \nabla_\eta\xi - [\xi, \eta] + \{\xi, \eta\}, \tag{11}$$

so we can insert this to rewrite the terms in which a derivative goes into Rho. Moreover, from section 1.4, we know that

$$Y^s(\xi, \eta) = \nabla_\xi(\mathsf{P}(\eta)) - \nabla_\eta(\mathsf{P}(\xi)) - \mathsf{P}([\xi, \eta]) + \mathsf{P}(\{\xi, \eta\}) + \{\mathsf{P}(\xi), \mathsf{P}(\eta)\}$$

can be recoverd from the pullback of the Cartan curvature and a component of P. Together with the above, we conclude that

$$\nabla\mathsf{P} = \mathcal{F}^1(\mathsf{P}) + \tfrac{1}{2}Y - \tfrac{1}{2}i_T\mathsf{P} - \tfrac{1}{2}\{\mathsf{P}, \mathsf{P}\}. \tag{12}$$

This proves the theorem in the case $k = 1$. To complete the proof we show by induction on k that $\nabla\mathcal{F}^k(\mathsf{P}) - \mathcal{F}^{k+1}(\mathsf{P})$ admits a universal expansion as claimed in theorem with $\ell \leq k$. Since, by induction, one Weyl derivative of any term allowed in step k leads to a term allowed in step $k+1$, this recursively implies that claim of the theorem.

The case $k = 0$ has been sorted out above, so we assume that $k > 0$ and our claim has been proved $\ell < k$.

Since $\mathcal{F}^{k-1}(\mathsf{P}) \in \Gamma(S^kT^*M \otimes T^*M)$ is symmetric in its first k-indices, we can compute $\mathcal{F}^k\mathsf{P}$ as $\frac{1}{k+1}$ times the sum over all cyclic permutations in the first $k+1$ entries of $\nabla\mathcal{F}^{k-1}(\mathsf{P}) \in \Gamma(T^*M \otimes S^kT^*M \otimes T^*M)$. Using abstract index notation $\mathcal{F}_{a_1\dots a_k b}$ for $\mathcal{F}^{k-1}\mathsf{P}$ we can write $\mathcal{F}^k(P)_{a_0\dots a_k b}$ as

$$\frac{1}{k+1}\sum_{i=0}^{k}\nabla_{a_i}\mathcal{F}_{a_0\dots\widehat{a_i}\dots a_k b},$$

where as usual the hat denotes omission and we observe we may arbitrarily permute the first k indices in $\mathcal{F}$. Using symmetry of $\mathcal{F}$ to rewrite $\nabla_{a_0}\mathcal{F}_{a_1\dots a_k b}$, we conclude that we can rewrite $\nabla\mathcal{F}^k(\mathsf{P}) - \mathcal{F}^{k+1}(\mathsf{P})$ as

$$\frac{1}{k+1}\sum_{i=1}^{k}\left(\nabla_{a_0}\mathcal{F}_{a_i a_1,\dots\widehat{a_i}\dots a_k b} - \nabla_{a_i}\mathcal{F}_{a_0 a_1\dots\widehat{a_i}\dots a_k b}\right).$$

If $k = 1$, then the terms $\mathcal{F}$ already are of the form $\nabla\mathsf{P}$. For $k > 1$, the induction hypothesis implies that, up to terms of the form we allow in our expansions, we may replace the occurrences of $\mathcal{F}$ by $\nabla\mathcal{F}^{k-2}(\mathsf{P})$. But then each term in our sum becomes twice an alternation of a double covariant derivative of $\mathcal{F}^{k-2}$. (Recall that $\mathcal{F}^0(\mathsf{P}) = \mathsf{P}$.)

But now for any tensor field t, the alternation of $\nabla^2 t$ maps $\xi, \eta \in \mathfrak{X}(M)$ to

$$\nabla_\xi \nabla_\eta t - \nabla_{\nabla_\xi \eta} t - \nabla_\eta \nabla_\xi t + \nabla_{\nabla_\eta \xi} t.$$

Inserting from (11) and rearranging terms, we see that this can be rewritten as

$$R(\xi, \eta) \bullet t - \nabla_{T(\xi,\eta)} t + \nabla_{\{\xi,\eta\}} t, \tag{13}$$

where in the first term the bullet denotes the tensorial action of the curvature. Now the last two terms in (13) are obtained from G_0-equivariant operations acting on $T \otimes \nabla t$ and of ∇t, respectively, and T can be obtained from the pullback of the Cartan curvature and from P. In our situation $t = \mathcal{F}^{k-2}(\mathsf{P})$, so by induction we can rewrite ∇t as a sum of terms of the allowed forms, so we see that these two summands altogether only produce terms of the allowed form.

For the first term in (13), we can rewrite the curvature R of $\nabla = \nabla^s$ according to section 1.4 as $W - (\partial \mathsf{P})_{\mathfrak{g}_0}$ where W is a component of the pullback of the Cartan curvature. But then the resulting terms can be obtained via G_0-equivariant maps from $W \otimes t$ respectively from $\mathsf{P} \otimes t$ and since t equals $\mathcal{F}^{k-2}(\mathsf{P})$, this leads to allowed terms only. □

3.4. Nearly invariant operators. Let us collect the information on affine invariants of the Weyl connections we have obtained so far. As discussed in Section 3.1, such an operator comes from a universal expression Ψ which involves bundle maps induced from G_0-equivariant linear maps and certain tensor products. If the operator acts on $\Gamma(\mathcal{V}M)$ these tensor products contain factors that are iterated covariant derivatives of either a section $\sigma \in \Gamma(\mathcal{V}M)$ or a component of the curvature of the involved connection (for which we will insert all Weyl connections). Now as discussed in Sections 1.4 and 2.6, we can universally decompose the curvature and torsion of the Weyl connections into components of the pullback of the Cartan connection and into components of the Rho-tensor. Since these decompositions are induced by G_0-equivariant maps, the same applies to iterated derivatives of (components of) the torsion and the curvature. Moreover, we can use Theorem 3.3 to rewrite iterated derivatives of the Rho-tensor via components of the pullback of the Cartan curvature and form-symmetrized iterated derivatives of the Rho-tensor and as before, this extends to components of the Rho-tensor. The upshot of this is that we may assume that all our factors in the tensor products are

- iterated covariant derivatives of $\sigma \in \Gamma(\mathcal{V}M)$
- iterated covariant derivatives of components of the pullback of the Cartan curvature
- components of form symmetrized iterated covariant derivatives of the Rho tensor.

Armed with this observation, we are ready to formulate and prove the main result of the paper.

3.5. THEOREM. *The nearly invariant operators are exactly the universal expansions obtained via $\Phi \mapsto \Phi^s$ from polynomial affine differential invariants of the natural covariant derivative D^- and its curvature and torsion on A.*

PROOF. Theorem 2.7 shows that affine differential invariants of D^- and the curvature and torsion of D induce nearly invariant operators.

To prove the converse inclusion, let us take a universal expression Ψ which gives rise to a nearly invariant operator which is formed in the way discussed above. Now let $\tilde{\Psi}^1$ be the universal expression obtained by removing from Ψ all terms in which the tensor product contains a component of a form symmetrized covariant derivative of the Rho tensor. For each summand in this expression count the total number of covariant derivatives in all factors showing up in the tensor product. Let k be the maximal number that occurs and let Ψ_1 be the sum of all terms in which the total number of covariant derivatives equals k. Then there is an obvious affine invariant Φ_1 of D^- on A that corresponds to Ψ_1 (in which we simply replace all covariant derivatives by D^-). Via $\Phi_1 \mapsto \Phi_1^s$, we obtain a nearly invariant operator defined on $\Gamma(\mathcal{V}M)$ and we subtract this from Ψ.

On the one hand, the result by construction is a nearly invariant operator $\tilde{\Psi}$. On the other hand, Theorem 2.5 implies that in Φ_1^s, we obtain exactly the same terms involving a total number of k covariant derivatives as in Ψ_s^1 while for all other terms, the total number of covariant derivatives that occur is strictly less than k. Otherwise put, the operator $\tilde{\Psi}$ has the property that in any term that does not contain a component of a form symmetrized iterated covariant derivative of the Rho tensor, there are less than k covariant derivatives in total.

This shows that we can iterate our procedure by applying the same construction to $\tilde{\Psi}$. After finitely many steps, we arrive at an expression for which the terms that do not involve any components of form symmetrized iterated covariant derivatives of Rho also do not contain any covariant derivatives and hence evidently are obtained from an affine invariant on A.

Hence we conclude that our original operator Ψ can be written as the sum of an operator of the form $\Phi \mapsto \Phi^s$ and a nearly invariant operator $\hat{\Psi}$ for which *any* term involves a component of a form symmetrized iterated covariant derivative of Rho. But now we can easily conclude that the nearly invariant operator $\hat{\Psi}$ vanishes identically. Indeed, take any Weyl structure s and consider $\Psi_s(\sigma)$ and a point $x \in M$. Then from Section 1.5 we know that we can find a Weyl structure $\tilde{s}$ which is normal at x and satisfies $\tilde{s}(x) = s(x)$. But by normality, any term in $\hat{\Psi}_{\tilde{s}}$ vanishes at x and hence $0 = \hat{\Psi}_{\tilde{s}}(\sigma)(x)$, and this equals $\hat{\Psi}_s(\sigma)(x)$ since $\hat{\Psi}$ is nearly invariant. This completes the proof. □

3.6. Remark. In our discussion, we have focused on the case of normal parabolic geometries, which equivalently encode some underlying structure and hence provides applications to invariants of that structure. The concept of Weyl structures as well as the basic calculus we develop here are available for general Cartan geometries and basically also the results we have proved above allow for extensions to this more general setting. The basic difference is that the relation between the pullback $s^*\kappa$ of the Cartan curvature, the curvature of the Weyl connection ∇^s determined by s and the Rho-tensor P^s becomes more complicated. So to extend the results, one would have to modify the definitions of affine invariants that are used in Theorems 2.7 and 3.5 appropriately. We do not go into details on this here.

References

[1] T. N. Bailey, M. G. Eastwood, and A. R. Gover, *Thomas's structure bundle for conformal, projective and related structures*, Rocky Mountain J. Math. **24** (1994), no. 4, 1191–1217, DOI 10.1216/rmjm/1181072333. MR1322223

[2] Thomas Branson, *Second order conformal covariants*, Proc. Amer. Math. Soc. **126** (1998), no. 4, 1031–1042, DOI 10.1090/S0002-9939-98-04092-1. MR1422849
[3] David M. J. Calderbank and Tammo Diemer, *Differential invariants and curved Bernstein-Gelfand-Gelfand sequences*, J. Reine Angew. Math. **537** (2001), 67–103, DOI 10.1515/crll.2001.059. MR1856258
23 (2005), 149–175.
[4] David M. J. Calderbank, Tammo Diemer, and Vladimír Souček, *Ricci-corrected derivatives and invariant differential operators*, Differential Geom. Appl. **23** (2005), no. 2, 149–175, DOI 10.1016/j.difgeo.2004.07.009. MR2158042
[5] A. Čap, *On canonical Cartan connections associated to filtered G-structures*, preprint `arXiv:1707.05627`, 46 pp.
[6] Andreas Čap and A. Rod Gover, *Tractor calculi for parabolic geometries*, Trans. Amer. Math. Soc. **354** (2002), no. 4, 1511–1548, DOI 10.1090/S0002-9947-01-02909-9. MR1873017
[7] A. Čap and T. Mettler, *Geometric theory of Weyl structures*, to appear in Commun. Contemp. Math., DOI: 10.1142/s0219199722500262.
[8] Andreas Čap and Hermann Schichl, *Parabolic geometries and canonical Cartan connections*, Hokkaido Math. J. **29** (2000), no. 3, 453–505, DOI 10.14492/hokmj/1350912986. MR1795487
[9] Andreas Čap and Jan Slovák, *Weyl structures for parabolic geometries*, Math. Scand. **93** (2003), no. 1, 53–90, DOI 10.7146/math.scand.a-14413. MR1997873
[10] A. Čap, J. Slovák, and V. Souček, *Invariant operators on manifolds with almost Hermitian symmetric structures. II. Normal Cartan connections*, Acta Math. Univ. Comenian. (N.S.) **66** (1997), no. 2, 203–220. MR1620484
[11] Andreas Čap, Jan Slovák, and Vladimír Souček, *Bernstein-Gelfand-Gelfand sequences*, Ann. of Math. (2) **154** (2001), no. 1, 97–113, DOI 10.2307/3062111. MR1847589
[12] Andreas Čap and Jan Slovák, *Parabolic geometries. I*, Mathematical Surveys and Monographs, vol. 154, American Mathematical Society, Providence, RI, 2009. Background and general theory, DOI 10.1090/surv/154. MR2532439
[13] E. Cartan, *Les espaces à connexion conforme*, Ann. Soc. Pol. Math. **2** (1923), 171–202.
[14] Michael Eastwood and Jan Slovák, *Semiholonomic Verma modules*, J. Algebra **197** (1997), no. 2, 424–448, DOI 10.1006/jabr.1997.7136. MR1483772
[15] Charles Fefferman, *Parabolic invariant theory in complex analysis*, Adv. in Math. **31** (1979), no. 2, 131–262, DOI 10.1016/0001-8708(79)90025-2. MR526424
[16] Charles Fefferman and C. Robin Graham, *The ambient metric*, Annals of Mathematics Studies, vol. 178, Princeton University Press, Princeton, NJ, 2012. MR2858236
[17] Paul Gauduchon, *Structures de Weyl-Einstein, espaces de twisteurs et variétés de type* $S^1 \times S^3$ (French), J. Reine Angew. Math. **469** (1995), 1–50, DOI 10.1515/crll.1995.469.1. MR1363825
[18] Marc Herzlich, *Parabolic geodesics as parallel curves in parabolic geometries*, Internat. J. Math. **24** (2013), no. 9, 1350067, 16, DOI 10.1142/S0129167X13500675. MR3109439
[19] Kengo Hirachi, *Normal form for pseudo-Einstein contact forms and intrinsic CR normal coordinates*, Complex Anal. Synerg. **8** (2022), no. 3, Paper No. 13, 7, DOI 10.1007/s40627-022-00101-7. MR4462884
[20] Ivan Kolář, Peter W. Michor, and Jan Slovák, *Natural operations in differential geometry*, Springer-Verlag, Berlin, 1993, DOI 10.1007/978-3-662-02950-3. MR1202431
[21] Tohru Morimoto, *Geometric structures on filtered manifolds*, Hokkaido Math. J. **22** (1993), no. 3, 263–347, DOI 10.14492/hokmj/1381413178. MR1245130
[22] R. W. Sharpe, *Differential geometry*, Graduate Texts in Mathematics, vol. 166, Springer-Verlag, New York, 1997. Cartan's generalization of Klein's Erlangen program; With a foreword by S. S. Chern. MR1453120
[23] J. Slovák, *Parabolic geometries*, Research Lecture Notes, Masaryk University, 1997, IGA Preprint (97/11), University of Adelaide, 70 pp.
[24] Noboru Tanaka, *On the equivalence problems associated with simple graded Lie algebras*, Hokkaido Math. J. **8** (1979), no. 1, 23–84, DOI 10.14492/hokmj/1381758416. MR533089
[25] T. Y. Thomas, *On conformal geometry*, Proc. N.A.S. **12** (1926), 352–359.
[26] T. Y. Thomas, *Conformal tensors*, Proc. N.A.S. **18** (1931), 103–189.
[27] Volkmar Wünsch, *On conformally invariant differential operators*, Math. Nachr. **129** (1986), 269–281, DOI 10.1002/mana.19861290123. MR864639

[28] Keizo Yamaguchi, *Differential systems associated with simple graded Lie algebras*, Progress in differential geometry, Adv. Stud. Pure Math., vol. 22, Math. Soc. Japan, Tokyo, 1993, pp. 413–494, DOI 10.2969/aspm/02210413. MR1274961

Faculty of Mathematics, University of Vienna, Oskar-Morgenstern-Platz 1, 1090 Wien, Austria
Email address: Andreas.Cap@univie.ac.at

Department of Mathematics and Statistics, Masaryk University, Kotlářská 2a, 611 37 Brno, Czech Republic
Email address: slovak@math.muni.cz

Contemporary Mathematics
Volume **788**, 2023
https://doi.org/10.1090/conm/788/15820

Coverings and pseudosymmetries of differential equations

Vladimir N. Chetverikov

ABSTRACT. Finite-dimensional coverings from systems of differential equations are investigated. This problem is of interest in view of its relationship with the computation of differential substitutions, nonlocal symmetries, recursion operators, and Bäcklund transformations. We show that the distribution specified by the fibers of a covering is determined by an integrable pseudosymmetry of the system. Conversely, every integrable pseudosymmetry of a system defines a covering from this system. The vertical component of the pseudosymmetry is a matrix analog of the evolution differentiation. The corresponding generating matrix satisfies a matrix analog of the linearization of the equation. A method for constructing coverings is given and demonstrated by the examples of the Laplace equation and the Kapitsa pendulum system.

1. Introduction

The concept of covering of differential equations was introduced in [**12**] as a generalization of the Wahlquist–Estabrook prolongation structures [**22**]. The notions of differential substitution, nonlocal symmetry, Bäcklund transformation, and recursion operator were formulated in the language of coverings [**2, 13**]. The problem of finding coverings over a given system turned out to be related to the description of integrable systems [**11**].

We study coverings from a given system of differential equations and use the Vinogradov's approach (see [**2, 10, 12, 13, 20**]). Namely, systems of differential equations are identified with infinite-dimensional submanifolds of spaces of infinite jets. Such submanifolds are called diffieties. Morphisms of the category of diffieties correspond to system transformations such that the independent and dependent variables of one systems are expressed via the independent and dependent variables and the derivatives of the dependent variables with respect to the independent variables of the other system. Isomorphisms of this category are called C-transformations [**2, 10**]. Coverings generalize C-transformations.

Differential substitutions (coverings of a special type) of evolution equations with one spatial variable were investigated in [**18**]. The set of vector fields defining the fibers of a covering was called a pseudosymmetry in view of its analogy with

2020 *Mathematics Subject Classification.* Primary 35A30, 93B27; Secondary 58J70, 58J72.

Key words and phrases. Coverings of differential equations, decompositions of systems, infinite jets, pseudosymmetries, multivector fields, the Schouten bracket.

This work was partially supported by the Ministry of Science and Higher Education of the Russian Federation, project no. 0705-2020-0047 and the Russian Foundation for Basic Research, project no. 20-07-00279.

©2023 American Mathematical Society

an infinitesimal symmetry. These results were generalized to arbitrary systems of differential equations in [**4**]. The defining equations for pseudosymmetries were derived. It was noted that a pseudosymmetry determines a covering only if it satisfies integrability conditions. These conditions were been formulated. Pseudosymmetries that satisfies them were called integrable.

In [**5**], to simplify the computations of multidimensional pseudosymmetries the exterior product of vector fields defining a pseudosymmetry were considered. Vinogradov's ideas to understand both multi-vector fields and differential forms as operators acting on the module of differential forms [**21**] were used. The definition of pseudosymmetry was rewritten in the language of the Schouten bracket of multivector fields and total derivatives with respect to the independent variables of the system. It was proved that any multivector pseudosymmetry is uniquely determined by its values on generators of the corresponding module of Cartan forms over the algebra of $\mathcal{C}$-differential operators. In addition, it was shown how to construct a new pseudosymmetry using an involutive closure of two pseudosymmetries.

The problem of the description of coverings from a given system is related to the decomposition problem. This explains the analogy between our results and the well-known results in [**9**], where a decomposition of an affine control system is associated with an affine distribution invariant with respect to that system. In our case, every covering from a given system is associated with an invariant integrable distribution on the diffiety of the system.

Note that the decompositions investigated in [**9**] correspond to a particular kind of coverings when the independent variable does not change, and the dependent variables are expressed only via the independent and dependent variables. Decompositions of a general kind were investigated in [**6**]. Unlike [**9**] in [**6**], C-transformations are used to transform systems into decomposed forms. Since in control theory C-transformations are called orbital equivalences, the decompositions examined in [**6**] are called orbital.

The present paper synthesizes and brings together the results of works [**4**–**6**]. Some proofs are simplified. There are methodological improvements. Moreover, the results of the work [**5**] are applied to control systems. Namely, it is shown that in the case of controllable systems, the module of Cartan q-forms over the algebra of $\mathcal{C}$-differential operators is free and a basis for this module is constructed. Thus, in order to construct a q-dimensional pseudosymmetry, it is enough to set its values on this basis, and arbitrary functions can be taken as such values. This result will be used in the future to construct multidimensional orbital decompositions of control systems.

Finally, we stress that the problem of constructing coverings from a given system is of interest as another tool for analyzing the internal structure of the system. Coverings, pseudosymmetries, and integrable pseudosymmetries are invariant under the group of C-transformations and can be used to classify systems of differential equations with respect to this group.

2. Coverings of differential equations

EXAMPLE 2.1. Consider a solution $u(x_1, x_2)$ of the Laplace equation $u_{x_1x_1} + u_{x_2x_2} = 0$ such that $u^2_{x_1x_1} + u^2_{x_1x_2} \neq 0$. By the implicit function theorem, the variables x_1, x_2 are expressed via the derivatives u_{x_1} and u_{x_2}. To each such solution $u(x_1, x_2)$ assign the function $v(z_1, z_2)$ such that $u = v(u_{x_1}, u_{x_2})$. It can be proved (see [**2**, Ch. 3, Example 6.1]) that function v also satisfies the Laplace equation

$$v_{z_1 z_1} + v_{z_2 z_2} = 0. \tag{2.1}$$

We get a mapping of solutions of the Laplace equation: $u(x_1, x_2) \mapsto v(z_1, z_2)$.

To construct the inverse mapping, one should find functions $w^1(z_1, z_2)$ and $w^2(z_1, z_2)$ such that

$$w^1_{z_1} = \frac{z_1 v_{z_1} - z_2 v_{z_2}}{z_1^2 + z_2^2} = -w^2_{z_2}, \qquad w^1_{z_2} = \frac{z_1 v_{z_2} + z_2 v_{z_1}}{z_1^2 + z_2^2} = w^2_{z_1}, \tag{2.2}$$

where the function $v(z_1, z_2)$ satisfies the Laplace equation (2.1) and the inequality $v^2_{z_1} + v^2_{z_2} \neq 0$. Using the implicit function theorem again, we express the variables z_1, z_2 via the functions $w^1(z_1, z_2)$ and $w^2(z_1, z_2)$ and find the function $v(z_1, z_2)$ from the relation $v = u(w^1, w^2)$.

The functions $w^1(z_1, z_2)$ and $w^2(z_1, z_2)$ are determined by their values at an initial point and by the function $v(z_1, z_2)$. Thus, to one function $v(z_1, z_2)$ there corresponds a set of function $u(x_1, x_2)$. In addition, the Laplace equation is equivalent to system (2.1)–(2.2). The mapping $u(x_1, x_2) \mapsto v(z_1, z_2)$ is a covering from the Laplace equation to the Laplace equation.

Consider an arbitrary system of differential equations for a vector function $u = (u^1, \ldots, u^m)$ of the variables $x = (x_1, \ldots, x_n)$:

$$G_\alpha(x, u, \ldots, u_\sigma, \ldots) = 0, \quad \alpha = 1, \ldots, r. \tag{2.3}$$

Here and in the following, $\sigma = (i_1, \ldots, i_n)$ is a multi-index, $|\sigma| = i_1 + \ldots + i_n$ is its length, and

$$u_\sigma = \left(\frac{\partial^{|\sigma|} u^1}{\partial x_1^{i_1} \ldots \partial x_n^{i_n}}, \ldots, \frac{\partial^{|\sigma|} u^m}{\partial x_1^{i_1} \ldots \partial x_n^{i_n}} \right).$$

The expression on the left-hand side in (2.3) means that the function G_α depends on x, u, and finitely many derivatives of the form u_σ. The maximum length $|\sigma|$ of multi-indices σ occurring in (2.3) coincides with the order of the system.

Suppose system (2.3) is equivalent to a system of the form

$$\Phi_\beta(z, v, \ldots, v_\sigma, \ldots) = 0, \quad \beta = 1, \ldots, r_1, \tag{2.4}$$

$$w^j_i = W^j_i(z, w, v, \ldots, v_\sigma, \ldots), \quad i = 1, \ldots, n, \ j = 1, \ldots, q, \tag{2.5}$$

where $z = (z_1, \ldots, z_n)$ is the set of independent variables and $v = (v^1, \ldots, v^{m_1})$, $w = (w^1, \ldots, w^q)$ are the sets of dependent variables of the system. Then one says that system (2.3) *covers* system (2.4). We also say that system (2.4)–(2.5) is a *decomposable form* of system (2.3).

3. Infinite jets

A detailed exposition of the material presented in this section can be found in [**2**, Ch. 3 and Ch. 4].

For $k \geq 0$, let $J^k(n, m)$ be the finite-dimensional space with coordinates

$$x_i, \ u^j, \ u^j_\sigma, \qquad i = 1, \ldots, n, \ j = 1, \ldots, m, \ |\sigma| \leq k. \tag{3.1}$$

If a vector function $s = (s^1(x), \ldots, s^m(x))^{\mathrm{T}}$ is k times differentiable in a neighborhood of a point $a = (a_1, \ldots, a_n) \in \mathbb{R}^n$, then the point of $J^k(n, m)$ with coordinates

$$x_i = a_i, \ u^j = s^j(a), \ u^j_\sigma = \frac{\partial^{|\sigma|} s^j(a)}{\partial x_1^{i_1} \cdots \partial x_n^{i_n}}, \qquad \sigma = (i_1, \ldots, i_n), \ |\sigma| \leq k,$$

is called the *k-jet* of s at a. The space $J^k(n,m)$ is called the *space of k-jets* with n independent and m dependent variables.

Relations (2.3) determine a surface $\mathcal{E}$ in $J^k(n,m)$, where k is the order of system (2.3). We shall identify system (2.3) with this surface. If $s = (s^1(x), \ldots, s^m(x))^{\mathrm{T}}$ is a solution of system (2.3), then s must satisfy all differential corollaries of this system, in particular, the equations

$$\frac{\partial G_\alpha}{\partial x_i} + \sum_{j,\sigma} u^j_{\sigma+1_i} \frac{\partial G_\alpha}{\partial u^j_\sigma} = 0, \quad \alpha = 1, \ldots, r, \; i = 1, \ldots, n, \tag{3.2}$$

where $1_i = (0, \ldots, 0, 1, 0, \ldots, 0)$, with 1 in the i-th place, $u^j_{(0,\ldots,0)} = u^j$. The union of systems (2.3) and (3.2) is called the *first prolongation of system* (2.3). Define the *l-th prolongation of a system* as the first prolongation of its $(l-1)$st prolongation. The corresponding surface in $J^{k+l}(n,m)$ will be denoted by $\mathcal{E}^{(l)}$. By definition, $\mathcal{E}^{(l)} = (\mathcal{E}^{(l-1)})^{(1)}$.

For any $s \geq 0$ and $k > s$, by $\pi_{k,s}$ denote the projection of $J^k(n,m)$ onto $J^s(n,m)$ taking the point with coordinates (3.1) to the point with the coordinates $x_i, u^j, u^j_\sigma, |\sigma| \leq s$. System $\mathcal{E} \subset J^k(n,m)$ is said to be *formally integrable* if for each positive integer l the surface $\mathcal{E}^{(l)}$ is a submanifold of $J^{k+l}(n,m)$ and the projection $\pi_{k+l,k+l-1} : \mathcal{E}^{(l)} \to \mathcal{E}^{(l-1)}$ is a vector bundle.

The space $J^\infty(n,m)$ of infinite jets is defined as the inverse limit of the chain of projections

$$J^0(n,m) \xleftarrow{\pi_{1,0}} J^1(n,m) \leftarrow \cdots \leftarrow J^k(n,m) \xleftarrow{\pi_{k+1,k}} J^{k+1}(n,m) \leftarrow \cdots.$$

Namely, an element of $J^\infty(n,m)$ is a sequence of points $\theta_k \in J^k(n,m)$, $k \geq 0$, such that $\pi_{k+1,k}(\theta_{k+1}) = \theta_k$ for any $k \geq 0$.

The *infinite jet of a function* s at a point a is defined as the sequence of k-jets of s at the point a and is denoted by $[s]^\infty_a$. Each point of $J^\infty(n,m)$ is the infinite jet of some function (see [**2**, Ch. 4, Sec. 1.1]). The coordinates (3.1) in the finite jet spaces generate the coordinates

$$x_i, \; u^j, \; u^j_\sigma, \qquad i = 1, \ldots, n, \; j = 1, \ldots, m, \; |\sigma| \geq 0. \tag{3.3}$$

in $J^\infty(n,m)$, which are called *canonical.*

The set $J^\infty(n,m)$ is equipped with the structure of an infinite-dimensional smooth manifold. Namely, one defines smooth (infinitely differentiable) functions, vector fields, and differential forms on $J^\infty(n,m)$. A *smooth function* on $J^\infty(n,m)$ is a function smoothly depending on a finite (but arbitrary) number of the canonical coordinates. The algebra of smooth functions on $J^\infty(n,m)$ will be denoted by $\mathcal{F}(J)$. Any derivation of this algebra is called a smooth *vector field* on $J^\infty(n,m)$.

For $i = 1, \ldots, n$, the vector field

$$D_i = \frac{\partial}{\partial x_i} + \sum_{j=1}^{m} \sum_{|\sigma| \geq 0} u^j_{\sigma+1_i} \frac{\partial}{\partial u^j_\sigma} \tag{3.4}$$

is called the *total derivative with respect to x_i on $J^\infty(n,m)$.*

A differential form on $J^\infty(n,m)$ is a differential form depending on finitely number of the canonical coordinates. Smooth functions, vector fields, and differential forms on $J^\infty(n,m)$ are related by the usual algebraic operations. In particular, the Lie derivative of a differential form ω along a vector field X will be denoted by $X\omega$.

The infinite prolongation $\mathcal{E}^\infty$ of a system $\mathcal{E} \subset J^k(n,m)$ is defined as the subset of $J^\infty(n,m)$ formed by the points $\theta = \{\theta_l\} \in J^\infty(n,m)$ such that for each positive integer l the point θ_{k+l} belongs to $\mathcal{E}^{(l)}$. The infinite prolongation of system (2.3) is given by the infinite system of equations

$$D_\sigma G_\alpha = 0, \qquad |\sigma| \geq 0, \quad \alpha = 1, \ldots, r,$$

where $D_\sigma = D_1^{i_1} \circ \ldots \circ D_n^{i_n}$ for $\sigma = (i_1, \ldots, i_n)$.

The structure of an infinite-dimensional smooth manifold on $J^\infty(n,m)$ inherites a similar structure on $\mathcal{E}^\infty$. In particular, a *smooth function (differential p-form) on* $\mathcal{E}^\infty$ is the restriction to $\mathcal{E}^\infty$ of some smooth function (differential p-form) on $J^\infty(n,m)$. Let $\mathcal{F}(\mathcal{E})$ denote the algebra of smooth functions on $\mathcal{E}^\infty$, and let $\Lambda^p(\mathcal{E}^\infty)$ denote the module of differential p-forms on $\mathcal{E}^\infty$. In the case of $\mathcal{E} \subset J^k(n,m)$, for coordinates in $\mathcal{E}^\infty$ one can take a part of the canonical coordinates in $J^\infty(n,m)$. Denote by $\mathcal{F}_l(\mathcal{E})$ the algebra of restrictions of elements of $\mathcal{F}(J)$ depending only on the coordinates $x_i, u^j, u^j_\sigma, |\sigma| \leq l$, to $\mathcal{E}^\infty$.

The derivations of the algebra $\mathcal{F}(\mathcal{E})$ are called *vector fields on* $\mathcal{E}^\infty$. It follows from the definition of $\mathcal{E}^\infty$ that the fields (3.4) are tangent to $\mathcal{E}^\infty$. The restriction of the field D_i to $\mathcal{E}^\infty$ will again be denoted by D_i. Let $\mathcal{D}(\mathcal{E}^\infty)$ denote the module of vector fields on $\mathcal{E}^\infty$.

The distribution generated by the fields $D_1, \ldots, D_n$ on $\mathcal{E}^\infty$ is called the *Cartan distribution* on $\mathcal{E}^\infty$ and is denoted by $\mathcal{C}(\mathcal{E})$. The plane of the Cartan distribution on $\mathcal{E}^\infty$ at a point $\theta \in \mathcal{E}^\infty$ will be denoted by $\mathcal{C}_\theta(\mathcal{E})$. The 1-forms on $\mathcal{E}^\infty$ vanishing on the Cartan distribution is called the *Cartan* 1*-forms*. For example, the 1-form $d_\mathcal{C} f = df - \sum_{i=1}^{n} D_i(f)\, dx_i$ is a Cartan 1-form for any $f \in \mathcal{F}(\mathcal{E})$. Denote by $\mathcal{C}\Lambda^1(\mathcal{E})$ the submodule of $\Lambda^1(\mathcal{E}^\infty)$ consisting of the Cartan 1-forms on $\mathcal{E}^\infty$. Denote by $\mathcal{C}^k\Lambda^k(\mathcal{E})$ the k-th power of the submodule $\mathcal{C}\Lambda^1(\mathcal{E})$, i.e., the submodule of $\Lambda^k(\mathcal{E}^\infty)$ generated by forms $\omega_1 \wedge \ldots \wedge \omega_k$, where $\omega_i \in \mathcal{C}\Lambda^1(\mathcal{E})$. It can be proved that $\omega \in \mathcal{C}^k\Lambda^k(\mathcal{E})$ if and only if $i_\xi(\omega_\theta) = 0$ for any $\xi \in \mathcal{C}_\theta(\mathcal{E})$ and $\theta \in \mathcal{E}^\infty$, where $i_\xi(\omega_\theta)$ denotes the interior product of the tangent vector ξ and the k-form ω_θ.

Any infinite jet of any solution of a system $\mathcal{E}$ lies in $\mathcal{E}^\infty$. The set of all infinite jets of a given solution s is an n-dimensional manifold, which is called the *graph of the solution* s in $\mathcal{E}^\infty$. The fields $D_1, \ldots, D_n$ are tangent to the graphs of solutions in $\mathcal{E}^\infty$. Moreover, any maximal integral manifold of the Cartan distribution on $\mathcal{E}^\infty$ is locally the graph of some solution in $\mathcal{E}^\infty$ (see [**2**, Ch. 4, Proposition 2.3]). Hence the infinite-dimensional manifold $\mathcal{E}^\infty$ and the Cartan distribution on $\mathcal{E}^\infty$ uniquely determine the system $\mathcal{E}$ and its solutions. The pair $(\mathcal{E}^\infty, \mathcal{C}(\mathcal{E}))$ is called the *diffiety* of the system $\mathcal{E}$. The space $J^\infty(n,m)$ with its Cartan distribution can be considered as the diffiety of the trivial system $0 = 0$.

4. Mappings of diffieties

A *smooth mapping* of a diffiety $(\mathcal{E}^\infty, \mathcal{C}(\mathcal{E}))$ into a diffiety $(\mathcal{S}^\infty, \mathcal{C}(\mathcal{S}))$ is a mapping

$$F : \mathcal{E}^\infty \longrightarrow \mathcal{S}^\infty \tag{4.1}$$

such that the induced mapping F^* takes smooth functions to smooth functions, i.e., $F^*\big(\mathcal{F}(\mathcal{S})\big) \subset \mathcal{F}(\mathcal{E})$, where $F^*(g) = g \circ F$. A bijective smooth mapping (4.1) is called a *diffeomorphism* if its inverse mapping is also smooth.

A diffeomorphism (4.1) is called a *$\mathcal{C}$-diffeomorphism* if it preserves the Cartan distribution, i.e.,

$$F_*(\mathcal{C}_\theta(\mathcal{E})) = \mathcal{C}_{F(\theta)}(\mathcal{S}), \qquad \forall \theta \in \mathcal{E}^\infty. \tag{4.2}$$

Since the n-dimensional integral manifolds of the Cartan distribution coincide locally with the graphs of solutions of the corresponding system, it follows from condition (4.2) that each $\mathcal{C}$-diffeomorphism takes the graphs of solutions of one system to the graphs of solutions of the other system. Thus, $\mathcal{C}$-diffeomorphic systems are equivalent systems.

A smooth mapping (4.1) is called a *covering* if the following conditions hold:

(1) the tangent mapping $F_{*,\theta}$ is a vector space epimorphism for any $\theta \in \mathcal{E}^\infty$;
(2) relation (4.2) holds;
(3) the dimension of the kernel of $F_{*,\theta}$ is the same for any point $\theta \in \mathcal{E}^\infty$.

The *dimension of a covering* F is the dimension of the kernel of $F_{*,\theta}$. For any point $\tilde{\theta} \in \mathcal{S}^\infty$, the set $F^{-1}(\tilde{\theta})$ is called the *fiber of the covering* F over the point $\tilde{\theta}$. If mapping (4.1) is a covering and $\mathcal{E}^\infty$ and $\mathcal{S}^\infty$ are the diffieties of systems $\mathcal{E}$ and $\mathcal{S}$ respectively, then one says that the system $\mathcal{E}$ *covers* the system $\mathcal{S}$, or that F is a *covering over the system* $\mathcal{S}$.

EXAMPLE 4.1. System (2.4)–(2.5) covers system (2.4). Moreover, for any finite-dimensional covering $\nu : \mathcal{E}^\infty \to \mathcal{S}^\infty$ and $\theta \in \mathcal{E}^\infty$, there are coordinates in a neighborhood of θ in which ν has this form. Indeed, let the system $\mathcal{S}$ have the form (2.4) and let $w = (w^1, \dots, w^q)$ be the coordinates in the fiber of ν in a neighborhood of the point θ. Then the derivatives $w^j_i = D_i(w^j)$ are functions of a finite number of coordinates x, w, u, $\dots$, u_σ, $\dots$. Hence the system $\mathcal{E}$ has locally the form (2.4)–(2.5) and the covering ν takes a point $(z, w, v, \dots, v_\sigma, \dots)$ to the point $(z, v, \dots, v_\sigma, \dots)$.

Coverings possess the following properties. A composition of coverings is a covering. Every $\mathcal{C}$-diffeomorphism is a covering of dimension zero. The dimension of a covering coincides with the dimension of any fiber of the covering. If a system $\mathcal{E}$ covers a system $\mathcal{S}$, then the number of independent variables of systems $\mathcal{E}$ and $\mathcal{S}$ is the same. The graph of each solution of the system $\mathcal{E}$ is taken by the covering (4.1) to the graph of a solution of the system $\mathcal{S}$. Conversely, for each solution s of the system $\mathcal{S}$ defined in a neighborhood of a point $a \in \mathbb{R}^n$ and for an arbitrary point θ of the fiber $F^{-1}([s]^\infty_a)$, there exists a unique solution $\tilde{s}$ of the system $\mathcal{E}$ such that $[\tilde{s}]^\infty_{\tilde{a}} = \theta$ and the covering (4.1) takes the graph of the solution $\tilde{s}$ to the graph of the solution s.

5. Pseudosymmetries of differential equations

Let us find the conditions for fibers of coverings. Consider system (2.4)–(2.5). Note that for $i = 1, \dots, n$ the total derivative with respect to x_i on the diffiety of system (2.4)–(2.5) has the form

$$D_i = \check{D}_i + \sum_{j=1}^{q} W^j_i(x, w, u, \dots, u_\sigma, \dots) \frac{\partial}{\partial w^j},$$

where $\check{D}_i$ is the total derivative with respect to x_i on the diffiety of system (2.4). It is easily seen that the column $X = \left(\partial/\partial w^1, \dots, \partial/\partial w^q\right)^{\mathrm{T}}$ of vector fields on the diffiety of system (2.4)–(2.5) satisfies the relations

$$[X, D_i] = A_i X, \qquad i = 1, \dots, n, \tag{5.1}$$

where $[X, D_i]$ is the column of commutators $[\partial/\partial w^s, D_i]$, $s = 1, \dots, q$, and $A_i X$ is the product of the function matrix $A_i = \left(\partial W_i^j/\partial w^s\right)_{s,j=1,\dots,q}$ and the column X.

Conditions (5.1) are not invariant with respect to $\mathcal{C}$–diffeomorphisms. Indeed, relations (5.1) can be transformed by a $\mathcal{C}$–diffeomorphism into relations of the form

$$[X, D_i] = A_i X + B_i D, \qquad i = 1, \dots, n, \tag{5.2}$$

where B_i is a $q \times n$ matrix of functions on the diffiety $\mathcal{E}^\infty$ of system (2.4)–(2.5) and D is the column of the vector fields $D_1, \dots, D_n$.

A columns X of vector fields on the diffiety $\mathcal{E}^\infty$ satisfying relations (5.2) is called a *pseudosymmetry of the system* $\mathcal{E}$ [**16, 18**]. The relations (5.2) and the Jacobi identity imply the relations

$$D_j(A_i) - D_i(A_j) + A_j A_i - A_i A_j = 0, \qquad \forall i, j, \tag{5.3}$$

$$D_j(B_i) - D_i(B_j) + A_j B_i - A_i B_j = 0, \qquad \forall i, j. \tag{5.4}$$

Equations (5.3) are known as the Maurer–Cartan equations (see [**15**] for solving these equations). From (5.3) it follows that the matrix differential operators $D_i + A_i$ and $D_j + A_j$ commute for any i and j.

If $q = 1$ and $A_i \equiv 0$ for all i, then equation (5.2) means that X defines a higher symmetry of the system. The following two theorems generalize Theorems 2.5 and 3.8 in [**2**, Ch. 4] characterized higher symmetries.

THEOREM 5.1 ([**4**]). *A column $X = (X_1, \dots, X_q)^{\mathrm{T}}$ of vector fields on $J^\infty(n, m)$ is a pseudosymmetry of $J^\infty(n, m)$ if and only if*

$$X = \mathfrak{D}_{\varphi, A} + MD, \tag{5.5}$$

where φ is a $q \times m$ matrix of arbitrary functions on $J^\infty(n, m)$, $A = (A_1, \dots, A_n)$ is a tuple of $q \times q$ matrices satisfying (5.3), M is a $q \times n$ matrix of arbitrary functions on $J^\infty(n, m)$, MD is the product of the matrix M and the column $D = (D_1, \dots, D_n)^{\mathrm{T}}$, and in the canonical coordinates in $J^\infty(n, m)$, the term $\mathfrak{D}_{\varphi, A}$ has the form

$$\mathfrak{D}_{\varphi, A} = \sum_\sigma (D + A)^\sigma(\varphi) \frac{\partial}{\partial u_\sigma}, \tag{5.6}$$

where $(D + A)^\sigma$, $\sigma = (i_1, \dots, i_n)$, is the composition $(D_1 + A_1)^{i_1} \circ \dots \circ (D_n + A_n)^{i_n}$ of $(q \times q)$ matrix differential operators acting on the function matrix φ, and the summand on the right-hand side in (5.6) is the product of the resulting matrix and the column $\dfrac{\partial}{\partial u_\sigma} = \left(\dfrac{\partial}{\partial u_\sigma^1}, \dots, \dfrac{\partial}{\partial u_\sigma^m}\right)^{\mathrm{T}}$.

PROOF. Using the relations (5.3) and straightforward computations, one can easily prove the relation

$$[\mathfrak{D}_{\varphi, A}, D_i](g) = A_i \mathfrak{D}_{\varphi, A}(g)$$

for any canonical coordinate function g on $J^\infty(n, m)$. Hence

$$[\mathfrak{D}_{\varphi, A}, D_i] = A_i \mathfrak{D}_{\varphi, A} \tag{5.7}$$

and the column $X = \mathfrak{D}_{\varphi, A} + MD$ satisfies the equality of the form (5.2) for $\mathcal{E}^\infty = J^\infty(n, m)$, where $B_i = -D_i(M)$.

Conversely, let a column X of vector fields satisfies (5.2). In the canonical coordinates (3.3) on $J^\infty(n,m)$, the Cartan distribution is defined by the set of 1-forms

$$d_{\mathcal{C}}u^j_\sigma = du^j_\sigma - \sum_{i=1}^n u^j_{\sigma+1_i}\,dx_i, \quad j=1,\dots,m, \quad |\sigma|\geq 0.$$

Set $\varphi^j_\sigma = X\rfloor d_{\mathcal{C}}u^j_\sigma$ and $M_s = X\rfloor dx_s$. Since $D_s\rfloor d_{\mathcal{C}}u^j_\sigma = 0$ and $D_s\rfloor dx_i$ is the zero function for $s\neq i$ and the unit function for $s=i$, it follows that the column $X-\sum_s M_sD_s$ has zero components for $\dfrac{\partial}{\partial x_1},\dots,\dfrac{\partial}{\partial x_n}$ and hence

$$X = \sum_{j,\sigma}\varphi^j_\sigma\frac{\partial}{\partial u^j_\sigma} + \sum_{s=1}^n M_sD_s$$

for some $\varphi^j_\sigma \in \mathcal{F}(J)$.

Using the infinitesimal analogue of Stokes' theorem (see [**19**, Ch. 3, Equation (1.10)]) and the relation $d(d_{\mathcal{C}}u^j_\sigma) = \sum\limits_{i=1}^n dx_i\wedge d_{\mathcal{C}}u^j_{\sigma+1_i}$, we get

$$[X,D_i]\rfloor d_{\mathcal{C}}u^j_\sigma = X\rfloor\Big(D_i\rfloor\sum_{s=1}^n dx_i\wedge d_{\mathcal{C}}u^j_{\sigma+1_i}\Big) + X(D_i\rfloor d_{\mathcal{C}}u^j_\sigma) - D_i(X\rfloor d_{\mathcal{C}}u^j_\sigma) =$$

$$= X\rfloor d_{\mathcal{C}}u^j_{\sigma+1_i} - D_i(X\rfloor d_{\mathcal{C}}u^j_\sigma) = \varphi^j_{\sigma+1_i} - D_i\varphi^j_\sigma.$$

The substitution of the right-hand side of relation (5.2) into the form $d_{\mathcal{C}}u^j_\sigma$ gives the relation

$$A_iX\rfloor d_{\mathcal{C}}u^j_\sigma + B_iD\rfloor d_{\mathcal{C}}u^j_\sigma = A_i\varphi^j_\sigma.$$

Consequently, $\varphi^j_{\sigma+1_i} = (D_i+A_i)\varphi^j_\sigma$, which proves equality (5.5), where $M = (M_1,\dots,M_n)$, $\varphi = (\varphi^1_\varnothing,\dots,\varphi^m_\varnothing)$, and $\varnothing$ is the empty multi-index. □

Matrices A_1, …, A_n and B_1, …, B_n satisfying (5.3) and (5.4) is called *coefficient matrices*, the matrix φ is called the *generating matrix* of the pseudosymmetry (5.5), and the corresponding column $\mathfrak{D}_{\varphi,A}$ is called the *evolution pseudosymmetry.*

THEOREM 5.2 ([**4**]). *Let a formally integrable system $\mathcal{E}$ be determined by equations* (2.3)*. Then a column $X = (X_1,\dots,X_q)^{\mathrm{T}}$ of vector fields is a pseudosymmetry of $\mathcal{E}$ if and only if X is the restriction of a column of the form* (5.5) *to $\mathcal{E}^\infty$, where the restrictions of the matrices A_1, …, A_n to $\mathcal{E}^\infty$ satisfy relations* (5.3)*, and the generating matrix $\varphi = (\varphi_{ij})$ satisfies*

$$\text{(5.8)}\qquad \sum_{j,\sigma}\frac{\partial G_\alpha}{\partial u^j_\sigma}(D+A)^\sigma(\varphi^j)\Big|_{\mathcal{E}^\infty} = 0, \qquad \alpha = 1,\dots,r,\ \varphi^j = (\varphi_{1j},\dots,\varphi_{qj})^{\mathrm{T}}.$$

PROOF. Let X be a column of vector fields on $J^\infty(n,m)$ of the form (5.5)–(5.6), let the restrictions of the matrices A_1, …, A_n to $\mathcal{E}^\infty$ satisfy relations (5.3),and let the generating matrix $\varphi = (\varphi_{ij})$ satisfy (5.8). Since the term MD is tangent to the infinite prolongation of any system, it follows that the tangency of the column X to the diffiety $\mathcal{E}^\infty$ is equivalent to the tangency of the column $\mathfrak{D}_{\varphi,A}$ to $\mathcal{E}^\infty$. The latter means that the functions $\mathfrak{D}_{\varphi,A}(G_\alpha)$, $\alpha=1$, …, r, are zero on $\mathcal{E}^\infty$, which is equivalent to system (5.8). Hence the restriction of X to $\mathcal{E}^\infty$ is well defined.

By arguing as in the proof of Theorem 5.1, but considering the restrictions of the fields $\ni_{\varphi,A}, D_i$ and the functions x_s, u^j_σ to $\mathcal{E}^\infty$ instead of these fields and functions themselves, we obtain the identity $[\ni_{\varphi,A}\big|_{\mathcal{E}^\infty}, D_i\big|_{\mathcal{E}^\infty}] = A_i\big|_{\mathcal{E}^\infty}\ni_{\varphi,A}\big|_{\mathcal{E}^\infty}$, which means that the restriction of X to $\mathcal{E}^\infty$ is a pseudosymmetry of system (2.3).

Conversely, let a column X of vector fields on $\mathcal{E}^\infty$ satisfies equality (5.2). By arguing as in the proof of Theorem 5.1, it is not hard to prove that the column X is determined by the functional matrices $\varphi^j = X\rfloor(d_{\mathcal{C}}u^j\big|_{\mathcal{E}^\infty})$ and $M_s = X\rfloor(dx_s\big|_{\mathcal{E}^\infty})$, where $X\rfloor(d_{\mathcal{C}}u^j_\sigma\big|_{\mathcal{E}^\infty}) = (D+A)^\sigma(\varphi^j)\big|_{\mathcal{E}^\infty}$. In contrast to $J^\infty(n,m)$, the forms $d_{\mathcal{C}}u^j_\sigma$ on $\mathcal{E}^\infty$ are related by the constraints

$$d_{\mathcal{C}}G_\alpha\big|_{\mathcal{E}^\infty} = \sum_{j,\sigma}\frac{\partial G_\alpha}{\partial u^j_\sigma}d_{\mathcal{C}}u^j_\sigma\big|_{\mathcal{E}^\infty} = 0, \qquad \alpha = 1,\dots,r.$$

Hence

$$X\rfloor d_{\mathcal{C}}G_\alpha\big|_{\mathcal{E}^\infty} = \sum_{j,\sigma}\frac{\partial G_\alpha}{\partial u^j_\sigma}(D+A)^\sigma(\varphi^j)\big|_{\mathcal{E}^\infty} = 0, \qquad \alpha = 1,\dots,r.$$

To complete the proof of Theorem 5.2, it suffices to note that for the coordinates on $\mathcal{E}^\infty$ one can take the restriction to $\mathcal{E}^\infty$ of part of the coordinates (3.3). Hence X is the restriction to $\mathcal{E}^\infty$ of the column (5.5). $\square$

6. Integrable pseudosymmetries

Along with conditions (5.2), the distribution generated by the fields $\partial/\partial w^1$, …, $\partial/\partial w^q$ satisfies the integrability condition. Note that the manifold $\mathcal{E}^\infty$ is infinite-dimensional. Hence a regular and involutive distribution on $\mathcal{E}^\infty$ can be not integrable. Conditions for the integrability of vector fields defining higher symmetries were obtained in [**3**]. Following [**3**], we give the following definition.

A column X of vector fields X_1, …, X_q on $\mathcal{E}^\infty$ is an *integrable pseudosymmetry* of the system $\mathcal{E}$ if the following conditions hold:

(i) the fields X_1, …, X_q generate an involutive distribution of dimension q that does not intersect with the Cartan distribution;
(ii) the column $X = (X_1,\dots,X_q)^{\mathrm{T}}$ is a pseudosymmetry of the system $\mathcal{E}$;
(iii) there exists a ring $\mathcal{K}$ of functions on $\mathcal{E}^\infty$ such that $\mathcal{F}_0(\mathcal{E}) \subset \mathcal{K} \subset \mathcal{F}_l(\mathcal{E})$ for some $l \geq 0$, all elements of the coefficient matrices A_1, …, A_n and B_1, …, B_n belong to $\mathcal{K}$, and $X_i(\mathcal{K}) \subset \mathcal{K}$ for any $i = 1$, …, q.

Note that condition (iii) allows to restrict the distribution generated by the fields X_1, …, X_q to a finite-dimensional manifold (see the proof of Theorem 6.6 below). Thus, from the conditions (i) and (iii) it follows that this distribution is integrable.

EXAMPLE 6.1. Any finite-dimensional Lie algebra of classical symmetries of a system forms an integrable pseudosymmetry of the system (see [**2**, Ch. 3] for classical symmetries).

EXAMPLE 6.2. The fields $\partial/\partial w^1$, …, $\partial/\partial w^q$ form an integrable pseudosymmetry of system (2.4)–(2.5). Indeed, $A_i = \left(\partial W^j_i/\partial w^s\right)_{s,j=1,\dots,q}$, $B_i = 0$, and $\mathcal{K}$ is the ring of smooth functions of t, w, u, …, u_σ, $|\sigma| \leq r$, where r is the order of system (2.5).

Let $\mathcal{E}$ be a formally integrable system, and let $\mathcal{K}$ be a function ring on the diffiety $\mathcal{E}^\infty$ such that $\mathcal{F}_0(\mathcal{E}) \subset \mathcal{K} \subset \mathcal{F}_l(\mathcal{E})$ for some $l \geq 0$. For $s > 0$, by $D^s\mathcal{K}$ we denote the ring generated by the functions $D_\sigma(f)$, where $f \in \mathcal{K}$ and $|\sigma| \leq s$. A point $\theta \in \mathcal{E}^\infty$ is called a *generic point of the ring* $\mathcal{K}$ [**3**] if the subspaces $\{df|_{\theta'} \in T^*_{\theta'} : f \in \mathcal{K}\}$ and $\{df|_{\theta'} \in T^*_{\theta'} : f \in D^l\mathcal{K}\}$ have constant dimension in some neighborhood of θ.

THEOREM 6.3. *Any $\mathcal{C}$-diffeomorphism maps integrable pseudosymmetries into integrable pseudosymmetries.*

PROOF. Let $X = (X_1, \dots, X_q)^{\mathrm{T}}$ be an integrable pseudosymmetry of a system $\mathcal{E}$ and let $F : \mathcal{E}^\infty \to \mathcal{S}^\infty$ be a $\mathcal{C}$-diffeomorphism. Since a diffeomorphism takes involutive distributions of dimension q into involutive distributions of dimension q, we see that the distribution generated by the fields $F_*(X_1), \dots, F_*(X_q)$ is involutive and has dimension q. In addition, it does not intersect with the Cartan distribution on $\mathcal{S}^\infty$, since F maps the Cartan distribution on $\mathcal{E}^\infty$ into the Cartan distribution on $\mathcal{S}^\infty$. Thus condition (i) is invariant under $\mathcal{C}$-diffeomorphisms.

Consider the distribution generated by the fields $X_1, \dots, X_q, D_1, \dots, D_n$, where $D_1, \dots, D_n$ are the total derivatives with respect to the independent variables of the system $\mathcal{E}$. From conditions (i) and (ii) and the definition of $D_1, \dots, D_n$ it follows that this distribution is involutive. Suppose at a point $\theta \in \mathcal{E}^\infty$ we have

$$c_1 X_{1,\theta} + \dots + c_q X_{q,\theta} + c_{q+1} D_{1,\theta} + \dots + c_{q+n} D_{n,\theta} = 0, \quad c_1, \dots, c_{q+n} \in \mathbb{R}.$$

Since the distribution generated by the fields $X_1, \dots, X_q$ does not intersect with the Cartan distribution (see (i)), it follows that $c_1 X_{1,\theta} + \dots + c_q X_{q,\theta} = 0$ and $c_{q+1} D_{1,\theta} + \dots + c_{q+n} D_{n,\theta} = 0$. Both sets of vectors $X_{1,\theta}, \dots, X_{q,\theta}$ and $D_{1,\theta}, \dots, D_{n,\theta}$ are linearly independent. Therefore $c_1 = \dots = c_{q+n} = 0$. Thus the vectors $X_{1,\theta}, \dots, X_{q,\theta}, D_{1,\theta}, \dots, D_{n,\theta}$ are linearly independent and the distribution generated by the fields $X_1, \dots, X_q, D_1, \dots, D_n$ has dimension $q + n$. As above, the distribution generated by the fields $F_*(X_1), \dots, F_*(X_q), F_*(D_1), \dots, F_*(D_n)$ is involutive and has dimension $q + n$. Since any $\mathcal{C}$-diffeomorphism preserves the Cartan distribution, it follows that this distribution coincides with the distribution generated by the fields $F_*(X_1), \dots, F_*(X_q), \tilde{D}_1, \dots, \tilde{D}_n$, where $\tilde{D}_1, \dots, \tilde{D}_n$ are the total derivatives with respect to the independent variables of the system $\mathcal{S}$. This implies condition (ii) for the fields $F_*(X_1), \dots, F_*(X_q)$ and, therefore, condition (ii) is also invariant under $\mathcal{C}$-diffeomorphisms.

Consider a ring $\mathcal{K}$ of functions on $\mathcal{E}^\infty$ satisfying condition (iii). It follows from smoothness of F that $F^*\big(\mathcal{F}_0(\mathcal{S})\big) \subset \mathcal{F}_{r_0}(\mathcal{E})$ for some $r_0 \geq 0$. On the other hand, for any $i = 1, \dots, n$ the Lie derivative along D_i takes $\mathcal{F}_{r_0}(\mathcal{E})$ to $\mathcal{F}_{r_0+1}(\mathcal{E})$. Since F^{-1} is a $\mathcal{C}$-diffeomorphism, and any $\mathcal{C}$-diffeomorphism preserves the Cartan distribution, we see that

$$(F^{-1})_*(\tilde{D}_i) = \sum_{j=1}^{n} a_{ij} D_j, \quad a_{ij} \in \mathcal{F}(\mathcal{E}).$$

Denote by l_0 the smallest integer such that $a_{ij} \in \mathcal{F}_{l_0}(\mathcal{E})$ for any $i, j = 1, \dots, n$. Then for any function $g \in \mathcal{F}_0(\mathcal{S})$ one has $F^*\big(\tilde{D}_i(g)\big) = (F^{-1})_*(\tilde{D}_i)\big(F^*(g)\big) = \sum_{j=1}^n a_{ij} D_j\big(F^*(g)\big) \in \mathcal{F}_r(\mathcal{S})$, where $r = \max(l_0, r_0 + 1)$. Any function of $\mathcal{F}_1(\mathcal{S})$ is a function of a finite set of functions of the form $g \in \mathcal{F}_0(\mathcal{S})$ or $\tilde{D}_i(g)$, $i = 1, \dots, n$. Hence, $F^*(\mathcal{F}_1(\mathcal{S})) \subset \mathcal{F}_r(\mathcal{E})$. Using elementary induction on s, one can easily prove the relation $F^*(\mathcal{F}_s(\mathcal{S})) \subset \mathcal{F}_{r+s-1}(\mathcal{E})$.

Similarly, for the $\mathcal{C}$-diffeomorphism F^{-1} there exists an integer $r_1 \geq 0$ such that for any $s \geq 0$ one has $(F^{-1})^*(\mathcal{F}_s(\mathcal{E})) \subset \mathcal{F}_{r_1+s}(\mathcal{S})$ and hence $\mathcal{F}_s(\mathcal{E}) \subset F^*(\mathcal{F}_{r_1+s}(\mathcal{S}))$. If we combine this with the definition of $D^s\mathcal{K}$, we get $F^*\big(\mathcal{F}_0(\mathcal{S})\big) \subset \mathcal{F}_{r_0}(\mathcal{E}) \subset D^{r_0}\mathcal{K} \subset \mathcal{F}_{r_0+l}(\mathcal{E}) \subset F^*\big(\mathcal{F}_{l_*}(\mathcal{S})\big)$, where $l_* = r_1 + r_0 + l$. Thus, $\mathcal{F}_0(\mathcal{S}) \subset (F^{-1})^*(D^{r_0}\mathcal{K}) \subset \mathcal{F}_{l_*}(\mathcal{S})$.

On the other hand, from (5.2) it follows that for any $g \in \mathcal{K}$ we have

$$X\big(D_i(g)\big) = D_i\big(X(g)\big) + A_i X(g) + B_i D(g), \quad i = 1, \ldots, n$$

and, therefore, $X_j\big(D_i(g)\big) \in D^1\mathcal{K}$ for any $j = 1, \ldots, q$, $i = 1, \ldots, n$. Hence $X_j(D^1\mathcal{K}) \subset D^1\mathcal{K}$ for $j = 1, \ldots, q$. Similarly, $X_j(D^s\mathcal{K}) \subset D^s\mathcal{K}$ and

$$F_*(X_j)\big((F^{-1})^*(D^s\mathcal{K})\big) \subset (F^{-1})^*(D^s\mathcal{K}), \qquad s \geq 0.$$

Thus condition (iii) holds for the fields $F_*(X_1), \ldots, F_*(X_q)$ and the ring $(F^{-1})^*(D^{r_0}\mathcal{K})$. This means that condition (iii) is invariant under $\mathcal{C}$-diffeomorphisms. □

We say that an *integrable pseudosymmetry* $X = (X_1, \ldots, X_q)^{\mathrm{T}}$ *determines a covering* if the fibers of the covering coincide with the maximal integral manifolds of the distribution generated by the fields $X_1, \ldots, X_q$. If this covering is of the form (4.1), then the system $\mathcal{S}$ is called the *quotient system* of the system $\mathcal{E}$ along X.

EXAMPLE 6.4. The covering of Example 2.1 is determined by a Lie algebra of classical symmetries. This Lie algebra is generated by symmetries $\partial/\partial x_1$ and $\partial/\partial x_2$ (see [**2**, Ch. 3, Example 6.1]).

THEOREM 6.5 ([**5**]). *Suppose a column* $X = (X_1, \ldots, X_q)^{\mathrm{T}}$ *is an integrable pseudosymmetry of a system* $\mathcal{E}$ *and* N *is a nonsingular* $q \times q$ *matrix of functions on* $\mathcal{E}^\infty$. *Then*

(a) *the product* NX *is also an integrable pseudosymmetry of the system* $\mathcal{E}$*;*
(b) *the integrable pseudosymmetries* X *and* NX *determine the same covering;*
(c) *if matrices* A_i *and* B_i *correspond to* X *(see* (5.2)*), then the matrices* $NA_iN^{-1} - D_i(N)N^{-1}$ *and* NB_i *correspond to* NX.

PROOF. The entries of the column NX are vector fields on $\mathcal{E}^\infty$ and determine the same distribution as the entries of the column X. Hence, this distribution is involutive and the statement (b) is proved.

Using the definition of commutator and (5.2), we get

$$[NX, D_i] = N[X, D_i] - D_i(N)X = NA_iX + NB_iD - \\ D_i(N)X = \big(NA_iN^{-1} - D_i(N)N^{-1}\big)NX + NB_iD.$$

This proves condition (ii) of the definition of integrable pseudosymmetry and the statement (c) of Theorem 6.5.

Let $\mathcal{K}$ be a ring of functions on $\mathcal{E}^\infty$ satisfying condition (iii) of the definition of integrable pseudosymmetry for the column X. Since the entries of the matrix N are elements of the ring $\mathcal{F}(\mathcal{E})$ and their number is finite, we see that there exist an integer $l_2 \geq 0$ such that any entry of N belongs to $\mathcal{F}_{l_2}(\mathcal{E})$. Therefore any entry of the matrix $D_i(N)$ belongs to $\mathcal{F}_{l_2+1}(\mathcal{E})$ for any $i = 1, \ldots, n$. From the definition of $D^s\mathcal{K}$ it follows that $\mathcal{F}_{l_2+1}(\mathcal{E}) \subset D^{l_2+1}\mathcal{K} \subset \mathcal{F}_{l_2+1+l}(\mathcal{E})$. Thus condition (iii) holds for NX and the ring $D^{l_2+1}\mathcal{K}$. This completes the proof of Theorem 6.5. □

A *regular point* of an integrable pseudosymmetry $X = (X_1, \dots, X_q)^{\mathrm{T}}$ is a generic point of the corresponding ring $\mathcal{K}$ (see condition (iii) of the definition of integrable pseudosymmetry).

THEOREM 6.6 ([4]). (I) *Let X be an integrable pseudosymmetry of a system $\mathcal{E}$ and let $\theta \in \mathcal{E}^\infty$ be a regular point of X. Then X determines a covering from a neighborhood of θ.*

(II) *If $\nu : \mathcal{E}^\infty \to \mathcal{S}^\infty$ is a finite-dimensional covering and $\theta \in \mathcal{E}^\infty$, then there is an integrable pseudosymmetry defining this covering in a neighborhood of the point θ.*

PROOF. The statement (II) follows from Examples 4.1 and 6.2 and Theorem 6.3, since any change of system variables (see Example 4.1) is a $\mathcal{C}$-diffeomorphism.

The proof of the statement (I) is divided into the following 8 steps.

STEP 1. We use the following theorem proved in [3] (see Theorem 1).

THEOREM 6.7. *Let $\mathcal{E}$ be a formally integrable system, and let $\mathcal{K}$ be a ring of functions on its diffiety $\mathcal{E}^\infty$ such that $\mathcal{F}_0(\mathcal{E}) \subset \mathcal{K} \subset \mathcal{F}_l(\mathcal{E})$ for some integer l. Then for any generic point $\theta \in \mathcal{E}^\infty$ of the ring $\mathcal{K}$, there exist a system $\mathcal{S}$ and a $\mathcal{C}$-diffeomorphism F of some neighborhood of the point θ to the diffiety $\mathcal{S}^\infty$ such that $\mathcal{K} \subset F^*\big(\mathcal{F}_0(\mathcal{S})\big)$ and any element of $F^*\big(\mathcal{F}_0(\mathcal{S})\big)$ is a function of elements of $\mathcal{K}$.*

Let $X = (X_1, \dots, X_q)^{\mathrm{T}}$ be an integrable pseudosymmetry of a system $\mathcal{E}$, $\mathcal{K}$ be the corresponding ring, and $\tilde{\theta} \in \mathcal{E}^\infty$ be a regular point of X. Then $\tilde{\theta}$ is a generic point of $\mathcal{K}$. Using Theorem 6.7, we construct $\mathcal{C}$-diffeomorphism F of some neighborhood $\mathcal{U}^\infty \subset \mathcal{E}^\infty$ of the point $\tilde{\theta}$ to the diffiety $\mathcal{S}^\infty$ of some system $\mathcal{S}$. From invariancy of integrable pseudosymmetries it follows that the fields $Z_1 = F_*(X_1)$, $\dots$, $Z_q = F_*(X_q)$ form an integrable pseudosymmetry of the system $\mathcal{S}$. Its ring $\mathcal{K}_Z$ is the ring $\mathcal{F}_0(\mathcal{S})$. Indeed, for any function $g \in \mathcal{F}_0(\mathcal{S})$ we have $F^*(g) \in \mathcal{K}$. Whence $X_i\big(F^*(g)\big) \in \mathcal{K}$ and hence $Z_i(g) = (F^{-1})^*\Big(X_i\big(F^*(g)\big)\Big) \in (F^{-1})^*(\mathcal{K}) \subset \mathcal{F}_0(\mathcal{S})$. Thus $Z_i\big(\mathcal{F}_0(\mathcal{S})\big) \subset \mathcal{F}_0(\mathcal{S})$ for any $i = 1, \dots, q$ and hence $\mathcal{K}_Z = \mathcal{F}_0(\mathcal{S})$. Moreover, the projection $\pi_{\infty,0}\big|_{\mathcal{S}^\infty}$ takes the fields $Z_1, \dots, Z_q$ to some fields $Z_1^0, \dots, Z_q^0$ on $\mathcal{S}^0 = \pi_{\infty,0}(\mathcal{S}^\infty)$.

STEP 2. Denote by $\tilde{D}_1, \dots, \tilde{D}_n$ the total derivatives with respect to the independent variables of the system $\mathcal{S}$. Any element of $\mathcal{F}_1(\mathcal{S})$ is a function of a finite set of elements of the form $g \in \mathcal{F}_0(\mathcal{S})$ or $\tilde{D}_i(g)$, $i = 1, \dots, n$. Besides, the column $Z = (Z_1, \dots, Z_q)^T$ satisfies relation (5.3). Therefore, for any $j = 1, \dots, q$ we have

$$Z_j\big(\tilde{D}_i(g)\big) = \tilde{D}_i\big(Z_j(g)\big) + \sum_{l=1}^{q} a_{ji}^l Z_l(g) + \sum_{k=1}^{n} b_{ji}^k \tilde{D}_k(g),$$

where a_{ji}^l and b_{ji}^k are elements of matrices A_i and B_i respectively. Thus, $Z_j\big(\tilde{D}_i(g)\big) \in \mathcal{F}_1(\mathcal{S})$ and hence $Z_i\big(\mathcal{F}_1(\mathcal{S})\big) \subset \mathcal{F}_1(\mathcal{S})$, i.e., the projection $\pi_{\infty,1}\big|_{\mathcal{S}^\infty}$ takes the fields $Z_1, \dots, Z_q$ to some fields $Z_1^1, \dots, Z_q^1$ on $\mathcal{S}^1$. By the same reasons, for any $s > 1$ the projection $\pi_{\infty,s}\big|_{\mathcal{S}^\infty}$ takes the fields $Z_1, \dots, Z_q$ to some fields $Z_1^s, \dots, Z_q^s$ on $\mathcal{S}^s = \pi_{\infty,s}(\mathcal{S}^\infty)$. Moreover, for $k > s$ the map $\pi_{k,s}\big|_{\mathcal{S}^k}$ takes the fields $Z_1^k, \dots, Z_q^k$ to the fields $Z_1^s, \dots, Z_q^s$.

STEP 3. Reasoning in the same way as in the proof of Theorem 6.3, it is not hard to prove that at the point $\theta = F(\tilde{\theta})$ the vectors $Z_{1,\theta}, \ldots, Z_{q,\theta}, \tilde{D}_{1,\theta}, \ldots, \tilde{D}_{n,\theta}$ are linearly independent. Therefore, there exists $k \geq 0$ such that the vectors

$$Z^k_{1,\theta_k}, \ \ldots, \ Z^k_{q,\theta_k}, \ \pi_{\infty,k,*}(\tilde{D}_{1,\theta}), \ \ldots, \ \pi_{\infty,k,*}(\tilde{D}_{n,\theta}), \tag{6.1}$$

where $\theta_k = \pi_{\infty,k}(\theta)$, are linearly independent. From the definition of linear independence it follows that vectors $Z^k_{1,\theta_k}, \ldots, Z^k_{q,\theta_k}$ are also linearly independent. From smoothness of the fields $Z^k_1, \ldots, Z^k_q$ it follows that these fields are linearly independent at each point of some neighborhood of the point $\theta_k \in \mathcal{S}^k$.

STEP 4. Since the map $\pi_{\infty,k}\big|_{\mathcal{S}^\infty}$ takes the fields $Z_1, \ldots, Z_q$ to the fields $Z^k_1, \ldots, Z^k_q$ on $\mathcal{S}^k$, we see that this map takes commutators of the fields $Z_1, \ldots, Z_q$ to the corresponding commutators of the fields $Z^k_1, \ldots, Z^k_q$. Taking into account condition (i), we see that the vector fields $Z^k_1, \ldots, Z^k_q$ generate regular involutive distribution in some neighborhood of the point θ_k. By the Frobenius theorem, this distribution is integrable and possesses a full set of first integrals $g_1, \ldots, g_p$. This means that if $g \in C^\infty(\mathcal{S}^k)$ is a first integral of this distribution, i.e., $Z^k_i(g) \equiv 0$ for any $i = 1, \ldots, q$, then g is a function of $g_1, \ldots, g_p$. Besides, the covectors $dg_{1,\theta_k}, \ldots, dg_{p,\theta_k}$ are linearly independent, i.e., the functions $g_1, \ldots, g_p$ are functionally independent.

STEP 5. It follows from the linear independence of the vectors (6.1) that, in a neighborhood of the point θ_k, the fields $Z^k_1, \ldots, Z^k_q$ have a set of common first integrals $z_1, \ldots, z_n$ such that the matrix

$$\big(\tilde{D}_i(z_j)(\theta)\big) = \big(\pi_{\infty,k,*}(\tilde{D}_{i,\theta})(z_j)\big)$$

is nonsingular.[1] It follows from the smoothness of the fields $\tilde{D}_1, \ldots, \tilde{D}_n$ and the functions $z_1, \ldots, z_n$ that the matrix $\big(\tilde{D}_i(z_j)(\theta)\big)$ is nonsingular at each point of some neighborhood of the point $\theta \in \mathcal{S}^\infty$. Consequently, the functions $z_1, \ldots, z_n$ can be taken for the new independent variables.

STEP 6. Since $z_1, \ldots, z_n$ are common first integrals of the fields $Z^k_1, \ldots, Z^k_q$ and hence of the fields $Z_1, \ldots, Z_q$, it follows from relations (5.2) for the fields Z_i that

$$B_i = [Z, D_{z_i}](z) = Z\big(D_{z_i}(z)\big) - D_{z_i}\big(Z(z)\big) = 0, \quad i = 1, \ldots, n, \quad z = (z_1, \ldots, z_n)^T.$$

Hence if g is a common first integral of the fields $Z_1, \ldots, Z_q$, then

$$Z\big(D_{z_i}(g)\big) = D_{z_i}\big(Z(g)\big) + A_i Z(g) = 0,$$

and hence $D_{z_i}(g)$ is a common first integral of these fields for each $i = 1, \ldots, n$ as well.

STEP 7. We supplement the set $z_1, \ldots, z_n$ with some functions $\phi^1, \ldots, \phi^s \in C^\infty(\mathcal{S}^k)$ to a maximal functionally independent set of common first integrals of the fields $Z^k_1, \ldots, Z^k_q$ in a neighborhood of the point $\theta_k \in \mathcal{S}^k$. As the vectors $Z^k_{1,\theta_k}, \ldots, Z^k_{q,\theta_k}$ are linearly independent, there exist smooth functions $w^1, \ldots, w^q$ on $\mathcal{S}^k$ such that the matrix $\big(Z^k_{i,\theta_k}(w_j)\big)$ is nonsingular. From smoothness of the fields $Z^k_1, \ldots, Z^k_q$ and the functions $w^1, \ldots, w^q$ it follows that the matrix $\big(Z^k_i(w_j)\big)$ is

[1] Here and in the following, we identify the functions $g \in \mathcal{F}_k(\mathcal{S})$ and $(\pi_{\infty,k})^*(g) \in \mathcal{F}(\mathcal{S})$.

nonsingular in some neighborhood of the point θ_k. Whence $z_1, \dots, z_n, \phi^1, \dots, \phi^s, w^1, \dots, w^q$ are local coordinates in a neighborhood of $\theta_k \in \mathcal{S}^k$.

Since $\pi_{\infty,k}$ takes the fields $Z_1, \dots, Z_q$ to the fields $Z_1^k, \dots, Z_q^k$, the functions[2] $z_1, \dots, z_n, \phi^1, \dots, \phi^s$ are common first integrals of fields $Z_1, \dots, Z_q$. Likewise, the functions $w^1, \dots, w^q$ possess the property $\det\big(Z_i(w_j)\big) \neq 0$ in a neighborhood of the point θ. The derivatives $D_{z_i}(\phi^j)$ are also common first integrals of fields $Z_1, \dots, Z_q$. On the other hand, $D_{z_i}(\phi^j)$ are functions on $\mathcal{S}^{k+1}$, and $\pi_{\infty,k+1}$ takes the fields $Z_1, \dots, Z_q$ to the fields $Z_1^{k+1}, \dots, Z_q^{k+1}$. Therefore, the functions $z_i, \phi^j, D_{z_i}(\phi^j)$ are common first integrals of the fields $Z_1^{k+1}, \dots, Z_q^{k+1}$.

The fields $Z_1^{k+1}, \dots, Z_q^{k+1}$ generate also an integrable distribution in some neighborhood of the point $\theta_{k+1} = \pi_{\infty,k+1}(\theta) \in \mathcal{S}^{k+1}$. Let $\{\psi^1, \dots, \psi^s\}$ be a full set of first integrals of this distribution. Since the functions $z_i, \phi^j, D_{z_i}(\phi^j)$ are also first integrals of this distribution, then they are functions of $\psi = (\psi^1, \dots, \psi^s)$, i.e., $z_i = z_i(\psi), \phi^j = \phi^j(\psi), D_{z_i}(\phi^j) = \phi_1^j(\psi)$. Since the functions z_i, ϕ^j are functionally independent, there exists a basis minor of the Jacobian matrix $\left(\frac{\partial(z_i,\phi^j,\phi_1^j)}{\partial(\psi^l)}\right)$ at the point θ_{k+1} such that the rows corresponding to z_i, ϕ^j are basis. The other basis rows correspond to $D_{z_i}(\phi^j), \psi^l$ for some i, j, l. Thus the functions $z_1, \dots, z_n, \phi^1, \dots, \phi^s, w^1, \dots, w^q$ and $D_{z_i}(\phi^j), \psi^l$ for some i, j, l constitute a coordinate system in a neighborhood of the point $\theta_{k+1} \in \mathcal{S}^{k+1}$. A similar coordinate system exists in a neighborhood of the point $\theta_{k+l} \in \mathcal{S}^{k+l}$ for any $l > 1$. As $\mathcal{S}^\infty$ is the diffiety of some system $\mathcal{S}$, there exists an integer l_0 such that coordinates $\tilde{\psi}^1, \dots, \tilde{\psi}^{s_0}$ in $\mathcal{S}^{k+l_0}$ and a part of their time derivatives $D_{z_i}(\tilde{\psi}^j)$ constitute a coordinate system in $\mathcal{S}^{k+l_0+1}$. Thus the functions $z_1, \dots, z_n, \phi^1, \dots, \phi^s, w^1, \dots, w^q, \psi^1, \dots, \psi^{s_*}$ and a part of derivatives $D_\sigma(\phi^j), D_\sigma(\psi^j)$ constitute a coordinate system in a neighborhood of the point $\theta \in \mathcal{S}^\infty$.

STEP 8. The derivatives $D_{z_i}(w^j)$ are functions on $\mathcal{S}^{k+1}$ and hence are functions of the coordinates in $\mathcal{S}^{k+1}$, i.e.,

$$D_{z_i}(w^j) = W_i^j\big(z, w, \phi, \psi, D_z(\phi)\big), \quad i = 1, \dots, n, \quad j = 1, \dots, q. \tag{6.2}$$

Take another equation of the system $\mathcal{S}$. Using relations (6.2), we eliminate time derivatives of w^j from this equation. We get

$$\Phi\big(z, w, \phi, \psi, \dots, D_\sigma(\phi), D_\sigma(\psi), \dots\big) = 0.$$

The derivative of this equation along a field Z_i is

$$\sum_{j=1}^{q} \Phi'_{w^j} Z_i(w^j) = 0.$$

Since the matrix $\big(Z_i(w^j)\big)_{i,j=1,\dots,q}$ is nonsingular in a neighborhood of the point θ, we see that $\Phi'_{w^j} \equiv 0$, $j = 1, \dots, q$, i.e., the function Φ is independent of w. Thus we have an equation for common first integrals of fields $Z_1, \dots, Z_q$:

$$\Phi\big(z, \phi, \psi, \dots, D_\sigma(\phi), D_\sigma(\psi), \dots\big) = 0. \tag{6.3}$$

We apply a similar argument to the remaining equations in the system $\mathcal{S}$. Finally, denote by $v^1, \dots, v^{s+s_*}$ the functions $\phi^1, \dots, \phi^s, \psi^1, \dots, \psi^{s_*}$. Then

[2] more precisely, their images under the map $(\pi_{\infty,k})^*$

system (6.3) is presented in the form of (2.4) and Equation (6.2) is presented in the form (2.5). Theorem 6.6 is proved. □

Let $X = (X_1, \dots, X_q)^{\mathrm{T}}$ be an integrable pseudosymmetry of a system $\mathcal{E}$ and let $D_1, \dots, D_n$ be the total derivatives with respect to independent variables of the system $\mathcal{E}$. To construct a covering determined by X, one needs to find functions $g_1, \dots, g_p$ such that

(1) $g_1, \dots, g_p \in \mathcal{K}$;
(2) $g_1, \dots, g_p$ are common first integrals of the vector fields $X_1, \dots, X_q$;
(3) $dg_1, \dots, dg_p$ are linearly independent at each point;
(4) the set $\{g_1, \dots, g_p\}$ is a maximal set satisfying 1)-3), i.e., if $g \in \mathcal{K}$ is a common first integral of the vector fields $X_1, \dots, X_q$, then g is a function of $g_1, \dots, g_p$.

Since the fields $X_1, \dots, X_q$ are linearly independent at a regular point θ, there exist functions $w_1, \dots, w_q \in \mathcal{F}(\mathcal{E})$ such that the matrix $\big(X_i(w_j)(\theta)\big)$ is nonsingular. Let us choose new independent variables $z_1, \dots, z_n$ among the functions $g_1, \dots, g_p$ such that the matrix $\big(D_i(z_j)(\theta)\big)$ is nonsingular. By ϕ denote the set of the functions $g_1, \dots, g_p$ that are not $z_1, \dots, z_n$. From the proof of Theorem 6.6 it follows that the system $\mathcal{E}$ is determined by the equations of the form (2.4)–(2.5), where v consists of ϕ and a part of derivatives $D_\sigma(\phi)$.

Methods for solving matrix equations of the form (5.8) are similar to methods for solving equations in higher symmetries (see [**2**]). Most often, there exists a variable v such that both sides of the equation are polynomials in v, but all components of the matrices φ and A are independent of v. Equating the coefficients at the same degrees of v, we obtain the consequences of this equation. Such a transition from an equation to its consequences will be called a *splitting procedure* with respect to v. If one of the new equations is simple enough and contains a derivative of the component of φ_i with respect to a variable w, then, solving this equation, we derive the dependence of φ_i on w. Substituting the expression for φ into (5.8) and applying the splitting procedure for the result with respect to another variable, we get new equations and repeat the procedure until the equation (5.8) is completely solved.

To find integrable pseudosymmetries, the splitting procedure should be combined with using Theorem 6.5 and condition (iii).

The following example demonstrates the technique of computing integrable pseudosymmetries and covering.

7. Example of calculating coverings from a control system

Consider the Kapitsa pendulum, which differs from the mathematical pendulum in that its suspension point can move along a vertical line. The equations of motion are (see for example [**7**]):

$$\dot{\alpha} = p + \frac{u}{l}\sin\alpha, \qquad \dot{q} = u \qquad \dot{p} = \left(\frac{g}{l} - \frac{u^2}{l^2}\cos\alpha\right)\sin\alpha - \frac{u}{l}p\cos\alpha, \tag{7.1}$$

where q is the vertical coordinate of the suspension point, u is its vertical velocity, p is a quantity proportional to the generalized momentum, α is the angle between the pendulum axis and the vertical axis, l is the pendulum length, g is the gravitational acceleration, and $\dot{\alpha}, \dot{q}, \dot{p}$ are the time derivatives of α, q, p.

Let us compute all the one-dimensional coverings from this system. By Theorem 6.6, there exists a one-dimensional integrable pseudosymmetry that determines such covering. We restrict ourselves to the case when the covering preserves the time function.

Let us first transform system (7.1). Assuming that the angle α lies between 0 and π, we introduce a new variable by the relation

$$y = q + g\frac{t^2}{2} - l\int_{\pi/2}^{\alpha} \frac{d\theta}{\sin\theta}.$$

It is easily shown that in terms of the variables y, α, system (7.1) has the form

$$\ddot{y} = a(\alpha)\,\dot{\alpha}^2, \quad a(\alpha) = l\,\frac{\cos\alpha}{\sin^2\alpha}. \tag{7.2}$$

Systems (7.1) and (7.2) are equivalent, therefore their decomposable forms coincide. Since we consider coverings preserving the time function, then $M \equiv 0$ in the representation (5.5). Thus to find desired coverings from system (7.1), it suffices to find all the one-dimensional integrable evolution pseudosymmetries of system (7.2).

Denote by $\mathcal{E}$ system (7.2). The ring $\mathcal{F}_s(\mathcal{E})$ consists of smooth functions depending on t, y, α, $\dot{y}$, $\dot{\alpha}$, $\ldots$, $\alpha^{(s+1)}$ for $s \geq 0$. By $\mathcal{F}_{-1}(\mathcal{E})$ denote the ring of smooth functions depending on $t, y, \alpha, \dot{y}$. Equation (5.8) for system (7.2) can be written as

$$(D_t + A)^2(\varphi_y) = 2a\,\dot{\alpha}\,(D_t + A)(\varphi_\alpha) + a'\,\dot{\alpha}^2\,\varphi_\alpha, \tag{7.3}$$

where $A, \varphi_y, \varphi_\alpha$ belong to $\mathcal{F}_s(\mathcal{E})$ for some $s \geq -1$.

Consider the three cases:

I) $\varphi_y \not\equiv 0,\ \varphi_\alpha \not\equiv 0$;
II) $\varphi_y \not\equiv 0,\ \varphi_\alpha \equiv 0$;
III) $\varphi_y \equiv 0,\ \varphi_\alpha \not\equiv 0$.

In case I) consider the points $\theta \in \mathcal{E}^\infty$ such that $\varphi_y(\theta) \neq 0$. From Theorem 6.5 it follows that the column $\psi_0 = (1, \psi = \varphi_\alpha/\varphi_y)^T$ is also a generating matrix of an integrable pseudosymmetry of system (7.2) in neighborhoods of these points. Equation (7.3) for this generating matrix is of the form

$$D_t(A) + A^2 = 2a(\alpha)\,\dot{\alpha}\,(D_t + A)(\psi) + a'(\alpha)\,\dot{\alpha}^2\,\psi. \tag{7.4}$$

Suppose that the function ψ depends on $\alpha^{(s+1)}$ for some $s \geq 0$ and is independent of $\alpha^{(i)}$ for any $i > s+1$, i.e., $\psi \in \mathcal{F}_s(\mathcal{E}) \setminus \mathcal{F}_{s-1}(\mathcal{E})$. Then from (7.4) it follows that $A \in \mathcal{F}_s(\mathcal{E}) \backslash \mathcal{F}_{s-1}(\mathcal{E})$ as well. Since the component of $\ni_{\psi_0,A}$ at $\dfrac{\partial}{\partial\alpha^{(k)}}$ is $(D_t + A)^k(\psi)$, the Lie derivative along $\ni_{\psi_0,A}$ takes $\mathcal{F}_k(\mathcal{E}) \setminus \mathcal{F}_{k-1}(\mathcal{E})$ to $\mathcal{F}_{s+k}(\mathcal{E}) \setminus \mathcal{F}_{s+k-1}(\mathcal{E})$. In this case, condition (iii) cannot hold. Therefore, ψ and A are functions of $t, y, \alpha, \dot{y}$ only.

Further, apply the splitting procedure for equation (7.4). Finding the coefficients at $\dot{\alpha}^3$ and $\dot{\alpha}^2$ in this equation, we obtain that ψ is independent of $\dot{y}$ and the function A is linear in $\dot{y}$. Calculating the coefficients at $\dot{y}^2, \dot{y}\dot{\alpha}, \dot{\alpha}, \dot{y}$, and 1 in (7.4), we get

$$\psi = \frac{\beta + c_1}{2\sqrt{a}(y + c_2)}, \quad A = \frac{\dot{y}}{y + c_2}, \quad \beta = \int_{\pi/2}^{\alpha} \sqrt{a(\theta)}\,d\theta, \tag{7.5}$$

where c_1, c_2 are arbitrary constants.

This pseudosymmetry determines the following decomposable form of system (7.2) and hence of system (7.1):

$$\dot{z}_1 = v - \frac{1}{2}z_1v^2, \qquad \dot{z}_2 = 1 - z_2v^2 \tag{7.6}$$
$$\dot{\zeta} = \zeta v^2$$

where

$$z_1 = \frac{\beta + c_1}{\sqrt{\ddot{y}}}, \quad z_2 = \frac{y + c_2}{\dot{y}}, \quad v = \dot{\alpha}\sqrt{\frac{a(\alpha)}{\dot{y}}}, \quad \zeta = \dot{y}.$$

Similarly, in the cases II) and III) we get the following pseudosymmetries:

II) $\varphi_y = 1,\ \varphi_l = 0,\ A = \frac{1}{t+c_3},\ c_3 = \text{const}$; or

IIa) $\varphi_y = 1,\ \varphi_\alpha = 0,\ A = 0$;

III) $\varphi_y = 0,\ \varphi_\alpha = 1,\ A = -\frac{a'\,\dot{\alpha}}{2a}$.

The corresponding decomposable forms are

II) $\begin{cases} \dot{z}_1 = -(t+c_3)a(z_2)\dot{z}_2^2 \\ \dot{\zeta} = \frac{\zeta - z_1}{t+c_3}, \end{cases}$ where $z_1 = y - \dot{y}(t+c_3),\quad z_2 = \alpha,\quad \zeta = y$;

IIa) $\begin{cases} \dot{z}_1 = a(z_2)\dot{z}_2^2 \\ \dot{\zeta} = z_1, \end{cases}$ where $z_1 = \dot{y},\quad z_2 = \alpha,\quad \zeta = y$;

III) $\dot{\zeta} = \sqrt{\frac{\ddot{z}}{a(\zeta)}}$, where $z = y,\quad \zeta = \alpha$.

Note that calculating the integral in (7.5), we obtain

$$\beta(\alpha) = \sqrt{l}\left(\frac{1}{2}\ln\frac{1+\sqrt{\cos\alpha}}{1-\sqrt{\cos\alpha}} - \operatorname{arctg}\sqrt{\cos\alpha}\right).$$

Considering the function β to be a new dependent variable, we transform system (7.2) for $\alpha \in (0, \pi/2)$ into the equation $\ddot{y} = \dot{\beta}^2$.

Decompositions can be used to solve boundary value problems and, in particular, for trajectory planning. As an example, consider the boundary value problem for system (7.1) with the conditions

$$\alpha(t_0) = \alpha_0, \quad q(t_0) = q_0, \quad \alpha(t_f) = \alpha_f, \quad q(t_f) = q_f. \tag{7.7}$$

Namely, it is required to find a solution of system (7.1) that satisfies conditions (7.7). This problem defines the following boundary conditions for system (7.2):

$$\alpha(t_0) = \alpha_0, \quad y(t_0) = y_0, \quad \alpha(t_f) = \alpha_f, \quad y(t_f) = y_f, \tag{7.8}$$

where

$$y_i = q_i + g\frac{t_i^2}{2} - l\int_{\pi/2}^{\alpha_i}\sqrt{a(\theta)}d\theta, \quad i = 0, f.$$

If $c_1 = -\beta(\alpha_f)$ and $c_2 = -y_f$, then the corresponding boundary conditions for system (7.6) is

$$z_1(t_f) = 0, \quad z_2(t_f) = 0, \quad z_2(t_0) = \frac{(y_0 - y_f)z_1(t_0)^2}{(\beta(\alpha_0) - \beta(\alpha_f))^2}.$$

Let $\big(z_1(t), z_2(t), v(t)\big)$ be a solution of the problem. Then conditions (7.8) define the initial condition

$$\zeta(t_0) = \frac{y_0 - y_f}{z_2(t_0)}$$

for the system

$$\dot{\zeta} = \zeta(v(t))^2,$$

whereas $\zeta(t_f)$ is arbitrary. Thus, the decomposition converts the boundary value problem into a Cauchy problem and a boundary value problem with fewer conditions than the original problem (3 instead of 4).

8. Commutation of pseudosymmetries

Let X and Y be two pseudosymmetries of a system $\mathcal{E}$. Denote by $[X \cup Y]$ the set of generators of the involutive closure of the union of the sets X and Y of vector fields.

THEOREM 8.1 ([**5**]). *If X and Y are two pseudosymmetries of a system $\mathcal{E}$ and the set $[X \cup Y]$ is finite, then this set also forms a pseudosymmetry of the system $\mathcal{E}$.*

PROOF. By definition, if $X = (X_1, \dots, X_q)^{\mathrm{T}}, Y = (Y_1, \dots, Y_r)^{\mathrm{T}}$ are pseudosymmetries of a system $\mathcal{E}$, then $[X_i, D_l]$ is a linear combination of the fields $X_1, \dots, X_q, D_1, \dots, D_n$, and $[Y_j, D_l]$ is a linear combination of the fields $Y_1, \dots, Y_r, D_1, \dots, D_n$ for any $i = 1, \dots, q$, $j = 1, \dots, r$, $l = 1, \dots, n$. The Jacobi identity for the fields X_i, Y_j, D_l has the form $\Big[[X_i, Y_j], D_l\Big] = -\Big[[Y_j, D_l], X_i\Big] + \Big[[X_i, D_l], Y_j\Big]$. Therefore $\Big[[X_i, Y_j], D_l\Big]$ is a linear combination of the fields $X_p, Y_s, [X_p, Y_s], D_k$ for any $i, p = 1, \dots, q$, $j, s = 1, \dots, r$, $l, k = 1, \dots, n$.

The commutators of the obtained vector fields possess a similar property. Thus, the commutator of any field of involutive closure with any field D_i is a linear combination of $D_1, \dots, D_n$ and fields of involutive closure. By the hypothesis of the theorem, the involutive closure has a finite set of generators. Therefore, this set is a pseudosymmetry of the system $\mathcal{E}$. $\square$

EXAMPLE 8.2. By $\ni_1$, $\ni_2$, $\ni_3$, and $\ni_4$ we denote the pseudosymmetries of system (7.2) established in Section 7 and corresponding to cases I), II), III), and IIa). It can readily be verified that one has the relations

$$[\ni_1, \ni_2] = \frac{1}{y + c_2}(\ni_1 - \ni_2), \quad [\ni_1, \ni_3] = \frac{h(\alpha)}{y + c_2}\ni_3, \quad h(\alpha) = \frac{(\beta + c_1)a'}{4a^{3/2}} - \frac{1}{2},$$
$$[\ni_1, \ni_4] = \frac{1}{y + c_2}(\ni_1 - \ni_4), \quad [\ni_2, \ni_3] = 0, \quad [\ni_2, \ni_4] = 0, \quad [\ni_3, \ni_4] = 0.$$

Consequently, any nonempty subset of the tuple $\{\ni_1, \ni_2, \ni_3, \ni_4\}$ generates an involutive distribution and, by Theorem 8.1, is a pseudosymmetry of system (7.1). This pseudosymmetry is integrable, because as a ring $\mathcal{K}$ we can take the ring of all functions depending on $t, y, \alpha, \dot{y}$. Thus, we obtain six two-dimensional, four three-dimensional, and one four-dimensional integrable pseudosymmetries and eleven coverings of the corresponding dimensions from system (7.1). Let us give three of them while indicating the corresponding pseudosymmetries:

$$\{\ni_1,\ni_3\}:\quad \begin{cases} \dot\zeta_1 = z(\zeta_1+c_2) \\ \dot\zeta_2 = \sqrt{\frac{(\dot z+z^2)(\zeta_1+c_2)}{a(\zeta_2)}}, \end{cases} \quad \text{where} \quad \begin{array}{l} z=\frac{\dot y}{y+c_2},\quad \zeta_1=y \\ \zeta_2=\alpha; \end{array}$$

$$\{\ni_1,\ni_2,\ni_3\}:\quad \begin{cases} \ddot\zeta_1 = \frac{\dot\zeta_1(t+c_3)-\zeta_1-c_2}{z} \\ \dot\zeta_2 = \sqrt{\frac{\dot\zeta_1(t+c_3)-\zeta_1-c_2}{za(\zeta_2)}}, \end{cases} \quad \text{where} \quad \begin{array}{l} z=\frac{\dot y(t+c_3)-y-c_2}{a(\alpha)\,\dot\alpha^2}, \\ \zeta_1=y,\quad \zeta_2=\alpha; \end{array}$$

$$\{\ni_1,\ni_2,\ni_3,\ni_4\}:\quad \begin{cases} \dddot\zeta_1 = z\ddot\zeta_1 \\ \dot\zeta_2 = \sqrt{\frac{\ddot\zeta_1}{a(\zeta_2)}}, \end{cases} \quad \text{where} \quad \begin{array}{l} z=\frac{\dddot y}{\ddot y},\quad \zeta_1=y \\ \zeta_2=\alpha. \end{array}$$

9. Multivector pseudosymmetries

The module $\bigwedge^r \mathcal{D}(\mathcal{E}^\infty)$ (for $r>0$) generated by elements of the form $X_1\wedge\cdots\wedge X_r$, $X_1,\ldots,X_r\in\mathcal{D}(\mathcal{E}^\infty)$, is called the *r-th exterior degree* of the module $\mathcal{D}(\mathcal{E}^\infty)$.

Elements of the module $\bigwedge^* \mathcal{D}(\mathcal{E}^\infty)=\bigoplus_{r>0}\bigwedge^r\mathcal{D}(\mathcal{E}^\infty)$ are called multivector fields on $\mathcal{E}^\infty$. An element $X\in\bigwedge^r\mathcal{D}(\mathcal{E}^\infty)$ we call *a homogeneous element of degree* $|X|=r$.

Define the *Schouten bracket* of two homogeneous elements $X=\bigwedge_{\alpha=1}^s X_\alpha$ and $Y=\bigwedge_{\beta=1}^k Y_\beta$ as

$$[X,Y]=\sum_{\alpha=1}^s\sum_{\beta=1}^k(-1)^{\alpha+\beta}[X_\alpha,Y_\beta]\wedge X_1\wedge\cdots\wedge\widehat{X_\alpha}\wedge\cdots\wedge X_s\wedge Y_1\wedge\cdots\wedge\widehat{Y_\beta}\wedge\cdots\wedge Y_k,$$

where, as usual, the symbol $\widehat{\ }$ over X_α and Y_β means that X_α and Y_β excluded from the product. Extend this bracket to $\bigwedge^*\mathcal{D}(\mathcal{E}^\infty)$ by linearity.

The following properties of the Schouten bracket are well known [**17**, **21**].

THEOREM 9.1. *Let X,Y,Z be homogeneous elements of the module $\bigwedge^*\mathcal{D}(\mathcal{E}^\infty)$. Then the following identities are valid:*

(1) $[X,Y]=-(-1)^{(|X|-1)(|Y|-1)}[Y,X]$ *(antisymmetry of Schouten bracket);*
(2) $(-1)^{(|X|-1)(|Z|-1)}[X,[Y,Z]]+(-1)^{(|Y|-1)(|X|-1)}[Y,[Z,X]]$
$+(-1)^{(|Z|-1)(|Y|-1)}[Z,[X,Y]]=0$ *(Jacobi identity for Schouten bracket);*
(3) $[X\wedge Y,Z]=(-1)^{|Y|(|Z|-1)}[X,Z]\wedge Y+X\wedge[Y,Z]$ *(Poisson identity).*

The Schouten bracket makes the module $\bigwedge^*\mathcal{D}(\mathcal{E}^\infty)$ into a Lie superalgebra if the degree of a homogeneous element $X\in\bigwedge^r\mathcal{D}(\mathcal{E}^\infty)$ is taken to be the integer $r-1$.

THEOREM 9.2 ([**5**]). *If the vector fields $X_1,\ldots,X_q$ form a pseudosymmetry of system $\mathcal{E}$, then their exterior product $X=X_1\wedge\cdots\wedge X_q$ satisfies the equalities*

$$[X,D_i]=a_{i,X}X+\sum_{j=1}^n Z_{ij,X}\wedge D_j,\qquad i=1,\ldots,n, \tag{9.1}$$

for some $a_{i,X}\in\mathcal{F}(\mathcal{E})$ and $Z_{ij,X}\in\bigwedge^{q-1}\mathcal{D}(\mathcal{E}^\infty)$.

PROOF. From the definitions of the Schouten bracket and pseudosymmetry it follows that

$$[X, D_i] = \sum_{\alpha=1}^{q}(-1)^{\alpha+1}[X_\alpha, D_i] \wedge X_1 \wedge \cdots \wedge \widehat{X_\alpha} \wedge \cdots \wedge X_q =$$
$$\sum_{\alpha=1}^{q}(-1)^{\alpha+1}\sum_{\beta=1}^{q} A_{i,\alpha\beta}X_\beta \wedge X_1 \wedge \cdots \wedge \widehat{X_\alpha} \wedge \cdots \wedge X_q +$$
$$\sum_{\alpha=1}^{q}(-1)^{\alpha+1}\sum_{j=1}^{n} B_{i,\alpha j}D_j \wedge X_1 \wedge \cdots \wedge \widehat{X_\alpha} \wedge \cdots \wedge X_q,$$

where $A_{i,\alpha\gamma}$ and $B_{i,\alpha\gamma}$ are $\alpha\gamma$–entries of the matrices A_i and B_i respectively. Using the properties of the exterior product, we get

$$[X, D_i] = \sum_{\alpha=1}^{q} A_{i,\alpha\alpha}X_1 \wedge \cdots \wedge X_q +$$
$$\sum_{j=1}^{n}\sum_{\alpha=1}^{q}(-1)^{\alpha+q}B_{i,\alpha j}X_1 \wedge \cdots \wedge \widehat{X_\alpha} \wedge \cdots \wedge X_q \wedge D_j$$

that is equality (9.1), where

$$a_{i,X} = \mathrm{tr}A_i = \sum_{\alpha=1}^{q} A_{i,\alpha\alpha}, \qquad Z_{ij,X} = \sum_{\alpha=1}^{q}(-1)^{\alpha+q}B_{i,\alpha j}X_1 \wedge \cdots \wedge \widehat{X_\alpha} \wedge \cdots \wedge X_q.$$

□

A homogeneous multivector field X on $\mathcal{E}^\infty$ satisfying the equality (9.1) will be called a *multivector pseudosymmetry* of system $\mathcal{E}$.

THEOREM 9.3 ([**5**]). *If X and Y are multivector pseudosymmetries of a system, then $X \wedge Y$ is also a multivector pseudosymmetry of this system.*

PROOF. From the Poisson identity and the definition of multivector pseudosymmetry, as well as the properties of the exterior product, it follows that

$$[X \wedge Y, D_i] = [X, D_i] \wedge Y + X \wedge [Y, D_i] =$$
$$(a_{i,X}X + \sum_{j=1}^{n} Z_{ij,X} \wedge D_j) \wedge Y +$$
$$X \wedge (a_{i,Y}Y + \sum_{j=1}^{n} Z_{ij,Y} \wedge D_j) =$$
$$(a_{i,X} + a_{i,Y})X \wedge Y + \sum_{j=1}^{n}(X \wedge Z_{ij,Y} - Z_{ij,X} \wedge Y) \wedge D_j.$$

We obtain the equality (9.1) for $X \wedge Y$, where

$$a_{i,X\wedge Y} = a_{i,X} + a_{i,Y}, \qquad Z_{ij,X\wedge Y} = X \wedge Z_{ij,Y} - Z_{ij,X} \wedge Y.$$

□

Note that the pseudosymmetry $X \wedge Y$ can be zero.

Now we prove one more property of the Schouten bracket. We define the *interior product* i_X of a multivector field $X = X_1 \wedge \cdots \wedge X_r \in \bigwedge^r \mathcal{D}(\mathcal{E}^\infty)$ and a differential form $\omega \in \Lambda^p(\mathcal{E}^\infty), p \geq r$, as $i_X(\omega) = (i_{X_r} \circ \cdots \circ i_{X_1})(\omega)$, where i_{X_i} is the interior product of the vector field X_i. Extend this operation to an arbitrary element of $\bigwedge^* \mathcal{D}(\mathcal{E}^\infty)$ by linearity.

THEOREM 9.4 ([**5**]). *The interior product i_X of a multivector field $X \in \bigwedge^* \mathcal{D}(\mathcal{E}^\infty)$ and the Lie derivative of differential forms with respect to a vector field $Z \in \mathcal{D}(\mathcal{E}^\infty)$ satisfy the equation*

$$i_X \circ Z - Z \circ i_X = i_{[X,Z]}. \tag{9.2}$$

PROOF. We use induction on $|X|$. The equality (9.2) is well known, if X is a vector field, i.e., when $|X| = 1$.

Suppose $X = X_1 \wedge X_2$, $|X_1| > 0, |X_2| > 0$. Then from the definition of the interior product it follows that $i_X = i_{X_2} \circ i_{X_1}$. By the inductive hypothesis,

$$i_X \circ Z = i_{X_2} \circ i_{X_1} \circ Z = i_{X_2} \circ (Z \circ i_{X_1} + i_{[X_1,Z]}).$$

Similarly,

$$Z \circ i_X = (i_{X_2} \circ Z - i_{[X_2,Z]}) \circ i_{X_1}.$$

If we combine this with the Poisson identity, we get

$$i_X \circ Z - Z \circ i_X = i_{X_2} \circ i_{[X_1,Z]} + i_{[X_2,Z]} \circ i_{X_1} = i_{[X_1 \wedge X_2, Z]}.$$

□

THEOREM 9.5 ([**5**]). *If X is a multivector pseudosymmetry of a system $\mathcal{E}$, $q = |X|$, $\omega \in \mathcal{C}^q\Lambda^q(\mathcal{E}^\infty)$, D_l is the total derivative with respect to x_l on $\mathcal{E}^\infty$, then*

$$i_X(D_l\omega) = (D_l + a_{l,X})(i_X\omega). \tag{9.3}$$

PROOF. Using antisymmetry of exterior product, we rewrite (9.1) as

$$[X, D_l] = a_{l,X}X + \sum_{j=1}^{n}(-1)^{q-1}D_j \wedge Z_{lj,X}.$$

Combining this with (9.2) and the definition of the interior product, we obtain

$$\begin{aligned}(i_X \circ D_l)(\omega) &= (D_l \circ i_X + i_{[X,D_l]})(\omega) = \\ &(D_l \circ i_X + a_{l,X}i_X + \sum_{j=1}^{n}(-1)^{q-1}i_{Z_{lj,X}} \circ i_{D_j})(\omega).\end{aligned}$$

This implies (9.3), since ω is a Cartan form and hence $i_{D_j}(\omega) = 0$. □

An operator of $\mathcal{F}(\mathcal{E})$ to $\mathcal{F}(\mathcal{E})$ of the form $g \mapsto \sum_{|\sigma| \leq k} a_\sigma D_\sigma(g), a_\sigma \in \mathcal{F}(\mathcal{E})$ is called a *scalar $\mathcal{C}$-differential operator* of order $\leq k$. The set $\mathcal{C}\,\mathrm{Dif}\,f(\mathcal{F}(\mathcal{E}))$ of all scalar $\mathcal{C}$-differential operators is a noncommutative ring with respect to the composition operation. The set $\mathcal{C}^q\Lambda^q(\mathcal{E}^\infty)$ is a left module over the ring $\mathcal{C}\,\mathrm{Dif}\,f(\mathcal{F}(\mathcal{E}))$. From Theorem 9.5 it follows that any multivector pseudosymmetry X of a system $\mathcal{E}$ is uniquely determined by its values on generators of the module $\mathcal{C}^q\Lambda^q(\mathcal{E}^\infty)$, $q = |X|$.

EXAMPLE 9.6. For the Laplace equation $u_{x_1x_1} + u_{x_2x_2} = 0$ generators of the module $\mathcal{C}^2\Lambda^2(\mathcal{E}^\infty)$ are

$$\omega_1^k = -d_{\mathcal{C}}u_{(k,0)} \wedge d_{\mathcal{C}}u_{(k,1)}, \quad \omega_2^k = d_{\mathcal{C}}u_{(k,0)} \wedge d_{\mathcal{C}}u_{(k+1,0)}, \qquad k \geq 0.$$

They are related by the equations

$$D_2(\omega_1^k) = D_1(\omega_2^k), \quad \Delta^k\big(D_1(\omega_1^0) + D_2(\omega_2^0)\big) = c_k\big(D_1(\omega_1^k) + D_2(\omega_2^k)\big),$$

where $\{c_k\}$ is some sequence of integers, $\Delta = D_1^2 + D_2^2$.

We now define the *Lie derivative* of $g \in \mathcal{F}(\mathcal{E})$ along a homogeneous element $X = X_1 \wedge \cdots \wedge X_q \in \bigwedge^q \mathcal{D}(\mathcal{E}^\infty)$ as

$$X(g) = \sum_{i=1}^{q}(-1)^{q-i}X_i(g)X_1 \wedge \cdots \wedge \widehat{X_i} \wedge \cdots \wedge X_q.$$

Extend this operation to an arbitrary element $X \in \bigwedge^* \mathcal{D}(\mathcal{E}^\infty)$ by linearity. A function $g \in \mathcal{F}(\mathcal{E})$ is called a *first integral of a multivector field* X, if $X(g) \equiv 0$. It is easy to see that $g \in \mathcal{F}(\mathcal{E})$ is a common first integral of vector fields $X_1, \ldots, X_q$ if and only if g is a first integral of the multivector field $X = X_1 \wedge \cdots \wedge X_q$. This fact can be used to compute coverings determined by a multivector pseudosymmetry.

EXAMPLE 9.7. Let X be the multivector pseudosymmetry of the Laplace equation such that

$$i_X(\omega_1^0) = x_2, \quad i_X(\omega_2^0) = x_1, \quad i_X(\omega_i^k) = 0, \quad a_{i,X} = 0, \quad Z_{ij,X} = 0$$

for any $k > 0, i, j = 1, 2$. Then

$$X = x_1\frac{\partial}{\partial u} \wedge \frac{\partial}{\partial u_{(1,0)}} - x_2\frac{\partial}{\partial u} \wedge \frac{\partial}{\partial u_{(0,1)}} + \frac{\partial}{\partial u} \wedge \frac{\partial}{\partial u_{(2,0)}}.$$

The corresponding decomposable form is

$$(z_2 - z_1)(z_2 v^1_{(1,0)} - z_1 v^1_{(0,1)}) = (z_2^2 + z_1^2)v^2 - (z_2 + z_1)v^1,$$
$$w^1_{(1,0)} = \frac{v^1 - z_1 w^2}{z_2 - z_1}, \qquad w^1_{(0,1)} = \frac{v^1 - z_2 w^2}{z_1 - z_2}.$$

The corresponding change of variables is

$$z_1 = x_1, \quad z_2 = x_2, \quad w^1 = u, \quad w^2 = u_{(1,0)} + u_{(0,1)},$$
$$v^1 = x_2 u_{(1,0)} + x_1 u_{(0,1)}, \quad v^2 = u_{(1,0)} + u_{(0,1)} + (x_2 - x_1)u_{(2,0)}.$$

EXAMPLE 9.8. In the situation of example 9.7, but when $i_X(\omega_2^1) = -1$ and $i_X(\omega_i^k) = 0$ for other k, i, j, we have

$$X = \frac{\partial}{\partial u} \wedge \frac{\partial}{\partial u_{(3,0)}} - \frac{\partial}{\partial u_{(1,0)}} \wedge \frac{\partial}{\partial u_{(2,0)}} + \frac{\partial}{\partial u_{(0,1)}} \wedge \frac{\partial}{\partial u_{(1,1)}}.$$

The decomposable form is

$$v^1_{(1,0)} = v^2_{(0,1)}, \quad v^1_{(0,1)} = -v^2_{(1,0)}, \quad w_{(2,1)} = v^1, \quad w_{(4,0)} = v^2.$$

The change of variables is

$$z_1 = x_1, \quad z_2 = x_2, \quad v^1 = u_{(2,1)}, \quad v^2 = u_{(4,0)}, \quad w = u.$$

This example demonstrates that the dimension of a covering can be greater than the degree of the corresponding multivector pseudosymmetry.

10. Generators of modules of Cartan q-forms for control systems

Recall that from Theorem 9.5 it follows that any multivector pseudosymmetry X of a system $\mathcal{E}$ is uniquely determined by its values on generators of the module $\mathcal{C}^q\Lambda^q(\mathcal{E}^\infty)$, $q = |X|$. Let us now find the generators of this module in the case of a control system

$$\dot{x} = f(t, x, u), \tag{10.1}$$

where $x = (x_1, \dots, x_n) \in \mathbb{R}^n$ is the state, $u = (u_1, \dots, u_m) \in \mathbb{R}^m$ is the control input. For system (10.1) the diffiety is an infinite-dimensional manifold $\mathcal{E}^\infty$ with coordinates

$$t,\ x_i,\ u_j^{(l)}, \qquad i = 1, \dots, n,\ j = 1, \dots, m,\ l \geq 0, \tag{10.2}$$

where $u_j^{(l)}$ denotes the variable corresponding to the lth order derivative of u_j with respect to t.

In coordinate system (10.2) the module $\mathcal{C}^1\Lambda(\mathcal{E})$ is generated by forms

$$d_\mathcal{C}x_1, \dots, d_\mathcal{C}x_n, d_\mathcal{C}u_1^{(0)}, \dots, d_\mathcal{C}u_m^{(0)}, d_\mathcal{C}u_1^{(1)}, \dots, d_\mathcal{C}u_m^{(1)}, d_\mathcal{C}u_1^{(2)}, \dots.$$

Denote by $\mathcal{H}_0$ the $\mathcal{F}(\mathcal{E})$-submodule in $\mathcal{C}^1\Lambda(\mathcal{E})$ generated by forms $d_\mathcal{C}x_1, \dots, d_\mathcal{C}x_n$. By definition, put

$$\mathcal{H}_{k+1} = \{\omega \in \mathcal{H}_k \mid D\omega \in \mathcal{H}_k\}, \quad k \geq 0.$$

The *dimension* of some submodule $\mathcal{H} \subset \mathcal{C}^1\Lambda(\mathcal{E})$ at a point $\theta \in \mathcal{E}^\infty$ is the dimension of the space of covectors $\{\omega_\theta \mid \omega \in \mathcal{H}\}$. A point $\theta \in \mathcal{E}^\infty$ is called *B-regular* if in a neighborhood of θ the rank of the matrix $\partial f/\partial u$ is equal to m and the dimensions of $\mathcal{H}_k$ and $\mathcal{H}_k + D(\mathcal{H}_k)$ are constant for any $k > 0$.

Note that the dimension of $\mathcal{H}_k$ at any point is finite and $\mathcal{H}_{k+1} \subset \mathcal{H}_k$. It follows that in a neighborhood of a B-regular point there exists an integer k^* such that $\mathcal{H}_{k+1} = \mathcal{H}_k = \mathcal{H}_{k^*}$ for $k \geq k^*$. The inequality $\mathcal{H}_{k^*} \neq \{0\}$ means that system (10.1) has non-trivial first integrals [**14**]. The only first integrals of a controllable system are the trivial ones [**8**]. Therefore, we consider only the case $\mathcal{H}_{k^*} = \{0\}$.

REMARK 10.1. In the autonomous case we can consider only functions, differential forms, and vector fields that are independent of t. In this case, $\mathcal{E}^\infty$ is a manifold with coordinates $(x, u^{(0)}, \dots, u^{(l)}, \dots)$ (without t), the vector field D does not contain the term $\partial/\partial t$, $d_\mathcal{C}g = dg$, $\mathcal{C}^1\Lambda(\mathcal{E})$ is identified with the module of all differential 1-forms on $\mathcal{E}^\infty$.

The following theorem was proved in [**1**] for autonomous systems.

THEOREM 10.2. *Let $\theta \in \mathcal{E}^\infty$ be a B-regular point for system* (10.1), *$\mathcal{H}_{k^*} = \{0\}$. Then in a neighborhood of θ there exist m forms $\omega_1, \dots, \omega_m \in \mathcal{H}_0$ such that the set $\{D^j(\omega_k) \mid k = 1, \dots, m, j \geq 0\}$ is a basis of the module $\mathcal{C}^1\Lambda(\mathcal{E})$.*

The set $(\omega_1, \dots, \omega_m)$ is called a *B-basis*. The proof of Theorem 10.2 is similar to the proof of the corresponding theorem from [**1**] and gives a **procedure for finding a B-basis**. Namely we have $\mathcal{H}_{k^*} = 0$ and $\mathcal{H}_{k^*-1} \neq 0$. Choose a basis $\omega_1, \dots, \omega_{m_1}$ ($m_1 \leq m$) of the module $\mathcal{H}_{k^*-1}$. Then the 1-forms $\omega_1, \dots, \omega_{m_1}$, $D(\omega_1), \dots, D(\omega_{m_1})$ are $\mathcal{F}(\mathcal{E})$-linear independent and lie in $\mathcal{H}_{k^*-2}$. To obtain a basis of the module $\mathcal{H}_{k^*-2}$ we add some 1-forms $\omega_{m_1+1}, \dots, \omega_{m_2}$ ($m_1 \leq m_2 \leq m$) to this collection. Repeating this construction for $\mathcal{H}_{k^*-3}, \dots, \mathcal{H}_0$, we get a B-basis $(\omega_1, \dots, \omega_m)$.

Note that there are two module structures in $\mathcal{C}^1\Lambda(\mathcal{E})$: over the algebra $\mathcal{F}(\mathcal{E})$ and over the algebra of scalar $\mathcal{C}$-differential operators. It follows from Theorem 10.2 that a B-basis is a basis of this module over the algebra of scalar $\mathcal{C}$-differential operators.

Let θ be a B-regular point for system (10.1), $\mathcal{H}_{k^*} = \{0\}$, and $(\omega_1, \dots, \omega_m)$ be a B-basis at θ. Introduce a linear (or total) order $\prec$ on the set of pairs (i,k), $i = 1, \dots, m$, $k \geq 0$, as follows:

$$(l,j) \prec (k,i) \iff l < k \quad \text{or} \quad l = k,\ j < i.$$

We say that a pair (k,i) is a *d-degree* of a Cartan 1-form ω, if

$$\omega = a_k^i D^i(\omega_k) + \sum_{(l,j)\prec(k,i)} a_l^j D^j(\omega_l), \quad a_k^i, a_l^j \in \mathcal{F}(\mathcal{E}), \quad a_k^i \not\equiv 0.$$

If $(l,j) \prec (k,i)$, then $(l,j+1) \prec (k,i+1)$. Hence the total derivative $D = D_t$ increases d-degree by the pair $(0,1)$.

Each Cartan q-form is a sum of monomials of the form

$$aD^{j_1}(\omega_{l_1}) \wedge \dots \wedge D^{j_q}(\omega_{l_q}), \quad a \in \mathcal{F}(\mathcal{E}). \tag{10.3}$$

Let us rearrange the factors in each monomial (10.3) such that $(l_1, j_1) \succ (l_2, j_2) \succ \dots \succ (l_q, j_q)$.

THEOREM 10.3. *Let θ be a B-regular point for system* (10.1), $\mathcal{H}_{k^*} = \{0\}$, *and $(\omega_1, \dots, \omega_m)$ be a B-basis in a neighborhood of θ. In a neighborhood of the point θ the module $\mathcal{C}^q\Lambda^q(\mathcal{E})$ over the algebra of scalar $\mathcal{C}$-differential operators is free. The set of the Cartan q-forms*

$$D^{j_2+1}(\omega_{l_1}) \wedge D^{j_2}(\omega_{l_1}) \wedge D^{j_3}(\omega_{l_3}) \wedge \dots \wedge D^{j_q}(\omega_{l_q}), \quad (l_1, j_2) \succ (l_3, j_3) \succ \dots \succ (l_q, j_q),$$

and

$$\omega_{l_1} \wedge D^{j_2}(\omega_{l_2}) \wedge \dots \wedge D^{j_q}(\omega_{l_q}), \quad l_1 > l_2, \quad (l_2, j_2) \succ \dots \succ (l_q, j_q),$$

forms a basis for this module.

PROOF. Let Ω be a Cartan q-form. It is a finite sum of monomials of the form (10.3). Let (i,k) be the maximum d-degree of 1-forms contained in Ω. Consider the monomials containing the 1-form $D^i(\omega_k)$. Each such monomial has the form

$$\Omega_s = a_s D^i(\omega_k) \wedge D^{j_2}(\omega_{l_2}) \wedge \dots \wedge D^{j_q}(\omega_{l_q}), \quad (k,i) \succ (l_2, j_2) \succ \dots \succ (l_q, j_q).$$

If $l_2 = k$, then $i > j_2$ and the q-form $\Omega_s - \Delta_s(\Omega_s^b)$, where

$$\Delta_s = a_s D^{i-j_2-1}, \quad \Omega_s^b = D^{j_2+1}(\omega_k) \wedge D^{j_2}(\omega_k) \wedge \dots \wedge D^{j_q}(\omega_{l_q}),$$

contains only 1-forms of d-degree less than (i,k). If $l_2 \neq k$, then $l_2 < k$ and the q-form $\Omega_s - \Delta_s(\Omega_s^b)$, where

$$\Delta_s = a_s D^i, \quad \Omega_s^b = \omega_k \wedge D^{j_2}(\omega_{l_2}) \wedge \dots \wedge D^{j_q}(\omega_{l_q}),$$

contains only 1-forms of d-degree less than (i,k). Thus $\Omega - \sum_s \Delta_s(\Omega_s^b)$ contains only 1-forms of d-degree less than (i,k).

Repeating this reasoning, we remove all 1-forms $D^{i-1}(\omega_k)$, $D^{i-2}(\omega_k)$, ..., ω_k from Ω. Repeating this reasoning again, we remove all monomials with ω_{k-1}, ω_{k-2}, ..., ω_1 from Ω. As a result, we get the representation $\Omega = \sum_s \Delta_s(\Omega_s^b)$, where Δ_s are scalar $\mathcal{C}$-differential operators, Ω_s^b are basic Cartan q-forms.

Assume

$$\sum_s \Delta_s(\Omega_s^b) = 0, \tag{10.4}$$

where Δ_s are scalar $\mathcal{C}$-differential operators, Ω_s^b are basic Cartan q-forms. Let (i,k) be the maximum d-degree of 1-forms contained in (10.4). Then $D^i(\omega_k)$ can be in (10.4) only in terms of the form

$$\sum_{j=1}^{i} a_j D^{i-j}\Big(D^j(\omega_k) \wedge D^{j-1}(\omega_k) \wedge D^{j_3}(\omega_{l_3}) \wedge \ldots \wedge D^{j_q}(\omega_{l_q})\Big) + \\ a_0 D^i\Big(\omega_k \wedge D^{\bar{j}_2}(\omega_{\bar{l}_2}) \wedge \ldots \wedge D^{\bar{j}_q}(\omega_{\bar{l}_q})\Big).$$

The last sum is equal to

$$D^i(\omega_k) \wedge \Big(\sum_{j=1}^{i} a_j D^{j-1}(\omega_k) \wedge D^{j_3}(\omega_{l_3}) \wedge \ldots \wedge D^{j_q}(\omega_{l_q}) + \\ a_0 D^{\bar{j}_2}(\omega_{\bar{l}_2}) \wedge \ldots \wedge D^{\bar{j}_q}(\omega_{\bar{l}_q})\Big) + \Omega_0$$

where Ω_0 does not contain $D^i(\omega_k)$. From (10.4) it follows that

$$\sum_{j=1}^{i} a_j D^{j-1}(\omega_k) \wedge D^{j_3}(\omega_{l_3}) \wedge \ldots \wedge D^{j_q}(\omega_{l_q}) + a_0 D^{\bar{j}_2}(\omega_{\bar{l}_2}) \wedge \ldots \wedge D^{\bar{j}_q}(\omega_{\bar{l}_q}) = 0.$$

But this is only possible when $a_j = 0$ for any $j = 0, 1, \ldots, i$. That is, the 1-form $D^i(\omega_k)$ cannot be contained in (10.4). This contradiction proves that equality (10.4) is impossible. □

Thus, any multivector pseudosymmetry of a controllable system is uniquely determined by its values on the basis forms from Theorem 10.3. Since the module $\mathcal{C}^q\Lambda^q(\mathcal{E})$ over the algebra of scalar $\mathcal{C}$-differential operators is free, these values are arbitrary and are not bound by any relations.

11. Conclusion

We have presented two approaches to constructing pseudosymmetries. The first approach is to solve equation (5.8) using the splitting procedure, Theorem 6.5, and condition (iii). The second approach is to find generators of the module of Cartan forms over the algebra of $\mathcal{C}$-differential operators and construct pseudosymmetries satisfying condition (iii). To construct a covering determined by the found integrable pseudosymmetry, it is sufficient to use the algorithm given after Theorem 6.6.

Note also that in the case of one dependent variable ($m = 1$), the ring $\mathcal{K}$ (see condition (iii)) coincides with $\mathcal{F}_l(\mathcal{E})$, where the integer l is determined by the matrices φ and A (cf. with example from section 7). In the case $m > 1$, we suggest first choosing $\mathcal{K}$ and then solving equation (5.8). Changing $\mathcal{K}$, we get different coverings.

References

[1] E. Aranda-Bricaire, C. H. Moog, and J.-B. Pomet, *Infinitesimal Brunovský form for nonlinear systems with applications to dynamic linearization*, Geometry in nonlinear control and differential inclusions (Warsaw, 1993), Banach Center Publ., vol. 32, Polish Acad. Sci. Inst. Math., Warsaw, 1995, pp. 19–33. MR1364417

[2] A. V. Bocharov, V. N. Chetverikov, S. V. Duzhin, N. G. Khor'kova, I. S. Krasil'shchik (ed.), A. V. Samokhin, Yu. N. Torkhov, A. M. Verbovetsky, and A. M. Vinogradov(ed.), *Symmetries and Conservation Laws for Differential Equations of Mathematical Physics*, Amer. Math. Soc., Providence, RI, 1999.

[3] V. N. Chetverikov, *On the structure of integrable $\mathcal{C}$-fields*, Differential Geom. Appl. **1** (1991), no. 4, 309–325, DOI 10.1016/0926-2245(91)90011-W. MR1244448

[4] V. N. Chetverikov, *Coverings and integrable pseudosymmetries of differential equations*, Differ. Equ. **53** (2017), no. 11, 1428–1439, DOI 10.1134/s0012266117110052. Translation of Differ. Uravn. **53** (2017), no. 11, 1461–1472. MR3741191

[5] Vladimir N. Chetverikov, *Coverings and multivector pseudosymmetries of differential equations*, Differential Geom. Appl. **74** (2021), Paper No. 101705, 13, DOI 10.1016/j.difgeo.2020.101705. MR4178991

[6] Vladimir N. Chetverikov, *Orbital decompositions and integrable pseudosymmetries of control systems*, Automatica J. IFAC **139** (2022), Paper No. 110189, 8, DOI 10.1016/j.automatica.2022.110189. MR4385156

[7] Michel Fliess, Jean Lévine, Philippe Martin, and Pierre Rouchon, *Flatness and defect of non-linear systems: introductory theory and examples*, Internat. J. Control **61** (1995), no. 6, 1327–1361, DOI 10.1080/00207179508921959. MR1613557

[8] Michel Fliess, Jean Lévine, Philippe Martin, François Ollivier, and Pierre Rouchon, *A remark on nonlinear accessibility conditions and infinite prolongations*, Systems Control Lett. **31** (1997), no. 2, 77–83, DOI 10.1016/S0167-6911(97)00028-5. MR1461803

[9] Alberto Isidori, *Nonlinear control systems*, 3rd ed., Communications and Control Engineering Series, Springer-Verlag, Berlin, 1995, DOI 10.1007/978-1-84628-615-5. MR1410988

[10] I. S. Krasil'shchik, V. V. Lychagin, and A. M. Vinogradov, *Geometry of Jet Spaces and Nonlinear Partial Differential Equations*, Gordon and Breach, New York, 1986.

[11] Joseph Krasil'shchik and Alexander Verbovetsky, *Geometry of jet spaces and integrable systems*, J. Geom. Phys. **61** (2011), no. 9, 1633–1674, DOI 10.1016/j.geomphys.2010.10.012. MR2813504

[12] I. S. Krasil′shchik and A. M. Vinogradov, *Nonlocal symmetries and the theory of coverings: an addendum to Vinogradov's "Local symmetries and conservation laws" [Acta Appl. Math. **2** (1984), no. 1, 21–78; MR0736872 (85m:58192)]*, Acta Appl. Math. **2** (1984), no. 1, 79–96, DOI 10.1007/BF01405492. MR736873

[13] I. S. Krasil′shchik and A. M. Vinogradov, *Nonlocal trends in the geometry of differential equations: symmetries, conservation laws, and Bäcklund transformations*, Acta Appl. Math. **15** (1989), no. 1-2, 161–209, DOI 10.1007/BF00131935. Symmetries of partial differential equations, Part I. MR1007347

[14] A. P. Krishchenko and V. N. Chetverikov, *Minimal realizations of nonlinear systems* (Russian, with Russian summary), Differ. Uravn. **46** (2010), no. 11, 1612–1622, DOI 10.1134/S001226611011008X; English transl., Differ. Equ. **46** (2010), no. 11, 1612–1623. MR2798757

[15] M. Marvan, *On zero-curvature representations of partial differential equations*, Differential geometry and its applications (Opava, 1992), Math. Publ., vol. 1, Silesian Univ. Opava, Opava, 1993, pp. 103–122. MR1255532

[16] M. Marvan, *Some local properties of Bäcklund transformations*, Acta Appl. Math. **54** (1998), no. 1, 1–25, DOI 10.1023/A:1006037726082. MR1656796

[17] Albert Nijenhuis, *Jacobi-type identities for bilinear differential concomitants of certain tensor fields. I, II*, Nederl. Akad. Wetensch. Proc. Ser. A. **58** = Indag. Math. **17** (1955), 390–397, 398–403. MR0074879

[18] V. V. Sokolov, *Pseudosymmetries and differential permutations* (Russian), Funktsional. Anal. i Prilozhen. **22** (1988), no. 2, 47–56, 96, DOI 10.1007/BF01077602; English transl., Funct. Anal. Appl. **22** (1988), no. 2, 121–129. MR947605

[19] Shlomo Sternberg, *Lectures on differential geometry*, Prentice-Hall, Inc., Englewood Cliffs, N.J., 1964. MR0193578

[20] A. M. Vinogradov, *The geometry of nonlinear differential equations* (Russian), Problems in geometry, Vol. 11 (Russian), Akad. Nauk SSSR, Vsesoyuz. Inst. Nauchn. i Tekhn. Informatsii, Moscow, 1980, pp. 89–134, 243. MR579929

[21] A. M. Vinogradov, *The union of the Schouten and Nijenhuis brackets, cohomology, and superdifferential operators* (Russian), Mat. Zametki **47** (1990), no. 6, 138–140. MR1074539

[22] H. D. Wahlquist and F. B. Estabrook, *Prolongation structures of nonlinear evolution equations*, J. Mathematical Phys. **16** (1975), 1–7, DOI 10.1063/1.522396. MR358111

Bauman Moscow State Technical University, 2-ja Baumanskaya str. 5, 105005, Moscow, Russia

Email address: chetverikov.vl@yandex.ru

Contemporary Mathematics
Volume **788**, 2023
https://doi.org/10.1090/conm/788/15821

Differential invariants of inviscid flows in pipes

Anna Duyunova

ABSTRACT. In this paper inviscid flows in the three-dimensional space are studied. A PDE system (the Euler system) describing such flows is considered. We also formulate this system in special coordinates corresponding to a model of a pipe. The Lie algebra of symmetries of this system is described. Then we give an algebra of differential invariants, in particular, in coordinate-free form. Finally, we present the Euler system describing flows along a circular helix, discuss its symmetries and invariants.

Introduction

In this paper, symmetries and differential invariants of a PDE system describing inviscid flows in a region of the three-dimensional space are studied. Specifically, we consider the Euler system describing inviscid flows in a constant gravitational field. We examine the two forms of this system, namely the standard form in the Cartesian coordinates [**1**] and the special curvilinear form. The latter is intended to model flows in a curved pipe.

With this paper, we continue the series of works devoted to studies of media flows in pipes. Before, only negligibly thin pipes, which were represented as curves, were considered (see, for example, [**5**]). Now we proceed to investigate flows in pipes of arbitrary forms, e. g., of varying radius. To this end, we employ the same method as we did before for flows in spheres and spherical layers [**2**]. The essence of the method is to write generalized hydrodynamics equations on an oriented Riemannian manifold (M, g). This approach was discussed and developed in [**3**] from a general point of view. It allows us to rewrite the Euler system in any coordinates.

The paper is organized as follows.

In Section 1 we give the generalized Euler system and briefly remind the geometric interpretation of thermodynamics [**7**]. The latter is of importance since the PDE system itself is not closed, namely, it needs two additional thermodynamic equations. Then we proceed to the symmetry Lie algebra of the Euler system written in Cartesian coordinates, or, in terms of generalized Euler equations, when $(M, g) = (\mathbb{R}^3, dx^2 + dy^2 + dz^2)$. Next, we give its symmetries. The symmetry Lie algebra of the PDEs has a subalgebra $\mathfrak{g}_{\mathrm{m}}$, which is generated by the time shift, the space translations, the (deformed) rotations and the Galilean transformations.

2020 *Mathematics Subject Classification.* Primary 76N99; Secondary 53A55, 76M60.
Key words and phrases. Euler equation, differential invariants, curvilinear coordinates.
The author was partially supported by Russian Science Foundation Grant Number 21-71-20034.

©2023 American Mathematical Society

These ten symmetries act trivially on thermodynamic quantities like the density or the temperature, that is why we call them *kinematic* symmetries. The corresponding differential invariants are also called kinematic. The answer to the question whether the kinematic symmetries are the only possible ones depends on thermodynamic state equations, which may not admit symmetries besides kinematic ones. As an example, we discuss the case of a thermodynamic state admitting one additional symmetry. Thus, the complete Euler system admits a larger eleven-dimensional symmetry algebra (see [**4**] for details). Such symmetries we call *Euler symmetries*, and the corresponding invariants are called *Euler invariants.* Later, we give the theorem describing the field of the kinematic invariants, invariant derivatives, and discuss their coordinate-free representation.

In Section 2 we present a curvilinear coordinate system in a tubular neighborhood $P \subset \mathbb{R}^3(X,Y,Z)$ of a smooth curve. This serves as a model of pipes. Note that the transformation we use is not a diffeomorphism. We rewrite the Euclidean metric in these curvilinear coordinates to obtain the Euler system using its generalized version. Besides, we give the Euler system when the smooth curve is a line. As a more specific example, we consider the case of a circular helix. For the case of this curve we calculate symmetries. It turns out that these symmetries cannot be obtained with a simple change of coordinates in the symmetries of the Euler system in Cartesian coordinates. Contrary to the symmetries, invariants and invariant derivatives can be obtained using their coordinate-free description.

1. The Euler system

The system of differential equations (the Euler system) describing flows on an oriented Riemannian manifold (M, g) consists of the following equations (see [**3**] for details):

$$\begin{cases} \rho(\mathbf{u}_t + \nabla_{\mathbf{u}}\mathbf{u}) = -\operatorname{grad} p + \mathbf{g}\rho, \\ \dfrac{\partial(\rho\,\Omega_g)}{\partial t} + \mathcal{L}_{\mathbf{u}}(\rho\,\Omega_g) = 0, \\ \rho\theta\,(s_t + \nabla_{\mathbf{u}} s) = \operatorname{div}(k \operatorname{grad} \theta), \end{cases} \tag{1}$$

where the vector field $\mathbf{u} = (u, v, w)$ is the flow velocity, p, ρ, s, θ are the pressure, density, specific entropy and the temperature of the medium respectively, k is the thermal conductivity, which is assumed to be a constant, and $\mathbf{g}$ is the gravitational acceleration.

Here ∇_X is the directional covariant derivative with respect to a vector field X, $\mathcal{L}_X$ is the Lie derivative along a vector field X, Ω_g is the volume form on M corresponding to the metric g. This metric also provides us with an isomorphism between TM and T^*M, which we also denote by g. Further, we omit any reference to M and simply write T or T^*.

It is easy to see that the system (1) is incomplete, namely, it contains $\dim M + 2$ equations for $\dim M + 4$ unknown functions. There are several ways to complete the system, and, usually, it is done with simple relations like $\rho = const$, $s = const$, or $\operatorname{curl} \mathbf{u} = 0$. Though such closures may seem reasonable and practical, they are also restrictive. A more general approach is to add two relations between thermodynamic quantities. To find them, we employ the same method as in the paper [**4**].

The idea of this method is based on interpretation of thermodynamic state equations as Legendrian, or Lagrangian, manifolds in contact, or symplectic, space, respectively.

It can be shown [**7**], that thermodynamic state can be defined as Lagrangian manifold L given by two functions, namely,

$$L = \{ F(p, \rho, s, \theta) = 0,\ G(p, \rho, s, \theta) = 0 \} \tag{2}$$

that satisfy the relation

$$[F, G] = 0 \mod \{F = 0,\, G = 0\},$$

where $[F, G]$ is the Poisson bracket with respect to the symplectic form

$$ds \wedge d\theta + \rho^{-2}\, d\rho \wedge dp.$$

The latter condition corresponds the first law of thermodynamics.

Moreover, the restriction of the quadratic differential form

$$d(\theta^{-1}) \cdot d\epsilon - \rho^{-2}\, d(p\theta^{-1}) \cdot d\rho$$

to the manifold of thermodynamic state is negative definite, here ϵ is the specific internal energy. This condition corresponds to the second law of thermodynamics [**7**].

Following [**4**], *by the Euler system* $\mathcal{E}$ *we mean the system* (1) *together with the two thermodynamic equations* (2).

First, we consider the Euler system in the three-dimensional Euclidean space. Though later in Section 2, we introduce a special curvilinear coordinates associated with an arbitrary curve. In this section to find symmetries and invariants we use the standard Cartesian coordinates.

1.1. Symmetry Lie algebra. To describe the Lie algebra of symmetries of the Euler system, we consider the Lie algebra $\mathfrak{g}$ generated by the following vector fields:

$$\begin{aligned}
&X_1 = \partial_t, \quad X_2 = \partial_x, \quad X_3 = \partial_y, \quad X_4 = \partial_z,\\
&X_5 = t\,\partial_x + \partial_u, \quad X_6 = t\,\partial_y + \partial_v, \quad X_7 = t\,\partial_z + \partial_w,\\
&X_8 = y\,\partial_x - x\,\partial_y + v\,\partial_u - u\,\partial_v,\\
&X_9 = \left(\frac{\mathrm{g}t^2}{2} + z\right)\partial_x - x\,\partial_z + (\mathrm{g}t + w)\,\partial_u - u\,\partial_v,\\
&X_{10} = \left(\frac{\mathrm{g}t^2}{2} + z\right)\partial_y - y\,\partial_z + (\mathrm{g}t + w)\,\partial_v - v\,\partial_w,\\
&X_{11} = \partial_p, \quad X_{12} = \partial_s, \quad X_{13} = \theta\,\partial_\theta, \quad X_{14} = p\partial_p + \rho\partial_\rho - s\partial_s,\\
&X_{15} = t\,\partial_t + x\,\partial_x + y\,\partial_y - \left(\frac{\mathrm{g}t^2}{2} - z\right)\partial_z - \mathrm{g}t\,\partial_w - s\,\partial_s,\\
&X_{16} = t\,\partial_t - \mathrm{g}t^2\partial_z - u\,\partial_u - v\,\partial_v - (2\mathrm{g}t + w)\,\partial_w - 2p\,\partial_p + s\,\partial_s.
\end{aligned}$$

Denote by $\mathfrak{h}$ the Lie algebra generated by the vector fields

$$Y_1 = \partial_s, \quad Y_2 = \partial_p, \quad Y_3 = \rho\,\partial_\rho, \quad Y_4 = s\,\partial_s, \quad Y_5 = p\,\partial_p, \quad Y_6 = \theta\,\partial_\theta.$$

Let $\vartheta : \mathfrak{g} \mapsto \mathfrak{h}$ be the following Lie algebra homomorphism

$$\vartheta(X) = X(\rho)\,\partial_\rho + X(s)\,\partial_s + X(p)\,\partial_p + X(\theta)\,\partial_\theta,$$

where $X \in \mathfrak{g}$.

Let also $\mathfrak{h}_t$ be the Lie subalgebra of the algebra $\mathfrak{h}$ that preserves the thermodynamic state (2).

THEOREM 1. *A Lie algebra of symmetries of the Euler system $\mathcal{E}$ coincides with* $\vartheta^{-1}(\mathfrak{h}_t)$.

The elements of the algebra $\mathfrak{g}_m = \ker \vartheta$ are called *kinematic symmetries.*

1.2. Thermodynamic states. Note that the pure thermodynamic part $\mathfrak{h}$ of symmetry algebra coincides with its counterpart from the case of the two-dimensional Euler system. Therefore, the symmetry classification of thermodynamic states is the same as in [**4**]. Here we show one example of the state equations, which have a one-dimensional symmetry algebra, that is, when the system $\mathcal{E}$ admits an eleven-dimensional symmetry algebra.

So, let $\dim \mathfrak{h}_t = 1$ and let $Z = \sum_{i=1}^{6} \lambda_i Y_i$ be a basis vector in this algebra.

If the coefficients $\lambda_1, \ldots, \lambda_6$ are in general position, the state equations admitting the symmetry Z are

$$p - C_1 \rho^{\frac{\lambda_5}{\lambda_3}} + \frac{\lambda_2}{\lambda_5} = 0, \quad \theta - C_2(\lambda_4 s + \lambda_1)^{\frac{\lambda_6}{\lambda_4}} = 0,$$

where C_1, C_2 are constants.

Moreover, the negative definiteness of the quadratic differential form

$$d(\theta^{-1}) \cdot d\epsilon - \rho^{-2}\, d(p\theta^{-1}) \cdot d\rho$$

on the Lagrangian surface leads to the relations

$$\frac{\lambda_4 s + \lambda_1}{\lambda_6} > 0, \quad \frac{C_1 \lambda_5}{\lambda_3} > 0$$

for all $s \in (-\infty, s_0]$.

1.3. Differential invariants. We consider two classes of differential invariants of the system $\mathcal{E}$: kinematic invariants, namely, functions that are invariant with respect to the prolonged action of $\mathfrak{g}_m$, and Euler invariants, that is, functions invariant with respect to the action of the Lie algebra $\vartheta^{-1}(\mathfrak{h}_t)$. This division makes calculations straightforward. At first, we find the kinematic invariants, then we use them to construct the Euler ones.

Here we present the main results of this subsection. All computation were made with Maple, and the corresponding files are available at `http://d-omega.org/appendices`.

1.3.1. *Kinematic invariants.* First of all, the functions ρ and θ generate all $\mathfrak{g}_m$-invariants of order zero.

Consider two vectors and a matrix

$$\operatorname{grad} \rho = \begin{pmatrix} \rho_x \\ \rho_y \\ \rho_z \end{pmatrix}, \quad \operatorname{grad} \theta = \begin{pmatrix} \theta_x \\ \theta_y \\ \theta_z \end{pmatrix}, \quad V = \begin{pmatrix} u_x & u_y & u_z \\ v_x & v_y & v_z \\ w_x & w_y & w_z \end{pmatrix},$$

and suppose that vectors $\operatorname{grad} \rho$ and $\operatorname{grad} \theta$ are linearly independent.

Denote by H a matrix made of column vectors

$$(\operatorname{grad} \rho \times \operatorname{grad} \theta, \operatorname{grad} \rho, \operatorname{grad} \theta)$$

with $\det \mathrm{H} \neq 0$, where $\operatorname{grad} \rho \times \operatorname{grad} \theta$ is the cross product. Then elements of the matrix $\mathrm{H}^{-1} V \mathrm{H}$ are nine functions, which are invariant with respect to the prolonged action of the Lie algebra $\mathfrak{g}_{\mathfrak{m}}$.

It is also easy to check that the functions

$$(\operatorname{grad} \rho)^2, \quad (\operatorname{grad} \theta)^2, \quad \operatorname{grad} \rho \cdot \operatorname{grad} \theta, \quad \theta_t + u\theta_x + v\theta_y + w\theta_z \tag{3}$$

are $\mathfrak{g}_{\mathfrak{m}}$-invariant.

We say now that a point $a \in \mathcal{E}_1$ and the corresponding $\mathfrak{g}_{\mathfrak{m}}$-orbit is *regular* if the invariants (3) and $(\mathrm{H}^{-1} V \mathrm{H})_{ij}$ are independent in a neighbourhood of this orbit and $\det \mathrm{H} \neq 0$.

Otherwise, we say that the point and the corresponding orbit is *singular*.

The Lie–Tresse theorem [**6**] gives us the following result.

THEOREM 2. *The field of the kinematic invariants of the system $\mathcal{E}$ is generated by the zero-order differential invariants ρ, θ, by the first-order differential invariants*

$$(\operatorname{grad} \rho)^2, \quad (\operatorname{grad} \theta)^2, \quad \operatorname{grad} \rho \cdot \operatorname{grad} \theta, \quad \theta_t + \theta_x u + \theta_y v + \theta_z w, \quad (\mathrm{H}^{-1} V \mathrm{H})_{ij}$$

and by the invariant derivatives

$$\nabla_1 = \frac{\mathrm{d}}{\mathrm{d}t} + u\frac{\mathrm{d}}{\mathrm{d}x} + v\frac{\mathrm{d}}{\mathrm{d}y} + w\frac{\mathrm{d}}{\mathrm{d}z},$$

$$\nabla_2 = \rho_x\frac{\mathrm{d}}{\mathrm{d}x} + \rho_y\frac{\mathrm{d}}{\mathrm{d}y} + \rho_z\frac{\mathrm{d}}{\mathrm{d}z}, \qquad \nabla_3 = \theta_x\frac{\mathrm{d}}{\mathrm{d}x} + \theta_y\frac{\mathrm{d}}{\mathrm{d}y} + \theta_z\frac{\mathrm{d}}{\mathrm{d}z},$$

$$\nabla_4 = (\rho_y\theta_z - \rho_z\theta_y)\frac{\mathrm{d}}{\mathrm{d}x} - (\rho_x\theta_z - \rho_z\theta_x)\frac{\mathrm{d}}{\mathrm{d}y} + (\rho_x\theta_y - \rho_y\theta_x)\frac{\mathrm{d}}{\mathrm{d}z},$$

where $\frac{\mathrm{d}}{\mathrm{d}t}$, $\frac{\mathrm{d}}{\mathrm{d}x}$, $\frac{\mathrm{d}}{\mathrm{d}y}$, $\frac{\mathrm{d}}{\mathrm{d}z}$ are the total derivatives with respect to t, x, y, z. This field separates regular orbits.

The Hilbert function $h(k)$ of $\mathfrak{g}_{\mathfrak{m}}$-invariants algebra (the number of independent invariants of pure order k) is

$$h(k) = 3k^2 + 8k + 5$$

for $k \geq 2$, and $h(1) = 13$, $h(0) = 2$.

The above constructions of invariants in Cartesian coordinates give us a hint on expressing these invariants in a coordinate-free form. We considered three invariant vector fields, namely, the gradients of the pressure and the temperature, and their cross product. Then we found coordinates of the operator V in this invariant basis thus obtaining invariant coordinates of V.

The operator V we used before is, in fact, the covariant differential of the flow velocity field $\mathbf{u}$, namely,

$$V = d_\nabla \mathbf{u} \in T \otimes T^*.$$

This operator corresponds to the rate-of-deformation tensor of the medium, see [**3**] for details.

Let us choose an invariant co-frame

$$\omega = (dp,\, d\theta,\, \star(dp \wedge d\theta)),$$

where $\star$ is the Hodge operator, and let $\omega' = g^{-1}(\omega)$ be its dual frame. Thus the invariants from Theorem 2 can be presented in a coordinate-free form. Namely, the inner products of the elements of ω, nine coordinates of the operator V in this co-frame and the material derivative of θ. The four invariant derivatives in a coordinate-free form are the horizontal lifts of the vector fields comprising the invariant frame ω'.

1.3.2. *Euler invariants.* Before (in Subsection 1.2) we considered thermodynamic equations admitting one thermodynamic symmetry. Here we find the Euler invariants for this case.

Let the equations of thermodynamic state admit one-dimensional symmetry algebra generated by the vector field

$$\xi_1 X_{11} + \xi_2 X_{12} + \xi_3 X_{13} + \xi_4 X_{14} + \xi_5 X_{15} + \xi_6 X_{16}.$$

Straightforward calculations show (see the aforementioned maple file) that for general ξ's the following theorem holds.

THEOREM 3. *The algebra of the Euler differential invariants for thermodynamic states admitting a one-dimensional symmetry algebra is generated by the differential invariants*

$$\theta\rho^{-\frac{\xi_3}{\xi_4}}, \quad \frac{\nabla_1\theta}{\theta J_{11}}, \quad (\operatorname{grad}\rho)^2\rho^{\frac{2\xi_5}{\xi_4}-2}, \quad \frac{\rho \operatorname{grad}\rho\cdot\operatorname{grad}\theta}{\theta(\operatorname{grad}\rho)^2}, \quad \frac{\theta\operatorname{grad}\rho\cdot\operatorname{grad}\theta}{\rho(\operatorname{grad}\theta)^2},$$

$$J_{11}\rho^{\frac{\xi_5+\xi_6}{\xi_4}}, \quad \frac{J_{21}\rho^{\frac{\xi_5}{\xi_4}}}{J_{11}\theta}, \quad \frac{J_{31}\theta}{J_{21}\rho}, \quad J_{12}\rho^{\frac{\xi_6}{\xi_4}}, \quad \frac{J_{22}}{J_{11}}, \quad \frac{J_{32}\theta}{J_{11}\rho}, \quad \frac{J_{13}\rho}{J_{12}\theta}, \quad \frac{J_{23}\rho}{J_{11}\theta}, \quad \frac{J_{33}}{J_{11}}$$

of the first order, here the elements of the matrix $\mathrm{H}^{-1}V\mathrm{H}$ *are denoted by* J_{ij}, *and by the invariant derivations*

$$\rho^{\frac{\xi_5+\xi_6}{\xi_4}}\nabla_1, \quad \rho^{\frac{2\xi_5}{\xi_4}-1}\nabla_2, \quad \rho^{\frac{2\xi_5-\xi_3}{\xi_4}}\nabla_3, \quad \rho^{\frac{3\xi_5-\xi_3}{\xi_4}-1}\nabla_4.$$

Due to the Lie–Tresse theorem [**6**], *this algebra separates the regular orbits.*

Other cases, including two-dimensional algebras of thermodynamic symmetries, can be analyzed similarly.

2. Flows in pipes

We propose a model of pipes as a tubular neighbourhood $P \subset \mathbb{R}^3(X, Y, Z)$ of a smooth curve l. Suppose that the smooth curve l is given in a parametric form

$$\{X = f_1(a),\, Y = f_2(a),\, Z = f_3(a)\}$$

with a being a natural parameter. Consider the plane that is normal to l at a point $l(a) = (f_1(a), f_2(a), f_3(a))$, and introduce a two-dimensional orthogonal coordinate system (O, x, y) in this plane, with $l(a)$ being an origin O, and the unit normal and the unit binormal vectors are vectors in the direction of the x-axis and y-axis, respectively.

Thus, in the neighbourhood of the curve, the curvilinear coordinates (a, x, y) can be transformed to the Cartesian coordinates (X, Y, Z) by the formulae

$$\begin{aligned} X =& f_1(a) + \frac{f_1''(a)}{|\kappa(a)|}x + b_1 y,\\ Y =& f_2(a) + \frac{f_2''(a)}{|\kappa(a)|}x + b_2 y,\\ Z =& f_3(a) + \frac{f_3''(a)}{|\kappa(a)|}x + b_3 y, \end{aligned}$$

where $\kappa(a)$ is the curvature and (b_1, b_2, b_3) is a unit binormal vector at the point $l(a)$. The Jacobian determinant of this coordinate transformation equals

$$1 - x|\kappa(a)|.$$

By a pipe of radius r, we mean the domain P the curve $l(a)$ is not greater than radius r of the pipe, we restrict the domain P of points satisfying the inequality

$$x \leq r < \frac{1}{|\kappa(a)|}.$$

The standard Euclidean metric $g = dX^2 + dY^2 + dZ^2$ in the curvilinear coordinates $(a, x, y) \subset P$ takes the form

$$g = \left((1 - x|\kappa(a)|)^2 + \tau^2(a)(x^2 + y^2)\right) da^2 + 2\tau(a)\,(x\,dy - y\,dx) \cdot da + dx^2 + dy^2,$$

where $\kappa(a)$ and $\tau(a)$ are the curvature and the torsion at the point $l(a)$ respectively. The associated with the metric g volume form has the form

$$\Omega_g = |1 - x|\kappa(a)||\; da \wedge dx \wedge dy.$$

Since the Euler system (1) contains the vector of gravitational acceleration $\mathbf{g} = -\mathrm{g}\partial_Z$, we have to rewrite the vector ∂_Z in the curvilinear coordinates (a, x, y):

$$\frac{f_3'(\,\partial_a + \tau(a)y\,\partial_x - \tau(a)x\,\partial_y)}{1 - x|\kappa(a)|} + \frac{f_3''\,\partial_x + \sqrt{\kappa^2(a)(1 - f_3'^2) - f_3''^2}\,\partial_y}{|\kappa(a)|}.$$

Now, we can rewrite the Euler system (1) for the Riemannian manifold (P, g). For convenience, we introduce the following notations:

$$\frac{\mathrm{d}}{\mathrm{d}t} = \partial_t + u\partial_a + v\partial_x + w\partial_y$$
$$\mu(a, x) = 1 - |\kappa(a)|\,x,$$

then the explicit form of the Euler system is

(4)

$$\begin{cases} \rho\dfrac{\mathrm{d}u}{\mathrm{d}t} = \dfrac{\rho u}{\mu}\left(|\kappa|\,(2v - \tau y u) + ux\,|\kappa|_a\right) - \dfrac{1}{\mu^2}(p_a + \tau(yp_x - xp_y)) - \dfrac{\mathrm{g}\rho f_3'}{\mu}, \\ \rho\dfrac{\mathrm{d}v}{\mathrm{d}t} = \rho u^2\left(\tau^2 x + \tau_a y - |\kappa|\left(\mu + \dfrac{y^2\tau^2}{\mu}\right) + \dfrac{xy\tau\,|\kappa|_a}{\mu}\right) + 2\tau\rho u\left(\dfrac{yv\,|\kappa|}{\mu} + w\right) - \\ \dfrac{\tau y}{\mu^2}p_a - \left(1 + \dfrac{\tau^2 y^2}{\mu^2}\right)p_x + \dfrac{\tau^2 xy}{\mu^2}p_y - \mathrm{g}\rho\left(\dfrac{f_3''}{|\kappa|} + \dfrac{\tau y f_3'}{\mu}\right), \\ \rho\dfrac{\mathrm{d}w}{\mathrm{d}t} = \rho u^2\left(\tau^2 y - \tau_a x + \dfrac{\tau x}{\mu}\left(|\kappa|\,\tau y - x\,|\kappa|_a\right)\right) - 2\tau\rho uv\left(1 + \dfrac{|\kappa|\,x}{\mu}\right) + \\ \dfrac{\tau x}{\mu^2}p_a + \dfrac{\tau^2 xy}{\mu^2}p_x - \left(1 + \dfrac{\tau^2 x^2}{\mu^2}\right)p_y - \mathrm{g}\rho\left(\dfrac{\sqrt{\kappa^2(1 - f_3'^2) - f_3''^2}}{|\kappa|} - \dfrac{f_3'\tau x}{\mu}\right), \\ \rho_t + (\rho u)_a + (\rho v)_x + (\rho w)_y = \dfrac{\rho}{\mu}\left(|\kappa|\,v + |\kappa|_a\,xu\right), \\ \dfrac{\rho\theta}{k}\dfrac{\mathrm{d}s}{\mathrm{d}t} = \dfrac{\theta_{aa}}{\mu^2} + \left(1 + \dfrac{\tau^2 y^2}{\mu^2}\right)\theta_{xx} + \left(1 + \dfrac{\tau^2 x^2}{\mu^2}\right)\theta_{yy} + \dfrac{2\tau}{\mu^2}\left(y\theta_{ax} - \tau xy\theta_{xy} - x\theta_{ay}\right) \\ + \dfrac{1}{\mu^3}\left((x\,|\kappa|_a + |\kappa|\,\tau y)\theta_a + \left(\tau^2(|\kappa|\,y^2 - \mu x) + |\kappa|_a\,\tau xy + \tau_a\mu y - |\kappa|\,\mu^2\right)\theta_x - \right. \\ \left.(y\tau^2(|\kappa|\,x + \mu) + x^2\tau\,|\kappa|_a + \mu x\tau_a)\theta_y\right). \end{cases}$$

The right hand sides of the Euler equations in this system contain three types of forces. The first and the second types are just the coordinates of the pressure gradient $\operatorname{grad} p$ and the gravitational force $\rho \mathrm{g}$, correspondingly. The force of the third type, which depends solely on the flow velocity and the curve invariants κ and τ, represents a fictitious force appearing because of the non-inertial nature of the coordinate frame (a, x, y).

Note that since the system (4) contains $|\kappa|$ and $|\kappa|_a = (\operatorname{sgn}\kappa)\kappa_a$, this system is discontinuous, and the curve $l(a)$ is thus divided into segments such that κ is either positive or negative, and on each segment the system (4) has a distinct form.

If $\kappa(a) = 0$ for all a, then the curve l is a line with a unit direction vector (m_1, m_2, m_3), and the Euler system has the form

$$
\begin{cases}
\rho \dfrac{\mathrm{d}u}{\mathrm{d}t} = -p_a - \mathrm{g}\rho m_3, \\
\rho \dfrac{\mathrm{d}v}{\mathrm{d}t} = -p_x, \\
\rho \dfrac{\mathrm{d}w}{\mathrm{d}t} = -p_y + \mathrm{g}\rho\sqrt{1 - m_3^2}, \\
\rho_t + (\rho u)_a + (\rho v)_x + (\rho w)_y = 0, \\
\dfrac{\rho\theta}{k}\dfrac{\mathrm{d}s}{\mathrm{d}t} = \theta_{aa} + \theta_{xx} + \theta_{yy}.
\end{cases}
$$

2.1. Helix pipes. Let us consider a circular helix:

$$
\begin{cases}
f_1(a) = \dfrac{\kappa}{\kappa^2 + \tau^2} \cos\sqrt{\kappa^2 + \tau^2}a, \\
f_2(a) = \dfrac{\kappa}{\kappa^2 + \tau^2} \sin\sqrt{\kappa^2 + \tau^2}a, \\
f_3(a) = \dfrac{\tau}{\sqrt{\kappa^2 + \tau^2}}a,
\end{cases}
$$

where $\kappa > 0$ and $\tau > 0$ are a constant curvature and torsion, respectively. Then the Euler equations has the form:

(5)

$$\begin{cases} \rho\dfrac{\mathrm{d}u}{\mathrm{d}t} = \dfrac{\rho u\kappa}{\mu}(2v - \tau yu) - \dfrac{1}{\mu^2}(p_a + \tau(yp_x - xp_y)) - \dfrac{\rho g\tau}{\mu\sqrt{\kappa^2+\tau^2}}, \\ \rho\dfrac{\mathrm{d}v}{\mathrm{d}t} = \rho u\left(\left(\tau^2 x - \kappa\mu - \dfrac{\kappa\tau^2 y^2}{\mu}\right)u + 2\tau\left(\dfrac{\kappa yv}{\mu} + w\right)\right) - \\ \dfrac{\tau y}{\mu^2}p_a - \left(1 + \dfrac{\tau^2 y^2}{\mu^2}\right)p_x + \dfrac{\tau^2 xy}{\mu^2}p_y - \dfrac{\rho g\tau^2 y}{\mu\sqrt{\kappa^2+\tau^2}}, \\ \rho\dfrac{\mathrm{d}w}{\mathrm{d}t} = \rho u\tau\,(u\tau y - 2v)\left(1 + \dfrac{\kappa x}{\mu}\right) + \dfrac{\tau x}{\mu^2}p_a + \dfrac{\tau^2 xy}{\mu^2}p_x - \left(1 + \dfrac{\tau^2 x^2}{\mu^2}\right)p_y + \\ \dfrac{\rho g}{\mu\sqrt{\kappa^2+\tau^2}}\left(\left(\kappa^2+\tau^2\right)x - \kappa\right), \\ \rho_t + (\rho u)_a + (\rho v)_x + (\rho w)_y = \dfrac{\kappa}{\mu}\rho v, \\ \dfrac{\rho\theta}{k}\dfrac{\mathrm{d}s}{\mathrm{d}t} = \dfrac{\theta_{aa}}{\mu^2} + \left(1 + \dfrac{\tau^2 y^2}{\mu^2}\right)\theta_{xx} + \left(1 + \dfrac{\tau^2 x^2}{\mu^2}\right)\theta_{yy} + \dfrac{2\tau}{\mu^2}\left(y\theta_{ax} - \tau xy\theta_{xy} - x\theta_{ay}\right) \\ + \dfrac{\kappa\tau y}{\mu^3}\theta_a + \dfrac{1}{\mu}\left(\dfrac{\kappa\tau^2 y^2}{\mu^2} - \dfrac{\tau^2 x}{\mu} - \kappa\right)\theta_x - \left(\dfrac{\kappa\tau^2 xy}{\mu^3} + \dfrac{\tau^2 y}{\mu^2}\right)\theta_y. \end{cases}$$

This system has a ten-dimensional kinematic symmetry algebra, generated by the vector fields $Z_1, \ldots, Z_{10}$, but these symmetries cannot be obtained from the symmetries of the Euler equations written in Cartesian coordinates. In particular, the lift of isometries of the metric g to 0-jets do not give symmetries of (5), which means that the symmetries of (5) are point symmetries. Though, there is a certain correspondence between the two algebras, for example, a similar role to the translations X_2, X_3, X_4 is played by three other symmetries, which involve the components of the flow velocity, i. e., u, v, w:

$$Z_1 = \frac{1}{\mu}\left(\partial_a + \tau y\,\partial_x - \frac{\lambda^2 x - \kappa}{\tau}\,\partial_y\right) + \frac{1}{\mu^2}\left(\kappa v\,\partial_u + (\kappa yv + \mu w)\,\partial_v - \tau v\,\partial_w\right),$$

$$Z_2 = \left(\lambda\cos(\lambda a) + \frac{\kappa\tau y\sin(\lambda a)}{\mu}\right)\partial_x + \left(\frac{\lambda u\cos(\lambda a)}{\mu} + \frac{\kappa v\sin(\lambda a)}{\mu^2}\right)(\kappa\partial_u - \tau\partial_w) +$$
$$\frac{\sin(\lambda a)}{\mu}(\kappa\partial_a - \tau\partial_y) + \left(\frac{\kappa\tau yu}{\mu}\lambda\cos(\lambda a) - \left(\lambda^2 u - \frac{\kappa^2\tau yv}{\mu^2} - \frac{\kappa\tau w}{\mu}\right)\sin(\lambda a)\right)\partial_v,$$

$$Z_3 = \left(-\lambda\sin(\lambda a) + \frac{\kappa\tau y\cos(\lambda a)}{\mu}\right)\partial_x - \left(\frac{\lambda u\sin(\lambda a)}{\mu} - \frac{\kappa v\cos(\lambda a)}{\mu^2}\right)(\kappa\partial_u - \tau\partial_w) +$$
$$\frac{\cos(\lambda a)}{\mu}(\kappa\partial_a - \tau\partial_y) - \left(\frac{\kappa\tau\lambda yu}{\mu}\sin(\lambda a) + \left(\lambda^2 u - \frac{\kappa^2\tau yv}{\mu^2} - \frac{\kappa\tau w}{\mu}\right)\cos(\lambda a)\right)\partial_v,$$

where $\lambda = \sqrt{\kappa^2 + \tau^2}$. The symmetries

$$Z_4 = tZ_1 + \frac{1}{\mu}\left(\partial_u + \tau y\,\partial_v - \frac{\lambda^2 x - \kappa}{\tau}\,\partial_w\right),$$

$$Z_5 = tZ_2 + \frac{\sin(\lambda a)}{\mu}(\kappa\partial_u - \tau\partial_w) + \left(\lambda\cos(\lambda a) + \frac{\kappa\tau y\sin(\lambda a)}{\mu}\right)\partial_v,$$

$$Z_6 = tZ_3 + \frac{\cos(\lambda a)}{\mu}(\kappa\partial_u - \tau\partial_w) + \left(-\lambda\sin(\lambda a) + \frac{\kappa\tau y\cos(\lambda a)}{\mu}\right)\partial_v$$

correspond to the Galilean symmetries X_5, X_6, X_7. Of course, there are the time shift ∂_t and an analogue of the rotation symmetries X_8, X_9, X_{10}, which have long expressions, and we do not give them here, but they can be found in the Maple files.

Furthermore, the Lie algebra of kinematic symmetries of (5) has the same structure as $\mathfrak{g}_{\mathfrak{m}}$, and each generator has its counterpart in $\mathfrak{g}_{\mathfrak{m}}$.

To construct the differential invariants for the system (5) explicitly, we can apply the coordinate-free presentation discussed before.

References

[1] G. K. Batchelor, *An introduction to fluid dynamics*, Second paperback edition, Cambridge Mathematical Library, Cambridge University Press, Cambridge, 1999. MR1744638

[2] Anna Duyunova, Valentin Lychagin, and Sergey Tychkov, *Differential invariants for spherical layer flows of viscid fluids*, J. Geom. Phys. **130** (2018), 288–292, DOI 10.1016/j.geomphys.2018.04.005. MR3807130

[3] Anna Duyunova, Valentin Lychagin, and Sergey Tychkov, *Continuum mechanics of media with inner structures*, Differential Geom. Appl. **74** (2021), Paper No. 101703, 10, DOI 10.1016/j.difgeo.2020.101703. MR4173971

[4] Anna Duyunova, Valentin V. Lychagin, and Sergey Tychkov, *Differential invariants for flows of fluids and gases*, Differential geometry, differential equations, and mathematical physics, Tutor. Sch. Workshops Math. Sci., Birkhäuser/Springer, Cham, [2021] ©2021, pp. 187–231, DOI 10.1007/978-3-030-63253-3_6. MR4238612

[5] Anna Duyunova, Valentin Lychagin, and Sergey Tychkov, *Symmetry classification of viscid flows on space curves*, J. Geom. Phys. **160** (2021), Paper No. 103997, 15, DOI 10.1016/j.geomphys.2020.103997. MR4171756

[6] Boris Kruglikov and Valentin Lychagin, *Global Lie-Tresse theorem*, Selecta Math. (N.S.) **22** (2016), no. 3, 1357–1411, DOI 10.1007/s00029-015-0220-z. MR3518554

[7] Valentin V. Lychagin, *Contact geometry, measurement, and thermodynamics*, Nonlinear PDEs, their geometry, and applications, Tutor. Sch. Workshops Math. Sci., Birkhäuser/Springer, Cham, 2019, pp. 3–52. MR3932297

V. A. Trapeznikov Institute of Control Sciences of RAS, Profsoyuznaya street, 65, Moscow, Russia, 117997

Email address: anna.duyunova@yahoo.com

Contemporary Mathematics
Volume **788**, 2023
https://doi.org/10.1090/conm/788/15822

Presymplectic gauge PDEs and Lagrangian BV formalism beyond jet-bundles

Maxim Grigoriev

Abstract. A gauge PDE is a geometrical object underlying what physicists call a local gauge field theory defined at the level of equations of motion (i.e. without specifying Lagrangian) in terms of Batalin-Vilkovisky (BV) formalism. This notion extends the BV formulation in terms of jet-bundles on the one hand and the geometrical approach to PDEs on the other hand. In this work we concentrate on gauge PDEs equipped with a compatible presymplectic structure and show that under some regularity conditions this data defines a jet-bundle BV formulation. More precisely, the BV jet-bundle arises as the symplectic quotient of the super jet-bundle of the initial gauge PDE. In this sense, presymplectic gauge PDEs give an invariant geometrical approach to Lagrangian gauge systems, which is not limited to jet-bundles. Furthermore, the presymplectic gauge PDE structure naturally descends to space-time submanifolds (in particular, boundaries, if any) and, in this respect, is quite similar to AKSZ sigma models which are long known to have this feature. We also introduce a notion of a weak presymplectic gauge PDE, where the nilpotency of the differential is replaced by a presymplectic analog of the BV master equation, and show that it still defines a local BV system. This allows one to encode BV systems in terms of finite-dimensional graded geometry, much like the AKSZ construction does in the case of topological models.

Contents

2020 *Mathematics Subject Classification.* Primary 35-XX, 70-XX, 49-XX; Secondary 53-XX, 83-XX.

Key words and phrases. Gauge theories, Batalin-Vilkovisky formalism, geometry of PDEs, AKSZ construction, supergeometry, presymplectic structures.

©2023 American Mathematical Society

1. Introduction

Modern theories of fundamental interactions such as Yang-Mills theory, Einstein gravity, and their modifications are local gauge field theories or, at least, can be described by such theories in certain regimes.

At the classical level such systems are partial differential equations (PDE), which in addition are required to be variational (Lagrangian) and to possess a family of variational symmetries locally parametrized by unconstrained space-time functions (gauge parameters). A powerful method of handling such systems is the Batalin-Vilkovisky (BV) formalism [**1–3**] (see also [**4, 5**] for a review) that extends usual Lagrangian formalism by incorporating the structure of the gauge symmetries into the extended configuration space which, in turn, naturally happens to be a differential graded symplectic manifold.

The version of the BV approach, which explicitly takes care of space-time locality, is also well known by now, see e.g. [**6, 7**] and [**8**] for a review and [**9–12**] for more recent developments. In this case, the BV configuration space is a jet-bundle, and in addition to the BV extension of the Lagrangian form and the BV symplectic structure, it comes equipped with the canonical Cartan distribution. This data also gives rise to a natural cohomology theory, the so-called local BRST cohomology, which happens to encode the invariant properties of the system and has direct physical meaning (global symmetries, consistent interactions, anomalies, counterterms etc.).

Although, in the geometrical theory of PDEs, the jet-bundle approach appears to be quite useful, it is not sufficiently flexible and is not invariant in the sense that one and the same PDE can be realised in terms of different jet-bundles. An invariant approach was developed by Vinogradov and his school and it deals with the infinitely-prolonged equations seen (in the simplest version) as bundles endowed with a Cartan distribution [**13–15**], see also [**16, 17**] for a review.

Similar ideas can be applied to local gauge theories described in terms of BV formalism. First of all, forgetting about symplectic structure gives a version of the jet-bundle BV approach applicable to not necessarily variational PDEs with a gauge structure [**9, 18**]. Furthermore, such a BV formulation can be extended beyond jet-bundles, giving a more flexible and invariant approach [**10**]. This approach was made more general and geometrical in [**19**], where the respective geometrical object was called a gauge PDE and its relation to the Vinogradov's notion of diffiety was made explicit. In particular, if a usual PDE is considered as a gauge PDE then the respective diffiety is obtained as its minimal model, i.e. by maximal equivalent reduction of the initial gauge PDE.

The notion of gauge PDE is not a straightforward merging of the notion of PDE and BV system (that would be just a PDE in the differential-graded category; this is known as a standard gauge PDE [**19**]). Instead, the Cartan distribution is somehow united with the gauge distribution encoded in the analog of the BV-BRST differential so that there is only one $\mathbb{Z}$-degree and one differential Q. This generalization turns out to be crucial in describing the frame-like formulation of gauge theories or, equivalently, gauge theories based on Cartan geometry. In particular, this approach includes, as a special case, so-called unfolded formalism [**20, 21**] which is based on representing equations of motion as a free differential algebra [**22**]. In fact, the main ingredients of the gauge PDE construction were initially proposed in [**9, 10**] as a tool to generate unfolded formulations and to relate them to the BV-BRST approach. In this context it is also worth mentioning the group-geometric approach to supergravity [**23**], which also employs free differential algebras.

In field theory applications we are mostly interested in Lagrangian systems. However, by its very definition Lagrangian is defined on a jet-bundle and hence cannot be directly encoded in the gauge PDE geometry. A natural question is to extend the gauge PDE approach to Lagrangian systems. It turns out that Lagrangian can often be (but not always is) encoded in the compatible graded presymplectic structure defined on the gauge PDE. In the case of a usual variational PDE, this presymplectic structure is precisely the presymplectic structure defined on the equation manifold [**24**–**27**] (see also [**28**–**30**] for more recent developments relevant in the present context). More precisely, given an equation manifold equipped with a compatible presymplectic structure, it defines a natural action functional [**31**, **32**], the so-called intrinsic action, defined on sections of the equation manifold (seen as a bundle over the space-time manifold). The structure of the intrinsic action originates from that of presymplectic AKSZ sigma-models [**33**] which, in turn, generalize AKSZ sigma models [**34**] (see also [**35**–**43**] for further developments), initially proposed as an elegant supergeometrical construction of the BV formulation of topological models.

The relation to AKSZ approach suggests that the intrinsic action construction can be extended to gauge PDEs. This was shown to be indeed the case in [**44**] in the case of Einstein gravity but the construction extends to generic diffeomorphism-invariant gauge theories. Extension to general gauge PDEs was recently proposed in [**45**]. It is also worth mentioning recent application of presymplectic AKSZ sigma models to constructing Lagrangians for higher spin gravity [**46**].

In this work we study presymplectic gauge PDEs, i.e. gauge PDEs equipped with compatible presymplectic structure. We apply the efficient technique of super jet-bundles to prove that a presymplectic gauge PDE indeed gives rise to a local BV system. Moreover, it defines a tower of symplectic structures of descending degree and a tower of compatible charges of which the top one is a Lagrangian itself. This is analogous to the case of AKSZ sigma models. In particular these symplectic structures and charges can be integrated over the submanifolds of the respective dimension. This shows that just like AKSZ models presymplectic gauge PDEs naturally induce the analogous structure on space-time submanifolds and/or boundaries.

We also introduce a concept of weak presymplectic gauge PDE, which is obtained by replacing the nilpoltency condition $Q^2 = 0$ with $\omega(Q,Q) = 0$ condition. Remarkably, this does not affect the entire construction and, again, the local BV system naturally emerges on a suitable symplectic quotient. This generalization appears useful in applications because it often allows one to replace genuine gauge PDE with a finite-dimensional weak analog which still encodes all the information on the system. The example of Yang-Mills theory in this formulation is explicitly constructed.

The paper is organized as follows: in Section **2** we recall how gauge systems are described in terms of graded geometry. This discussion is extended to local gauge theories in Section **3**. This is done in terms of so-called gauge PDEs, which are also briefly recalled. The structure of presymplectic gauge PDEs is studied in Section **4**. In Section **5** we introduce a weak analog of presymplectic gauge PDE and show that it still encodes a local BV system. The construction is illustrated with the example of Yang-Mills theory. Finally, in Section **6** we demostrate that presymplectic gauge PDE structure naturally descents to space-time submanifolds and boundaries and again consider Yang-Mills theory as an example.

2. Preliminaries

2.1. Q-manifolds. What we are after is a geometric setup underlying classical gauge systems. If we temporarily disregard additional structures such as space-time locality and Lagrangian, the data of a gauge system consists of a manifold of field configurations together with a submanifold singled out by the equations of motion, a gauge distribution that is involutive on the submanifold, and the data defining gauge for gauge symmetries and their higher analogs. All these can be conveniently packed into the structure of Q-manifold:

DEFINITION 2.1. Q-manifold (M, Q) is a $\mathbb{Z}$-graded supermanifold equipped with a homological vector field Q, i.e. a vector field of degree 1 satisfying $Q^2 = 0$, $\mathrm{gh}(Q) = 1$, $|Q| = 1$, where $\mathrm{gh}(\cdot)$ denotes $\mathbb{Z}$-degree (often called ghost degree) and $|\cdot|$ Grassmann parity.

It is probably more appropriate to call it dg manifold but we follow the tradition initiated in [**47, 48**]. For simplicity, we restrict ourselves to the case where fermionic fields and/or fermionic gauge parameters are not present so that $\mathbb{Z}_2$-degree (Grassmann parity) is determined by $\mathbb{Z}$-degree, i.e. given by $\mathbb{Z}$-degree mod 2.

2.2. Q-manifolds as gauge systems. It is instructive to see how the information about the underlying gauge system can be extracted from the data of a Q-manifold. To be more precise, by a gauge system we temporarily mean a system whose space-time manifold is 0-dimensional. The considerations still apply to generic systems but then one should disregard all the space-time related issues such as locality, space-time geometry, etc. and is forced to work with, in general, infinite-dimensional manifolds of field configurations.

Given a Q-manifold (M, Q), its body (which can be seen as a submanifold obtained by setting to zero all the coordinates of nonvanishing degree) is a space of "field configurations". The submanifold singled out by the equations of motion (its points are "solutions" of the underlying system) is the intersection of the zero-locus of Q with the body of M. It is useful to give a more invariant characterization. Namely, the body of M can be seen as the space of maps from a point (considered as a Q-manifold with the trivial $\mathbb{Z}$-grading and the trivial Q-structure) to M. Then solutions (points where Q-vanishes) can be identified as Q-maps, i.e. maps satisfying $\sigma^* \circ Q = 0$. In the same language the gauge transformations (the gauge distribution) can be defined as follows: $\delta_\xi \sigma^* = \xi^*_\sigma Q$, where ξ^*_σ is a "gauge parameter map" $\mathcal{C}^\infty(M) \to \mathbb{R}$ satisfying $\mathrm{gh}(\xi^*_\sigma) = -1$ and $\xi^*_\sigma(fg) = \xi^*_\sigma(f)\sigma^*(g) + (-1)^{|f|}\xi^*_\sigma(f)\sigma^*(g)$. In a similar way one can define gauge for gauge symmetries.

It is useful to represent gauge transformations in terms of vector fields defined on M. Given a vector field Y of degree -1 it defines an infinitesimal symmetry $V = [Q, Y]$ of the Q-manifold. Because $\mathrm{gh}(V) = 0$ it induces an infinitesimal transformation of the body of M, which is a gauge symmetry. Indeed, taking $\xi_\sigma = \sigma^* \circ Y$ one finds that the associated gauge symmetry is precisely the one determined by V.

It is also instructive to characterise solutions and gauge transformations in coordinate terms. For simplicity let us restrict ourselves to the situation where only coordinates $\mathcal{P}_a, \phi^i, C^\alpha$ of degree $-1, 0, 1$ respectively are present. Generic Q of degree 1 then reads as:

$$Q = E_a(\phi)\frac{\partial}{\partial \mathcal{P}_a} + C^\alpha R^i_\alpha(\phi)\frac{\partial}{\partial \phi^i} + \text{terms quadratic in } C, \mathcal{P}\,. \tag{2.1}$$

For a given map σ one finds that σ^* acts trivially on $C, \mathcal{P}$ so that the only nontrivial condition is $\sigma^*(Q\mathcal{P}_a) = 0$, giving $\sigma^*(E_a) = 0$. In other words, solutions are points where $E_a(\phi)$ vanish so that relations $E_a(\phi) = 0$ are indeed interpreted as equations of motion.

Let us turn to gauge symmetries, a generic gauge parameter map ξ^*_σ of degree -1 is determined by $\xi^*_\sigma(c^\alpha) = \epsilon^\alpha \in \mathbb{R}$. The gauge variation of $z^i = \sigma^*\phi^i$ is then given by

$$\delta_{\xi_\sigma} z^i = \xi^*_\sigma(Q\phi^i) = R^i_\alpha(\phi^i)\epsilon^\alpha \tag{2.2}$$

In other words vector fields $R^i_\alpha(\phi)\frac{\partial}{\partial \phi^i}$ define a gauge distribution. It is involutive on the surface $E_a = 0$ as a consequence of $Q^2 = 0$.

A natural question is when two gauge systems should be considered equivalent. If the gauge invariance is not present it is natural to call systems equivalent if they have isomorphic spaces of solutions. Situation is slightly more tricky when gauge symmetry is present as some of the variables can be gauged away by passing to the space of gauge orbits and one should compare the quotients. A subtlety is that gauge symmetries may act in a nontrivial way as not all gauge transformations can be represented as shift transformations of the form $\delta z = \epsilon$, with z being a global coordinate. These vague considerations lead to the following notion of equivalence for Q-manifolds:

Definition 2.2 ([**19**]). 1. A contractible Q-manifold is that of the form $(T[1]W, \mathrm{d_W})$ where W is a graded vector space considered as a graded manifolds and $\mathrm{d_W}$ is the de Rham differential on W seen as a Q-structure on $T[1]W$.

2. A Q-manifold (N, q) is called an equivalent reduction of (M, Q) if (M, Q) is a locally-trivial Q-bundle over (N, q) which admits a global Q-section and whose fibers are contractible Q-manifolds. The equivalent reduction generates the equivalence relation for Q-manifolds.

Recall that Q-bundle $\pi : (M, Q) \to (N, q)$ is a locally- trivial bundle of graded manifolds such that $\pi^* \circ q = Q \circ \pi^*$ [**49**]. A Q-bundle π is called locally-trivial if it is locally isomorphic, as a Q-manifold, to the product of the (N, q) and a typical fiber (F, q'), i.e. $Q = q + q'$ in the adapted coordinates.

The above notion of the equivalent reduction is a globalization of the well-known, in the context of local BRST cohomology, concept of eliminating contractible pairs, see e.g. [**7**,**50**]. In the field theory context the appropriate extension of the above equivalence was put forward in [**9**], which in turn is a nonlagrangian version of the equivalence through the elimination of so-called generalised auxiliary fields [**51**,**52**].

3. Local gauge theories as gauge PDEs

3.1. Gauge PDE.

DEFINITION 3.1. Gauge PDE $(E, Q, T[1]X)$ is a $\mathbb{Z}$-graded fiber bundle $\pi : E \to T[1]X$ whose total space is equipped with a homological vector field Q of degree 1 satisfying: $Q \circ \pi^* = \pi^* \circ \mathrm{d_X}$ and $\mathrm{d_X}$ is the de Rham differential on X seen as a homological vector field on $T[1]X$. Moreover, $(E, Q, T[1]X)$ should be equivalent to a non-negatively graded gauge PDE.

The last condition means that the surface singled out by the equations of motion encoded in Q (see below) is itself a smooth fiber bundle and that the role of all the negative degree variables is to provide a resolution of this surface. In the usual jet-space BV formulation this condition corresponds to the acyclicity of the Koszul-Tate differential contained in Q. One often also imposes some extra conditions of a technical nature to exclude pathological examples.

We assume that X is a real manifold (space-time manifold) while E (and $T[1]X$) is a $\mathbb{Z}$-graded one with the degree denoted by $\mathrm{gh}(\cdot)$. Note that gauge PDEs are special cases of Q-bundles [**49**]. However, gauge PDEs are necessarily $\mathbb{Z}$-graded and are typically infinite-dimensional. Further details and references can be found in [**19**] (see also [**10, 53, 54**] for the earlier and less-geometrical version).

By definitionm, a field configuration σ (i.e. a section $\sigma : T[1]X \to E$) is a solution if

$$d_X \circ \sigma^* = \sigma^* \circ Q\,. \tag{3.1}$$

The infinitesimal gauge transformation of the configuration σ is given by

$$\delta\sigma^* = d_X \circ \xi^*_\sigma + \xi^*_\sigma \circ Q \tag{3.2}$$

where $\xi^*_\sigma : \mathcal{C}^\infty(E) \to \mathcal{C}^\infty(T[1]X)$ is a gauge parameter map, which has a degree -1 and satisfies

$$\xi^*_\sigma(fg) = (\xi^*_\sigma(f)\sigma^*(g) + (-1)^{|f|}\sigma^*(f)\xi^*_\sigma(g)\,, \tag{3.3}$$

as well as $\xi^*_\sigma(\pi^*\alpha) = 0$ for any $\alpha \in \mathcal{C}^\infty(T[1]X)$. In a similar way one defines gauge for gauge symmetries. It can be also useful to parameterize gauge parameters in terms of ghost degree -1 vector fields Y on E as follows: $\xi_\sigma = \sigma^* \circ Y$, where Y is understood as a derivation of the algebra of functions on E. It is easy to check that condition (3.3) is indeed satisfied.

The notion of equivalence naturally extends from Q-manifolds to Q-bundles [**19**]. An equivalent reduction of a gauge PDE $(E, Q, T[1]X)$ is a Q-subbundle $(G, Q_G, T[1]X)$ of E such that the initial gauge PDE is a locally-trivial Q-bundle over G with a contractible fiber. Moreover, the projection $E \to G$ is also a Q-map of Q-bundles over $T[1]X$.

3.2. Examples.

3.2.1. *PDE.* The first example is the usual PDE. Let $E_X \to X$ be an equation manifold seen as a bundle over the space-time X. Algebra of horizontal forms on E_X can be identified with the algebra of functions on E which is E_X pulled back to $T[1]X$ by the canonical projection $T[1]X \to X$. Under this identification the horizontal differential becomes a homological vector field on E, which by some abuse of conventions we also denote by $\mathrm{d_h}$. In a suitable local coordinate system x^a, θ^a, ψ^A, where x^a, θ^a are standard coordinates on $T[1]X$ pulled back to E, and ψ^A originate from local coordinates on a typical fiber, one has:

$$\mathrm{d_h} = \theta^a D_a\,, \qquad D_a = \frac{\partial}{\partial x^a} - \Gamma^A_a(x, \psi)\frac{\partial}{\partial \psi^A}\,, \tag{3.4}$$

where D_a are components of the total derivative on the equation manifold. Note that this is a generic form of the Q-structure on E if the typical fiber is a real manifold (so that $\mathrm{gh}(\psi^A) = 0$). Using these coordinates it is easy to check that solutions of $(E, \mathrm{d_h}, T[1]X)$ are indeed the covariantly-constant (with respect to the Cartan distribution seen as an Ehresmann connection) sections in agreement with the standard picture, see e.g. [**16, 17**]. Note that a closely related representation of PDEs is known under the name of unfolded formulation [**20, 21**].

3.2.2. *AKSZ sigma model.* Let X be a space-time manifold and (F, q) a generic Q-manifold then take $(E, Q, T[1]X)$ to be a globally-trivial Q-bundle $(E, Q) = (F, q) \times (T[1]X, \mathrm{d_X})$ over $T[1]X$. This construction is known as AKSZ sigma model and originates from [**34**], where its Lagrangian version (in which case F is equipped with a symplectic form of degree $\dim X - 1$) was put forward as a tool to generate elegant BV formulations for topological models.

A typical example is to take $F = \mathfrak{g}[1]$ and $q = d_{CE}$, where $\mathfrak{g}[1]$ is a Lie algebra and d_{CE} its Chevalley-Eilenberg differential. A section σ is determined by the 1-form $A^\alpha_a(x)\theta^a = \sigma^*(c^\alpha)$, where c^α are linear coordinates on $\mathfrak{g}[1]$ and x^a local coordinates on X. The gauge PDE equations of motion then imply that $\mathrm{d_X} A + \frac{1}{2}[A, A] = 0$ and the gauge transformations are $\delta A = \mathrm{d_X}\epsilon + [A, \epsilon]$.

3.2.3. *BV-formulation at the level of equations of motion.* The BV approach to local gauge theories was initially designed for Lagrangian systems in which case the system is defined in terms of a certain extension S_{BV} of the classical action while the Q structure emerges as a Hamiltonian vector field generated by S_{BV} and the odd symplectic structure. The version of BV for not necessarily Lagrangian systems deals with Q structure from the very start while the symplectic structure and hence S_{BV} is disregarded or not specified, see [**9, 18**] for further details.

In the setting of local gauge theories the BV formulation at the level of equations of motion consists of the following data: a fiber bundle $\mathcal{E} \to X$, where X is the space-time manifolds and the the typical fiber is a $\mathbb{Z}$-graded manifold. The associated jet-bundle $J^\infty(\mathcal{E})$ is again a graded manifold and is equipped with a vertical evolutionary vector field s, called BRST differential, satisfying $\mathrm{gh}(s) = 1$ and $s^2 = 0$. Recall that evolutionary vector fields are those preserving the contact ideal. This condition can be also written as $[D_a, s] = 0$, where $D_a = \frac{\partial}{\partial x^a} + \dots$ denote total derivatives on $J^\infty(\mathcal{E})$, i.e. the horizontal lifts of the coordinate vector field $\frac{\partial}{\partial x^a}$ on X.

This data encodes the information on equations of motion, gauge symmetries as well as gauge for gauge symmetries of the underlying system. Indeed, a section $\sigma : X \to \mathcal{E}$ is called a solution if its prolongation $\sigma_{\mathrm{pr}} : X \to J^\infty(\mathcal{E})$ satisfies $\sigma^*_{\mathrm{pr}} \circ s = 0$. While gauge transformations are defined as $\delta\sigma^*_{\mathrm{pr}} = \xi^*_{\mathrm{pr}} \circ s$, where ξ^* is a ghost degree -1 map $\mathcal{C}^\infty(J^\infty(\mathcal{E})) \to \mathcal{C}^\infty(X)$ satisfying $\xi^*(fg) = \xi^*(f)\sigma^*(g) + (-1)^{|f|}\xi^*(f)\sigma^*(g)$ for all $f, g \in \mathcal{C}^\infty(J^\infty(\mathcal{E}))$. In a similar way one defines gauge for gauge symmetries.

It is not difficult to construct a gauge PDE out of this data. Namely, take as $E \to T[1]X$ a pull back of $J^\infty(\mathcal{E})$ by the canonical projection $T[1]X \to X$. Functions on E can be identified with the algebra of horizontal forms on $J^\infty(\mathcal{E})$. In particular, E is equipped with the horizontal differential $\mathrm{d_h}$ which can be written as $\theta^a D_a$, where θ^a are coordinates on the fibers of $T[1]X$. As a Q-structure of the total space one takes $Q = \mathrm{d_h} + s$. It is easy to check that it satisfies $\pi^* \circ \mathrm{d_X} = Q \circ \pi^*$, where $\mathrm{d_X}$ is the de Rham differential on X seen as a vector field on $T[1]X$. It was proved in [**10**] (see also [**9**] for the earlier but less general statement and [**19**] for a more geometrical exposition) that, at least locally, the gauge system encoded in $(E, Q, T[1]X)$ is an equivalent reformulation of the starting point BV-system.

4. Presymplectic gauge PDEs

In this section we demonstrate how generic Lagrangian local gauge field theories can be described by gauge PDEs equipped with compatible presymplectic structures.

4.1. Lagrangian BV system. We begin by recalling the Lagrangian BV formulation of local gauge field theories in terms of jet-bundles, for a review see e.g. [**8**] and references therein. As in Section **3.2.3** we assume that $J^\infty(\mathcal{E})$ is equipped with a ghost grading and an evolutionary vector field s, satisfying $s^2 = 0, \mathrm{gh}(s) = 1$. Now let, in addition, $\mathcal{E}$ be equipped with a closed $(n+2)$-form of degree -1 and of the following structure:

$$\omega^{\mathcal{E}} = (dx)^n \omega_{AB}(x, \psi^A) d\psi^A d\psi^B \,, \tag{4.1}$$

with ω_{AB} invertible. Here x^a are local coordinates on X pulled back to $\mathcal{E}$ and ψ^A the fiber coordinates. Although the requirement is formulated in coordinate terms it is actually invariant.[1]

The above form defines an $(n, 2)$ form $\overset{n}{\omega}$ on the associated jet-bundle $J^\infty(\mathcal{E})$, which is just a pullback of $\omega^{\mathcal{E}}$ to $J^\infty(\mathcal{E})$. It terms of standard local coordinates on $J^\infty(E)$ induced by x^a, ψ^A it can be written as

$$\overset{n}{\omega} = (dx)^n \omega_{AB}(x, \psi^A) \mathrm{d_v}\psi^A \mathrm{d_v}\psi^B \,, \tag{4.2}$$

where $\mathrm{d_v}$ is the vertical differential. Recall that on a jet-bundle de Rham differential decomposes as $d = \mathrm{d_h} + \mathrm{d_v}$, see e.g. [**27, 55, 56**] for more details on jet-bundles. If in addition $\overset{n}{\omega}$ is s-invariant modulo a total derivative, i.e.

$$L_s \overset{n}{\omega} + \mathrm{d_h} \overset{n-1}{\omega} = 0 \,, \tag{4.3}$$

for some $\overset{n-1}{\omega} \in \bigwedge^{n-1,2}(J^\infty(\mathcal{E}))$ we say that $(J^\infty(\mathcal{E}), s, \overset{n}{\omega})$ defines a local BV system (or local BV formulation).

To see that the above data indeed encodes BV master-action let us restrict to local analysis so that $\mathrm{d_h}$ is acyclic in horizontal form degree $< n$ and d (as well as $\mathrm{d_v}$) is acyclic in (vertical) form degree > 0. Using $\mathrm{d_v}\overset{n}{\omega} = 0$ and the fact that $\mathrm{d_v}$ mod $\mathrm{d_h}$ cohomology is trivial in this degree, one finds:

$$i_s \overset{n}{\omega} = -\mathrm{d_v} \overset{n}{\mathcal{L}} - \mathrm{d_h} \overset{n-1}{\chi} \,, \tag{4.4}$$

for some $\overset{n}{\mathcal{L}} \in \bigwedge^{n,0}(J^\infty(\mathcal{E}))$ and some $\overset{n-1}{\chi} \in \bigwedge^{n-1,1}(J^\infty(\mathcal{E}))$. Applying $\mathrm{d_v}$ to both sides one finds $L_s \overset{n}{\omega} + \mathrm{d_h}\mathrm{d_v} \overset{n-1}{\chi} = 0$ so that $\overset{n-1}{\omega}$ can be assumed $\mathrm{d_v}$-closed (as we do in what follows). Note that $\overset{n}{\mathcal{L}}$ is defined up to a total derivative, indeed $\overset{n}{\mathcal{L}} + \mathrm{d_h} \overset{n-1}{\epsilon}$ and $\overset{n-1}{\chi} + \mathrm{d_v} \overset{n-1}{\epsilon}$ is again a solution to (4.4) for any $\overset{n-1}{\epsilon} \in \bigwedge^{n-1,0}(J^\infty(\mathcal{E}))$.

Furthermore, considering $\mathrm{d_v}(i_s i_s \overset{n}{\omega})$ one finds (see Appendix for details):

$$\frac{1}{2} i_s i_s \overset{n}{\omega} = \mathrm{d_h} \overset{n-1}{\mathcal{L}} \,, \tag{4.5}$$

[1] The alternative is to define a vertical 2-form with values in densities on X

for some $\overset{n-1}{\mathcal{L}} \in \bigwedge^{n-1,0}(J^\infty(\mathcal{E}))$. Relations (4.4) and (4.5) are an equivalent reformulation of the standard axioms of local BV formulation, where $\mathcal{L}_{BV} \equiv \overset{n}{\mathcal{L}}$ is the integrand of the BV master action and s the associated BRST differential. In particular, (4.5) is equivalent to the BV master equation $\frac{1}{2}(L_{BV}, L_{BV}) = \mathrm{d_h}(\cdot)$, where the bracket is the odd Poisson bracket determined by $\overset{n}{\omega}$. Note that the odd Poisson bracket on local $(n,0)$-forms is not unique and is defined up to a total derivative. See e.g. [**57**] for a more detailed discussion of this point.

In what follows we also need the inverse statement:

PROPOSITION 4.1. *Let $J^\infty(\mathcal{E})$ be equipped with a symplectic structure $\overset{n}{\omega}$ determined (as explained above) by a nondegenerate $\omega^{\mathcal{E}}$ and a vertical evolutionary vector field s, $\mathrm{gh}(s) = 1$ (not necessarily nilpotent) such that equations* (4.4) *and* (4.5) *are satisfied for some $\overset{n}{\mathcal{L}}$, $\overset{n-1}{\chi}$, and $\overset{n-1}{\mathcal{L}}$. It follows that $\overset{n}{\mathcal{L}}, \overset{n}{\omega}$ define a local BV system on $J^\infty(\mathcal{E})$.*

PROOF. Let s' be an evolutionary vertical vector field determined by $i_{s'}\overset{n}{\omega} + \delta^{EL}\overset{n}{\mathcal{L}} = 0$, where δ^{EL} is the Euler-Lagrange differential. This is the standard definition of a Hamiltonian vector field with Hamiltonian $\mathcal{L}_{BV}$. It is the standard fact (see e.g. [**57**]) that nilpotency of s' is guaranteed if (4.5), with s replaced by s', is satisfied for some $\overset{n-1}{\mathcal{L}}$. It is indeed satisfied as $i_s\omega - i_{s'}\omega = \mathrm{d_h}(\cdot)$ thanks to (4.4) and the definition of s'. □

4.2. Descent-completed BV systems. The above Lagrangian BV formulation can be extended in such a way that the symplectic structure $\overset{n}{\omega}$ is completed by forms of lower horizontal degree to a cocycle of the total BRST differential $s + \mathrm{d_h}$. To avoid discussion of possible obstructions we work locally.

Using $\mathrm{d_v}\overset{n-1}{\omega} = 0$ one finds that (4.3) implies $\mathrm{d_h} L_s \overset{n-1}{\omega} = 0$. Acyclicity of $\mathrm{d_h}$ in horizontal form degree $< n$ then implies the next term in the descent equation:

$$L_s \overset{n-1}{\omega} + \mathrm{d_h}\overset{n-2}{\omega} = 0\,. \tag{4.6}$$

Repeating the same arguments as above gives

$$i_s \overset{n-1}{\omega} = -\mathrm{d_v}\overset{n-1}{\mathcal{L}} - \mathrm{d_h}\overset{n-2}{\chi} \tag{4.7}$$

for some $\overset{n-1}{\mathcal{L}} \in \bigwedge^{n-1,0}(J^\infty(\mathcal{E}))$ and $\overset{n-2}{\chi} \in \bigwedge^{n-2,1}(J^\infty(\mathcal{E}))$, which also implies that $\overset{n-2}{\omega}$ can be assumed $\mathrm{d_v}$-exact as we do in what follows. Continuing in the same way one finds $L_s\overset{n-k}{\omega} + \mathrm{d_h}\overset{n-k-1}{\omega} = 0$ for all $k > 1$, which in turn implies the existence of $\overset{n-k}{\mathcal{L}} \in \bigwedge^{n-k,0}(J^\infty(\mathcal{E}))$ and $\overset{n-k-1}{\chi} \in \bigwedge^{n-k-1,1}(J^\infty(\mathcal{E}))$ such that

$$\begin{aligned} \omega^{\mathrm{tot}} &= \overset{n}{\omega} + \overset{n-1}{\omega} + \ldots + \overset{1}{\omega} + \overset{0}{\omega}\,, \qquad \mathcal{L}^{\mathrm{tot}} = \overset{n}{\mathcal{L}} + \overset{n-1}{\mathcal{L}} + \ldots + \overset{1}{\mathcal{L}} + \overset{0}{\mathcal{L}}\,, \\ \chi^{\mathrm{tot}} &= \overset{n}{\chi} + \overset{n-1}{\chi} + \ldots + \overset{1}{\chi} + \overset{0}{\chi}\,, \end{aligned} \tag{4.8}$$

with $\overset{n}{\chi}$ defined such that $\mathrm{d_v}\overset{n}{\chi} = \overset{n}{\omega}$, satisfy

$$(\mathrm{d_h} + L_s)\omega^{\mathrm{tot}} = 0\,, \qquad i_s\omega^{\mathrm{tot}} = -\mathrm{d_v}\mathcal{L}^{\mathrm{tot}} - \mathrm{d_h}\chi^{\mathrm{tot}}\,, \qquad \omega^{\mathrm{tot}} = \mathrm{d_v}\chi^{\mathrm{tot}}\,, \tag{4.9}$$

and

$$\frac{1}{2} i_s i_s \omega^{\mathrm{tot}} = \mathrm{d_h}\mathcal{L}^{\mathrm{tot}}\,. \tag{4.10}$$

Note that $\omega^{\text{tot}}, \mathcal{L}^{\text{tot}}$, and χ^{tot} are inhomogeneous in the horizontal form degree.

A BV jet-bundle equipped with $\omega^{\text{tot}}, \mathcal{L}^{\text{tot}}, \chi^{\text{tot}}$ satisfying (4.9) and (4.10) is refereed to in what follows as a descent-completed local BV system. Although, as we have just recalled, a usual local BV system can be extended, at least locally, to a descent completed one, such an extension is in general not guaranteed and/or is not unique. In this sense it contains extra data with respect to the local BV formulation reviewed in Section **4.1**. For instance, $\overset{n-1}{\mathcal{L}}$ and $\overset{n-1}{\omega}$ correspond to, respectively, the BRST charge and the symplectic structure of the associated Hamiltonian formulation, see e.g. [**12, 58, 59**].

In one or another version descent-completed BV systems are known in the literature. In particular, the total BRST differential $s + \mathrm{d_h}$ was actively employed from the early days of the local BRST cohomology, see e.g. [**7, 50, 60**]. Furthermore, the symplectic structure equivalent to ω^{tot} and the equivalent of $\mathcal{L}^{\text{tot}}$ appeared already in [**53, 54**] (see also a related approach of [**33**] and its developments [**31, 44**]) within a slightly different formalism. In this approach the essential part of $\mathcal{L}^{\text{tot}}$ is constructed as a descent completion of the Lagrangian. Detailed study of the descent completion of (pre)symplectic structures in the jet-bundle BV approach was put forward in [**12, 59**]. The notion of descent-completed [2] BV was fully explicit in [**61, 62**] (see also [**58**], where a related structure have appeared in a slightly different framework).

4.3. Presymplectic BV systems. The usual Lagrangian BV formulation recalled in Section **4.1** has a straightforward generalization to the case where the symplectic structure $\omega^{\mathcal{E}}$ introduced in (4.1), is allowed to have a nonvanishing kernel on the vertical tangent vectors but its rank is required to be constant.

It turns out that almost all the considerations of Section **4.1** remain true. In particular, $\omega^{\mathcal{E}}$ determines $\mathrm{d_v}$-closed $\overset{n}{\omega} \in \bigwedge^{n,2}(J^\infty(\mathcal{E}))$. If BRST differential s preserves $\overset{n}{\omega}$ modulo boundary terms, i.e. equation (4.3) is satisfied, one finds $\overset{n}{\mathcal{L}} \in \bigwedge^{n,0}(J^\infty(\mathcal{E}))$ satisfying (4.4) and (4.5) for some $\overset{n-1}{\chi}$ and $\overset{n-1}{\mathcal{L}}$.

Let us restrict to local analysis and take a quotient of $\mathcal{E}$ modulo the distribution of vertical vector fields that are in the kernel of $\omega^{\mathcal{E}}$. This can be performed fiberwise, giving a new bundle $\mathcal{G}$ together with a projection $p : \mathcal{E} \to \mathcal{G}$. $\omega^{\mathcal{E}}$ induces a nondegenerate $n+2$ form $\omega^{\mathcal{G}}$ on $\mathcal{G}$, which has the same structure (4.1). Recall that locally a symplectic quotient can be realised as a submanifold and in the case at hand as a subbundle (because the distribution is vertical). To simplify the analysis we realise $\mathcal{G}$ as a subbundle of $\mathcal{E}$ and hence $J^\infty(\mathcal{G})$ is also a subbundle of $J^\infty(\mathcal{E})$ in a natural way. Pulling back equation (4.4) to $J^\infty(\mathcal{G})$ one finds that it remains true with s replaced by its projection to $J^\infty(\mathcal{G})$. At the same time it is easy to see that the projection of s to $J^\infty(\mathcal{G})$ is evolutionary. It then follows that we arrived at the local BV system on $J^\infty(\mathcal{G})$ with the symplectic structure $\omega^{\mathcal{G}}$ and the BV master action being a pullback of $\overset{n}{\mathcal{L}}$ to $J^\infty(\mathcal{G})$.

The above construction can be also understood in field-theoretical terms. Suppose that R is a vertical evolutionary vector field from the kernel of $\overset{n}{\omega}$, i.e. $i_R\overset{n}{\omega} = 0$. It follows it defines a gauge symmetry of the BV Lagrangian, Indeed, equation (4.4)

[2] called "Lax" in [**61**]

then implies

$$L_R \overset{n}{\mathcal{L}} + (-1)^{|R|} \mathrm{d_h} i_R \overset{n-1}{\chi} = 0\,. \tag{4.11}$$

From this perspective the passage to $\mathcal{G}$ can be seen as gauge-fixing these symmetries. Note that one can choose R to be a prolongation of a kernel vertical vector field on $\mathcal{E}$ in which case the respective gauge transformations are algebraic.

4.4. Presymplectic gauge PDEs. Let us first introduce vertical forms on $(E, Q, T[1]X)$ as the quotient of all local forms by the ideal $\mathcal{I}$ generated by elements of the form $\pi^*\alpha$, where α is a form of positive degree on $T[1]X$ and π is the projection $\pi : E \to T[1]X$. In coordinate terms this means that forms proportional to dx^a or $d\theta^a$ are to be considered equivalent to zero. An equivalence class of a representative $\rho \in \bigwedge^\bullet(E)$ is denoted by $[\rho]$. The de Rham differential is clearly well defined on the quotient as well as the Lie derivative along the vector fields that project to the base, e.g. vertical vector fields.

The following definition is taken from [**45**] though less general versions were in [**31, 44**] while the examples appeared already in [**33**]:

DEFINITION 4.2. A compatible presymplectic structure on gauge PDE $(E, Q, T[1]X)$ is a vertical two form $[\omega]$ of degree $n-1$, $n = \dim X$ satisfying $[d\omega] = 0$, $[L_Q\omega] = 0$. A gauge PDE equipped with a compatible presymplectic structure is called presymplectic gauge PDE.

In what follows we find it useful to work in terms of a fixed representative ω of $[\omega]$.

It turns out that under certain regularity conditions a presymplectic gauge PDE defines a local BV system (and in fact a descent-completed BV system as well). To construct it consider a super-jet bundle $SJ^\infty(E)$ of $\pi : E \to T[1]X$, i.e. a bundle of jets of super-sections. It is again a graded manifold which is a bundle over E and, of course, over $T[1]X$. We refer to [**63**] for general discussion of super-jet bundles and to [**19**] for precisely this setup.

Let V be a vector field on E. Then its prolongation is defined as a unique vector field $\bar{V}$ on $SJ^\infty(E)$ that projects to E by $\Pi : SJ^\infty(E) \to E$ and whose vertical component is evolutionary. Applying this to Q gives its prolongation $\bar{Q}$ which we decompose as $\bar{Q} = D + s$, where D is horizontal and s vertical. Because prolongation is a canonical operation one has $\mathrm{gh}(\bar{Q}) = 1$ and $(\bar{Q})^2 = 0$. Furthermore, because $\bar{Q}$ projects to $\mathrm{d_X}$ under $\pi \circ \Pi$, $SJ^\infty(E)$ equipped with $\bar{Q}$ and seen as a bundle over $(T[1]X, \mathrm{d_X})$ is again a gauge PDE. Moreover, it follows D is a horizontal lift of $\mathrm{d_X}$, and hence, $D = \theta^a D_a$, where D_a is a total derivative (horizontal lift of $\frac{\partial}{\partial x^a}$). If u^A are fiber coordinates of E, then $su^A = Qu^A - \theta^a D_a u^A$. The action of s on the remaining coordinates is determined by $[D_a, s] = 0 = [D^\theta_a, s]$, where D^θ_a is a horizontal lift of $\frac{\partial}{\partial\theta^a}$ and the commutator always denote the graded commutator.

In what follows we need to employ a useful coordinate system on $SJ^\infty(E)$. Given an adapted coordinate system x^a, θ^a, u^A the associated coordinates are x^a, θ^a, $\psi^A_{a_1\ldots a_l|b_l\ldots b_k}$ which are defined as follows: $\psi^A_{a_1\ldots a_l|b_1\ldots b_k} = D_{a_1}\ldots D_{a_l}\psi^A_{|b_1\ldots b_k}$, $D^\theta_a = \frac{\partial}{\partial\theta^a}$, and $\psi^A_{|b_1\ldots b_k}$ coincides with $D^\theta_{b_1}\ldots D^\theta_{b_k}u^A$ at $\theta^a = 0$. In particular, $u^A = \psi^A + \theta^a\psi^A_{|a} + \frac{1}{2}\theta^a\theta^b\psi^A_{|ba} + \ldots$, for further details see e.g. [**19**]. Note that in this coordinate system $[D^\theta_0, s] = [\frac{\partial}{\partial\theta^a}, s] = 0$, i.e. components of s are θ-independent while $D = \theta^a D_a$.

The 2-form ω also admits a canonical lift $\bar{\omega}$ to $SJ^\infty(E)$, namely $\bar{\omega} = \Pi^*(\omega)$. In terms of $\psi^A_{.|.}$-coordinates, $\bar{\omega}$ is given by $\omega_{AB}(\psi^A(\theta))d\psi^A(\theta)d\psi^B(\theta)$ and can be decomposed into its θ-homogeneous components:

$$\bar{\omega} = \overset{n}{\bar{\omega}} + \overset{n-1}{\bar{\omega}} + \ldots + \overset{0}{\bar{\omega}}\,. \tag{4.12}$$

Just like presymplectic structure ω on E we consider $\bar{\omega}$ on $SJ^\infty(E)$ as a representative of an equivalence class modulo the ideal generated by $dx^a, d\theta^a$ pulled back from $T[1]X$ to $SJ^\infty(E)$. By some abuse of notations we keep denoting the ideal by $\mathcal{I}$. However, in contrast to $E \to T[1]X$, super-jet bundle $SJ^\infty(E)$ comes equipped with a canonical Cartan distribution, i.e. the horizontal distribution generated by D_a, D^θ_a, so that one can pick a vertical component $\bar{\omega}^v$ as a unique representative of the equivalence class. Note that $\bar{\omega}^v$ only depends on the equivalence class $[\omega]$ of ω.

A vertical evolutionary vector field on $SJ^\infty(E)$ preserves the subspaces of horizontal and vertical forms. The same applies to $D = \theta^a D_a$. Because prolongation is canonical $L_Q\omega \in \mathcal{I}$ implies $L_{\bar{Q}}\bar{\omega} \in \mathcal{I}$. Taking a purely vertical part of both sides gives $L_{\bar{Q}}\bar{\omega}^v = 0$. Decomposing $\bar{\omega}^v$ into the θ-homogeneous components we get the usual descent equation:

$$L_s\overset{n}{\bar{\omega}}{}^v + L_D\overset{n-1}{\bar{\omega}}{}^v = 0\,, \quad L_s\overset{n-1}{\bar{\omega}}{}^v + L_D\overset{n-2}{\bar{\omega}}{}^v = 0\,, \quad \ldots\,, \tag{4.13}$$

which says that $\overset{k}{\bar{\omega}}{}^v$ is s-invariant modulo a total derivative. Indeed, the algebra of vertical forms on $SJ^\infty(E)$ can be identified with the algebra of all local forms on $SJ^\infty(E)_0$ which is, by definition, $SJ^\infty(E)$ restricted to the zero section of $T[1]X$ (i.e. in standard coordinates $SJ^\infty(E)_0$ is a submanifold in $SJ^\infty(E)$ singled out by $\theta^a = 0$). Moreover, $SJ^\infty(E)_0$ can be seen as a jet-bundle $J^\infty(\bar{E})$ of another bundle $\bar{E} \to X$. Under this identification θ^a goes to dx^a while L_D goes to $\mathrm{d_h}$ so that the above equations coincide with the first equation in (4.9). In the coordinate terms, this identification simply amounts to disregarding $dx^a, d\theta$ (which anyway do not enter vertical forms) and then replacing θ^a with dx^a.

In this way we arrive at the presymplectic BV system whose defining data consist of a vertical evolutionary vector field s and a 2-form $\overset{n}{\bar{\omega}}$ of θ-degree n, ghost degree $n-1$, and satisfying $L_s\overset{n}{\bar{\omega}} + L_D\overset{n-1}{\bar{\omega}} = 0$. Indeed, the analysis of Section **4.1** applies to a degenerate $\overset{n}{\bar{\omega}}$ as well, and hence there exists a θ-homogeneity n local function $\mathcal{L}_{BV}$ satisfying:

$$\mathrm{d_v}\mathcal{L}_{BV} + i_s\overset{n}{\bar{\omega}}{}^v = L_D(\cdot)\,, \qquad i_si_s\overset{n}{\bar{\omega}}{}^v = D(\cdot)\,. \tag{4.14}$$

Note that $\mathcal{L}_{BV}$ is not unique as it is defined modulo D-exact terms. Note also that $s, \bar{\omega}^v$ contains extra data with respect to just BV-symplectic form and s. Namely, the descent completion of $\overset{n}{\bar{\omega}}{}^v$.

To arrive at a genuine local BV system it is again convenient to identify $SJ^\infty(E)_0$ as $J^\infty(\bar{E})$. Considered as an $(n,2)$ form on $J^\infty(\bar{E})$, form $\overset{n}{\bar{\omega}}{}^v$ is a pull-back from $\bar{E}$ (does not involve derivatives) and hence we are in the setting of Section **4.3**. It follows that if $\overset{n}{\bar{\omega}}{}^v$ is regular this defines, at least locally, a local BV system on the associated symplectic quotient.

Of course, this BV system captures only part of the initial information. For instance, descent completion of $\overset{n}{\bar{\omega}}{}^v$ is not encoded in this BV system. In this respect, the presymplectic gauge PDE provides a more detailed description. Furthermore, the data of presymplectic gauge PDE, as defined above, determines $\mathcal{L}_{BV}$ only modulo $\mathrm{d_h}$-exact terms (boundary terms). As we discuss in the next Section, to fix this ambiguity one needs to choose a presymplectic potential $\chi \in \bigwedge^1(E)$ such that $\omega - d\chi \in \mathcal{I}$.

It is important to stress that the local BV system encoded in $(E, Q, T[1]X, \omega)$ again defines a gauge PDE. Indeed, forgetting the BV symplectic structure and applying the procedure explained in Section **3.2.3** one arrives at the gauge PDE. This gauge PDE is in general not equivalent to the initial $(E, Q, T[1]X, \omega)$ with ω forgotten but is weaker (has more inequivalent solutions).

It is also important to note that a descent-completed BV system discussed in Section **4.2** gives examples of presymplectic gauge PDE, with Q and ω being respectively $\mathrm{d_h} + s$ and ω^{tot}. In this way we conclude that any descent-completed BV system can be seen as a presymplectic gauge PDE of a special type. As we are going to see in the next section, the inverse statement is also true in the sense that, at lest locally, a presymplectic gauge PDE determines a descent-completed BV formulation, provided one has fixed a symplectic potential χ such that $\omega - d\chi \in \mathcal{I}$.

4.5. AKSZ-like representation of the Lagrangian. Let us make an explicit contact with the AKSZ-like representation of the Lagrangian $\mathcal{L}_{BV}$ or, more precisely, the respective BV master-action. Given a presymplectic gauge PDE $(E, T[1]X, Q, [\omega])$ let us pick a particular representative ω of $[\omega]$ and a 1-form χ (symplectic potential) such that $\omega = d\chi$. In addition, we impose the following extra technical condition $i_Q L_Q \omega = 0$. It follows, there exists function $\mathcal{L}$ such that $i_Q\omega + d\mathcal{L} \in \mathcal{I}$ so that the full set of relations read as:

$$Q^2 = 0\,, \qquad \omega = d\chi\,, \qquad i_Q\omega + d\mathcal{L} \in \mathcal{I}\,, \qquad i_Q L_Q \omega \in \mathcal{I}\,. \tag{4.15}$$

In what follows we refer to the above system as a descent-completed presymplectic gauge PDE and denote it by $(E, Q, T[1]X, \chi, \mathcal{L})$. Note that as shown in the Appendix these relations also imply $i_Q i_Q \omega = 0$ and $Q\mathcal{L} = 0$.

Let us pick a local trivialization x^a, θ^a, ψ^A of E and decompose Q as

$$Q = \mathrm{d_X} + q\,, \qquad \mathrm{d_X} = \theta^a \frac{\partial}{\partial x^a}\,, \quad q = q^A(\psi, x, \theta)\frac{\partial}{\partial \psi^A}\,. \tag{4.16}$$

Of course, the decomposition depends on the choice of trivialization and, in general, is defined only locally. Consider the prolongation of Q to $SJ^\infty(E)$ for $\mathrm{d_X}$ and q separately. In the local coordinates $\psi^A_{\cdot|\cdot}$ employed in Section **4.4** the prolongation of $\mathrm{d_X}$ reads as:

$$(\mathrm{d_X})^{\mathrm{pr}} = D + d^F\,, \qquad D = \theta^a D_a\,, \tag{4.17}$$

where d^F is a vertical part of $(\mathrm{d_X})^{\mathrm{pr}}$ defined in Section **4.4**.

Let Π denote a canonical projection of $SJ^\infty(E)$, considered as a bundle over E, i.e. $\Pi : SJ^\infty(E) \to E$. Differential forms and vector fields admit a natural lift from E to $SJ^\infty(E)$: for differential forms the lift is just a pullback by Π while for vector fields the lift is given by prolongation. Because by definition prolongation of a vector field V on E is a vector field V^{pr} such that it projects to V by Π and whose vertical part is evolutionary, one has:

$$i_{V^{\mathrm{pr}}}(\Pi^*(\alpha)) = \Pi^*(i_V\alpha)\,, \qquad \alpha \in \bigwedge{}^l(E)\,. \tag{4.18}$$

In particular:

$$i_D\bar{\omega} + i_{d^F}\bar{\omega} = \Pi^*(i_{\mathrm{d_X}}\omega)\,, \qquad \bar{\omega} = \Pi^*\omega\,, \tag{4.19}$$

$$i_{\bar{q}}\bar{\omega} + \Pi^*(i_{\mathrm{d_X}}\omega) + d\bar{\mathcal{L}} \in \mathcal{I}\,, \qquad \bar{q} = q^{\mathrm{pr}}\,, \quad \bar{\mathcal{L}} = \Pi^*\mathcal{L}\,. \tag{4.20}$$

Furthermore, $\omega = d\chi$ implies $\bar{\omega} = d\bar{\chi}$ so that

$$i_D\bar{\omega} = i_D(d\bar{\chi}) = L_D\bar{\chi} + d(i_D\bar{\chi})\,. \tag{4.21}$$

Combining the above relations one finds:

$$i_{d^F}\bar{\omega} + i_{\bar{q}}\bar{\omega} = -d(\bar{\mathcal{L}} + i_D\bar{\chi}) - L_D\bar{\chi} + \mathcal{I}\,. \tag{4.22}$$

Taking the vertical part and using $\bar{q} + d^F = s$ one gets

$$i_s\bar{\omega}^v = -\mathrm{d_v}(\bar{\mathcal{L}} + i_D\bar{\chi}) - L_D\bar{\chi}^v\,, \tag{4.23}$$

which is nothing but the second relation (4.9) with $\mathcal{L}^{\mathrm{tot}} = \bar{\mathcal{L}} + i_D\bar{\chi}$, $\omega^{\mathrm{tot}} = \bar{\omega}^v$, and $\chi^{\mathrm{tot}} = \bar{\chi}^v$. The remaining relations in (4.9) are consequences of the above one and $\bar{\omega}^v = \mathrm{d_v}\bar{\chi}^v$.

Let us check that (4.10) is also satisfied. As demonstrated in the Appendix $i_Qi_Q\omega = 0$ as well as $Q\mathcal{L} = 0$ so that $i_{\bar{Q}}i_{\bar{Q}}\bar{\omega} = \Pi^*(i_Qi_Q\omega) = 0$ and hence:

$$i_si_s\bar{\omega} = i_{\bar{Q}-D}i_{\bar{Q}-D}\bar{\omega} = -2i_{\bar{Q}}i_D\bar{\omega} + i_Di_D\bar{\omega}\,. \tag{4.24}$$

Using $\omega = d\chi$ one finds

$$i_Di_D\bar{\omega} = i_Di_Dd\bar{\chi} = i_DL_D\bar{\chi} + i_Ddi_D\bar{\chi} = L_D(i_D\bar{\chi}) + D(i_D\bar{\chi}) = 2D(i_D\bar{\chi})\,, \tag{4.25}$$

where we made use of $D^2 = 0$. Furthermore, applying i_Q to both sides of $i_Q\omega + d\mathcal{L} = \alpha$ one finds that $i_Q\alpha = 0$ thanks to $i_Qi_Q\omega = 0 = Q\mathcal{L}$. Because α is a 1-form from $\mathcal{I}$ one has $i_q\alpha = 0$ so that $i_{\mathrm{d_X}}\alpha = 0$ and

$$i_{\bar{Q}}i_D\bar{\omega} = -i_D(d\bar{\mathcal{L}} + \Pi^*(\alpha)) = -D\bar{\mathcal{L}} - \Pi^*(i_{\mathrm{d_X}}\alpha) = -D\bar{\mathcal{L}}\,, \tag{4.26}$$

where we made use of $i_D\Pi^*(\alpha) = \Pi^*(i_{\mathrm{d_X}}\alpha)$. All in all this gives:

$$\frac{1}{2}i_si_s\bar{\omega} = D(i_D\bar{\chi} + \bar{\mathcal{L}})\,. \tag{4.27}$$

We conclude that all the relations (4.9) and (4.10) are satisfied and hence:

PROPOSITION 4.3. *Decent completed gauge PDE $(E, Q, T[1]X, \chi, \mathcal{L})$ defines a structure of the descent-completed BV formulation on $SJ^\infty(E)_0$ with possibly degenerate (in the sense of Section* **4.3***) presymplectic structure.*

Here we do not address the issue of how to perform a symplectic reduction in this setup.

Finally, extracting the θ-homogeneity n component of (4.23) gives:

$$i_s\overset{n}{\bar{\omega}}{}^v = -\mathrm{d_v}(\overset{n}{\bar{\mathcal{L}}} + i_D\overset{n-1}{\bar{\chi}}) - L_D\overset{n-1}{\bar{\chi}}{}^v\,. \tag{4.28}$$

It follows $\overset{n}{\bar{\mathcal{L}}} + i_D\overset{n-1}{\bar{\chi}}$ can be taken as a BV Lagrangian $\mathcal{L}_{BV}$ on $SJ^\infty(E)$. The respective BV action can be written as:

$$S_{BV}[\widehat{\sigma}] = \int_{T[1]X} \widehat{\sigma}^*(\chi)(d_X) + \widehat{\sigma}^*(\mathcal{L})\,, \tag{4.29}$$

where $\widehat{\sigma}$ is a supersection of E.

Of course, S_{BV} does not depend on the fields parameterizing the kernel of the presymplectic structure and interpreting it as a standard BV action requires passing to the symplectic quotient or imposing additional gauge conditions which set these fields to particular configurations. With these additional gauge conditions the BV action can be used in the path integral just like the conventional BV-AKSZ action.

The above BV action is a generalization of that for AKSZ sigma-models [**34**], see also [**33, 44**] for a version applicable to diffeomorphism-invariant systems and [**31, 32**] for PDEs equipped with a symplectic current.[3] Note also that in the case $n = 1$ and an invertible ω this construction was already in [**36**].

To conclude, presymplectic gauge PDEs give a rather flexible way to encode Lagrangian BV formulation of local gauge systems without employing jet-bundles. It can also be thought of as a generalisation of the BV-AKSZ formalism to the case of not necessarily topological and not necessarily diffeomorphims-invariant systems.

5. Weak presymplectic gauge PDEs

5.1. Weak presymplectic Q-manifolds. Despite presymplectic gauge PDEs being a rather flexible and invariant way to describe Lagrangian gauge systems, they are not very concise and explicit. For instance, for a nonilinear PDE it is often impossible to find an explicit expression for the underlying Q-structure and one only has the existence statement [**10**]. An alternative would be to work in terms of the usual Lagrangian BV formalism, where the BV Lagrangian and the symplectic structure come from finite jets and can often be given explicitly, but in this way one is stuck to jet-bundles and hence looses flexibility.

It turns out that the data of a presymplectic gauge PDE can be encoded in a more flexible and often finite-dimensional bundle over $T[1]X$, where condition $Q^2 = 0$ is relaxed in a controlled way. Let us first illustrate the idea in the case of a toy model of "weak presymplectic gauge PDE" with $\dim X = 0$.

PROPOSITION 5.1. *Suppose we are given with a graded manifold E equipped with a vector field Q, $\mathrm{gh}(Q) = 1$ and a closed 2-form ω, $\mathrm{gh}(\omega) = l$ satisfying*

$$d\omega = 0\,, \qquad L_Q\omega = 0\,, \qquad i_Q i_Q\omega = 0\,, \tag{5.1}$$

for some function $\mathcal{L}$. If in addition ω is regular, its symplectic quotient, possibly locally defined, is naturally a symplectic Q-manifold.

PROOF. Let $\mathcal{K} \subset TE$ be a kernel distribution of ω. For any $K \in \mathcal{K}$ one has $[K, Q] \in \mathcal{K}$ thanks to $L_Q\omega = 0$. For any function f satisfying $\mathcal{K}f = 0$ one has $\mathcal{K}Qf = 0$ so that Q induces a vector field on the symplectic quotient. Let $p : E \to E'$ denote a projection to the symplectic quotient and Q' and ω' denote the respective structures induced on the symplectic quotient. It follows $\omega = p^*\omega'$ and $p^* \circ Q' = Q \circ p^*$, i.e. Q projects to Q'. It then follows that $i_{Q'} i_{Q'}\omega' = 0$, $L_{Q'}\omega' = 0$, and also $i_{Q'}\omega' + d\mathcal{L}' = 0$ for some $\mathcal{L}'$. Because ω' is symplectic, one concludes that $(Q')^2 = 0$. In particular, Hamiltonian $\mathcal{L}'$ of Q' satisfies master equation $\{\mathcal{L}', \mathcal{L}'\} = 0$, where the bracket denotes the (graded) Poisson structure determined by ω'. ☐

5.2. Weak presymplectic gauge PDEs. The notion and the statement can be generalized to the case where $\dim X > 0$ by replacing $Q^2 = 0$ with in general weaker condition $i_Q i_Q\omega = 0$ and keeping the remaining conditions intact. This gives a more flexible version of the presymplectic BV-AKSZ formalism developed in the previous section. More precisely, a weak presymplectic gauge PDE is a $\mathbb{Z}$-graded fiber bundle over $T[1]X$, equipped with a vector field Q satisfying $\mathrm{gh}(Q) = 1$,

[3] In fact the BV action of Lagrangian formulation proposed in [**53, 54**] also has the same structure.

$Q \circ \pi^* = \pi^* \circ \mathrm{d_X}$ and a 2-form ω of degree $n-1$, $n = \dim X$ satisfying:

$$d\omega = 0, \qquad L_Q\omega \in \mathcal{I}, \qquad i_Q i_Q \omega = 0, \qquad i_Q L_Q \omega \in \mathcal{I}. \tag{5.2}$$

In addition ω is assumed regular (in the same sense as in Section **4.4**, i.e. that the induced $\overset{n}{\omega}{}^v$ is regular on $SJ^\infty(E)$). This definition can be made even more flexible but it is enough to illustrate the idea. Note that the above relations imply $i_{Q^2}\omega \in \mathcal{I}$.

In order to see that such system still defines a local BV system, let us observe that in the analysis of Section **4.5** we only employed $Q^2 = 0$ to derive $i_Q i_Q \omega = 0$ and $Q\mathcal{L} = 0$. Now $i_Q i_Q \omega = 0$ by assumption while $Q\mathcal{L} = 0$ by the same reasoning as in the Appendix. Picking χ and $\mathcal{L}$ such that $\omega = d\chi$ and $i_Q\omega + d\mathcal{L} \in \mathcal{I}$ and repeating the considerations of Section **4.5** in this more general situation we arrive at equations (4.23) and (4.27) and hence to a descent-completed presymplectic BV system. Because we assumed $\overset{n}{\omega}{}^v$ regular this in turn defines a local BV system.

A remarkable feature of such an extension of the presymplectic BV-AKSZ formalism is that it is often enough to consider finite-dimensional E and there is no need to work with infinite-dimensional manifolds. It does not mean that one should necessarily relax $Q^2 = 0$ to achieve that, as in some cases, including Maxwell field and Einstein gravity, there exist finite-dimensional presymplectic gauge PDEs encoding the full-scale BV formulation [**44, 45**] (without BV extension similar examples were already in [**33**]).

5.3. Example. It is not difficult to give a nontrivial example of the weak presymplectic AKSZ system. Consider the following weak presymplectic Q-manifold: coordinates are x^a, θ^a, C, F^{ab}, where θ^a, C are of degree 1 and coordinates C, F^{ab} take values in a reductive real Lie algebra. The weak Q-structure is determined by:

$$Qx^a = \theta^a\,, \quad QC = -\frac{1}{2}[C,C] + \frac{1}{2}\theta^a\theta^b F_{ab}\,, \quad QF_{ab} = [F_{ab}, C]\,. \tag{5.3}$$

Note that Q is not nilpotent: $Q^2 F_{ab}$ is proportional to $[F_{ab}, \theta^c\theta^d F_{cd}]$. Of course, this Q-manifold is just a truncation of the minimal BRST complex of Yang-Mills theory, see e.g. [**50**].

As a presymplectic potential we take [**33**]:

$$\chi = \mathrm{Tr}(\theta^{(n-2)F^{ab}}_{ab} dC)\,, \qquad (\theta)^{(n-k)}_{a_1\ldots a_k} = \frac{1}{(n-k)!}\epsilon_{a_1\ldots a_k b_1\ldots b_{n-k}}\theta^{b_1}\ldots\theta^{b_{n-k}}\,. \tag{5.4}$$

where the indexes of F are raised and lowered by the constant Minkowski metric η_{ab}. It is easy to check that $i_Q i_Q \omega = 0$, where $\omega = d\chi$. The "covariant Hamiltonian" $\mathcal{L}$ determined through $i_Q\omega + d\mathcal{L} \in \mathcal{I}$ reads as

$$\mathcal{L} = \frac{1}{2}\,\mathrm{Tr}(F^{ab}[C,C])\theta^{(n-2)}_{ab} - \frac{1}{2}\,\mathrm{Tr}(F_{ab}F^{ab})\theta^{(n)}\,. \tag{5.5}$$

The BV-AKSZ action (4.29) takes the form of the usual first order formulation of YM action:

$$S_{BV} = \int d^n x (F^{ab}(\partial_a A_b - \partial_b A_a + [A_a, A_b]) - \frac{1}{2}F_{ab}F^{ab}) + \ldots\,, \tag{5.6}$$

where we only listed explicitly the terms not involving variable of nonvanishing ghost degree, $A_a(x)\theta^a = \sigma^*(C)$, and by some abuse of notations $F^{ab}(x) = \sigma^*(F^{ab})$. Passing to the simplectic quotient gives the respective BV field-antifield space.

The last step is completely analogous to the one in the case of the Maxwell field considered in [**45**], to which we refer for further details.

6. Induced gauge PDEs on submanifolds and boundaries

As we have seen, presymplectic gauge PDEs can be considered as a far-going generalization of AKSZ sigma models. It is well-known that for a given AKSZ model one can replace the space-time manifold with its submanifold, or add extra dimension(s) without spoiling consistency. A typical example is given by $n = 1$ AKSZ sigma model whose restriction to a point (i.e. just the target space) turns out to be the respective BFV phase space [**36**]. More generally, the Hamiltonian BFV description of generic AKSZ models can be obtained by restricting the model to a codimension 1 space-like submanifold [**64**]. This extends to a rather general gauge theories by representing them as AKSZ sigma models using so-called parent formulation approach [**53, 54**]. In fact the Lagrangian parent formulation of these Refs. is a special class of presymplectic gauge PDEs. In the non-Lagrangian setup, gauge PDEs give a natural tool to study induced theories on submanifolds and (asymptotic) boundaries, as was demonstrated in [**65–67**] (see also [**68, 69**] for further applications) in the context of AdS/CFT correspondence for higher-spin gauge theories. Let us also mention an alternative approach to BV systems on manifolds with boundaries [**58, 61, 70, 71**] (see also Refs. therein), which operates in the usual BV setup and does not explicitly employ (presymplectic) gauge PDEs.

The simplest situation is where $(E, Q, T[1]X)$ is a gauge PDE and we specify a submanifold $\Sigma \subset X$ (e.g. Σ can be a boundary of X) of its space-time manifold X. There is a naturally defined induced gauge PDE structure on $T[1]\Sigma$. Indeed, $T[1]\Sigma$ can be seen as a Q-submanifold of $T[1]X$, i.e. $\mathrm{d_X}$ is tangent to $T[1]\Sigma$. Introducing local coordinates x^i, z^α adapted to the embedding so that Σ is singled out by $z^\alpha = 0$, submanifold $T[1]\Sigma \subset T[1]X$ is singled out by $z^\alpha = 0, \mathrm{d_X} z^\alpha = 0$.

Let $E_\Sigma \to T[1]\Sigma$ be a restriction of E to $T[1]\Sigma \subset T[1]X$. It is easy to see that Q is tangent to E_Σ, seen as a submanifold of E, and hence defines a Q-structure Q_Σ on E_Σ, which projects to d_Σ and hence defines a new gauge PDE $(E_\Sigma, Q_\Sigma, T[1]\Sigma)$. In the case where Σ is a boundary of X this can be regarded as a theory of boundary values of the fields described by $(E, Q, T[1]X)$. This is precisely the procedure applied in [**65, 66**] to study boundary values of higher-spin fields in the AdS/CFT setup.

Let us now turn to presymplectic gauge PDEs. The pullback ω_Σ of presymplectic structure ω to E_Σ gives a presymplectic structure in $(E_\Sigma, Q_\Sigma, T[1]\Sigma)$. However, the ghost degree of ω_Σ is $\dim X - 1$ and not $\dim \Sigma - 1$ so that it is not a usual presymplectic gauge PDE. For Σ of codim 1 we are in fact dealing with a version of Hamiltonian BFV system, as was initially observed in [**36, 64**] in the AKSZ setup.

We refrain from giving here a general framework for presymplectic gauge PDEs induced on submanifolds and boundaries but consider instead a weak presymplectic gauge PDE, describing Yang-Mills system and presented in Section **5.3**. For simplicity we assume that X is a 4-dimensional Minkowski space and take Σ to be the hyperplain $x^0 = 0$. We use x^i, θ^i, $i = 1, 2, 3$ to denote standard coordinates on $T[1]\Sigma$. The induced symplectic potential reads explicitly as:

$$\chi_\Sigma = \mathrm{Tr}(F^{0i}\epsilon_{ijk}\theta^j\theta^k\, dC)\,. \tag{6.1}$$

The form $\overset{3}{\omega}{}^{v}_{\Sigma}$ induced by $\omega_\Sigma = d\chi_\Sigma$ on $SJ^\infty(E_\Sigma)$ (see Section **4.4**) reads explicitly as

$$\overset{3}{\omega}{}^{v}_{\Sigma} = 2\,\mathrm{Tr}\left(d\overset{0}{F}{}^{0i} dA_i + d\overset{1}{F}{}^{0i}_{i} d\overset{0}{C}\right)\theta^{(3)}\,, \tag{6.2}$$

where we introduced coordinates $\psi^A(\theta)$ via

$$\begin{aligned} F^{ab}(\theta) &= \overset{0}{F}{}^{ab} + \overset{1}{F}{}^{ab}_{j}\theta^j + \frac{1}{2}\overset{2}{F}{}^{ab}_{jk}\theta^j\theta^k\,,\\ C(\theta) &= \overset{0}{C} + A_j\theta^j + \frac{1}{2}\overset{2}{C}_{jk}\theta^j\theta^k\,. \end{aligned} \tag{6.3}$$

This gives a presymplectic structure $\omega^{\bar{E}_\Sigma}$ on the fibers of the bundle $\bar{E}_\Sigma \to X$, defined in such a way that $SJ^\infty(E_\Sigma)_0 = J^\infty(\bar{E}_\Sigma)$. Furthermore, all the coordinates save for $\mathcal{C} = \overset{0}{C}$, A_i, $\pi^i = \overset{0}{F}{}^{0i}$ and $\mathcal{P} = \overset{1}{F}{}^{0i}_{i}$ are in the kernel and the symplectic reduction gives us the reduced bundle $\bar{\mathcal{E}}_\Sigma$, equipped with the canonical symplectic structure

$$\omega^{\bar{\mathcal{E}}_\Sigma} = 2\,\mathrm{Tr}(d\pi^i dA_i + d\mathcal{C}d\mathcal{P})(dx)^3\,, \tag{6.4}$$

where we switched to the conventions of Section **4.1**. It is easy to see that the space of sections of $\bar{\mathcal{E}}_\Sigma$ coincides with the phase space of the constrained Hamiltonian system formulation of Yang-Mills theory. Indeed, π^i are momenta conjugate to A_i and $\mathcal{C}$ and $\mathcal{P}$ are canonically conjugated ghosts and ghost momenta. Furthermore, (6.4) induces a canonical symplectic structure on the space of sections of $\bar{\mathcal{E}}_\Sigma$.

Applying (4.29) in the present situation (and having in mind that the degree of ω_Σ is different) we get

$$\Omega[A,\pi,\mathcal{C},\mathcal{P}] = \int_{T[1]\Sigma}(\pi^*(\chi_\Sigma)(d_\Sigma)+\pi^*(\mathcal{L})) = -2\int d^3x\,\mathrm{Tr}(\pi^i(\partial_i\mathcal{C}+[A_i,\mathcal{C}])-\mathcal{P}[\mathcal{C},\mathcal{C}])\,, \tag{6.5}$$

which is proportional to the standard expression for the BFV-BRST charge of YM theory. Obtaining BRST-invariant Hamiltonian is not so straightforward in this case because it is not 1-st order in spatial derivatives and hence its derivation involves extra steps. In fact, "covariant Hamiltonian" $\mathcal{L}$ defined on the initial gauge PDE encodes both the BFV-BRST charge and the covariant analog of the usual Hamiltonian.

Appendix A. Technical details

Descent "covariant Hamiltonians". In the setting of Section **4.1** consider $\mathrm{d_v}(i_s i_s \overset{n}{\omega})$. We have

$$\mathrm{d_v}(i_s i_s\overset{n}{\omega}) = -2i_s L_s\overset{n}{\omega} = 2i_s\mathrm{d_h}\overset{n-1}{\omega} = 2\mathrm{d_h}i_s\overset{n-1}{\omega}\,. \tag{A.1}$$

Observe that $i_s\overset{n-1}{\omega}$ satisfies: $\mathrm{d_v}i_s\overset{n-1}{\omega} = -L_s\overset{n-1}{\omega} = \mathrm{d_h}\overset{n-2}{\omega}$ where in the last equality we made use of the descent equation. It follows from triviality of $\mathrm{d_v}$ mod $\mathrm{d_h}$ cohomology in this degree that $i_s\overset{n-1}{\omega} = -\mathrm{d_v}\overset{n-1}{\mathcal{L}} - \mathrm{d_h}\overset{n-1}{\chi}$ for some $(n-1,0)$-form $\overset{n-1}{\mathcal{L}}$ and $(n-1,1)$ form $\overset{n-1}{\chi}$. Using this in (A.1) gives

$$\mathrm{d_v}(i_s i_s\overset{n}{\omega}) = 2\mathrm{d_v}\mathrm{d_h}\overset{n-1}{\mathcal{L}} \tag{A.2}$$

which in turn implies $(\frac{1}{2}i_s i_s \overset{n}{\omega} - \mathrm{d_h} \overset{n-1}{\mathcal{L}}) = \pi^* f$ for some $f(x, dx)$. However because $i_s i_s \overset{n}{\omega}$ is of degree $n+1$ such f should vanish, giving (4.5).

Checking $i_Q i_Q \omega = 0$. In the setup of definition **4.2** let ω be a fixed representative of the vertical form $[\omega]$. Assume in addition that the representative is such that

$$i_Q L_Q \omega \in \mathcal{I}\,, \qquad i_Q d\omega \in \mathcal{I}\,, \qquad i_Q i_Q d\omega \in \mathcal{I}\,. \tag{A.3}$$

Let $\mathcal{L}$ be a covarinat Hamiltonian defined through $i_Q\omega + d\mathcal{L} \in \mathcal{I}$. One has

$$d(Q\mathcal{L}) = d(i_Q d\mathcal{L}) = L_Q(d\mathcal{L}) = L_Q(-i_Q\omega + \mathcal{I}) \in \mathcal{I}\,. \tag{A.4}$$

Because $d(Q\mathcal{L})$ is a 1-form $d(Q\mathcal{L}) \in \mathcal{I}$ implies $d(Q\mathcal{L}) = d\pi^*(f)$, $f \in \mathcal{C}^\infty(T[1]X)$. However, $\mathrm{gh}(Q\mathcal{L}) = n+1$ but on $T[1]X$ there are no functions of degree $n+1$ so that $Q\mathcal{L} = 0$.

Furthermore,

$$d(i_Q i_Q \omega) = i_Q d i_Q \omega - L_Q i_Q \omega = i_Q i_Q d\omega - i_Q L_Q \omega - i_{[Q,Q]}\omega - i_Q L_Q \omega \in \mathcal{I}\,, \tag{A.5}$$

where we made use of (A.3) and $Q^2 = 0$. Just like in deriving $Q\mathcal{L} = 0$ we use $\mathrm{gh}(i_Q i_Q \omega) = n+1$ to see that, again, $i_Q i_Q \omega = 0$. Note that if we impose $i_Q i_Q \omega = 0$ in place of $Q^2 = 0$ the last relation implies $i_{Q^2}\omega \in \mathcal{I}$.

Acknowledgments

I wish to thank I. Dneprov, K. Druzhkov, J. Frias, M. Henneaux, I. Krasil'shchik, A. Sharapov, E. Skvortsov, A. Verbovetsky for fruitful discussions. I am garateful to A. Kotov for the collaboration at the early stage of this project. Part of this work was done when the author participated in the thematic program "Higher Structures and Field Theory" at the Erwin Schrödinger International Institute for Mathematics and Physics, Vienna, Austria. This work was partially completed while the author participated in the workshop "Higher Spin Gravity and its Applications" supported by the Asia Pacific Center for Theoretical Physics, Pohang, Korea.

References

[1] I. A. Batalin and G. A. Vilkovisky, *Gauge algebra and quantization*, Phys. Lett. B **102** (1981), no. 1, 27–31, DOI 10.1016/0370-2693(81)90205-7. MR616572

[2] I. A. Batalin and G. A. Vilkovisky, *Closure of the gauge algebra, generalized Lie equations, and Feynman rules*, Group theoretical methods in physics, Vol. 1–3 (Zvenigorod, 1982), Harwood Academic Publ., Chur, 1985, pp. 187–205. MR989345

[3] I. A. Batalin and G. A. Vilkovisky, *Closure of the gauge algebra, generalized Lie equations and Feynman rules*, Nuclear Phys. B **234** (1984), no. 1, 106–124, DOI 10.1016/0550-3213(84)90227-X. MR736479

[4] Marc Henneaux and Claudio Teitelboim, *Quantization of gauge systems*, Princeton University Press, Princeton, NJ, 1992. MR1191617

[5] Joaquim Gomis, Jordi París, and Stuart Samuel, *Antibracket, antifields and gauge-theory quantization*, Phys. Rep. **259** (1995), no. 1-2, 145, DOI 10.1016/0370-1573(94)00112-G. MR1345328

[6] Glenn Barnich, Friedemann Brandt, and Marc Henneaux, *Local BRST cohomology in the antifield formalism. I. General theorems*, Comm. Math. Phys. **174** (1995), no. 1, 57–91. MR1372800

[7] Glenn Barnich, Friedemann Brandt, and Marc Henneaux, *Local BRST cohomology in Einstein-Yang-Mills theory*, Nuclear Phys. B **455** (1995), no. 1-2, 357–408, DOI 10.1016/0550-3213(95)00471-4. MR1361836

[8] Glenn Barnich, Friedemann Brandt, and Marc Henneaux, *Local BRST cohomology in gauge theories*, Phys. Rep. **338** (2000), no. 5, 439–569, DOI 10.1016/S0370-1573(00)00049-1. MR1792979

[9] G. Barnich, M. Grigoriev, A. Semikhatov, and I. Tipunin, *Parent field theory and unfolding in BRST first-quantized terms*, Comm. Math. Phys. **260** (2005), no. 1, 147–181, DOI 10.1007/s00220-005-1408-4. MR2175993

[10] Glenn Barnich and Maxim Grigoriev, *First order parent formulation for generic gauge field theories*, J. High Energy Phys. **1** (2011), 122, 36, DOI 10.1007/JHEP01(2011)122. MR2792251

[11] D. S. Kaparulin, S. L. Lyakhovich, and A. A. Sharapov, *Local BRST cohomology in (non-)Lagrangian field theory*, J. High Energy Phys. **9** (2011), 006, 34, DOI 10.1007/JHEP09(2011)006. MR2889892

[12] Alexey A. Sharapov, *Variational tricomplex, global symmetries and conservation laws of gauge systems*, SIGMA Symmetry Integrability Geom. Methods Appl. **12** (2016), Paper No. 098, 24, DOI 10.3842/SIGMA.2016.098. MR3553349

[13] A. M. Vinogradov, *On the algebro-geometric foundations of Lagrangian field theory* (Russian), Dokl. Akad. Nauk SSSR **236** (1977), no. 2, 284–287. MR0501142

[14] A. Vinogradov, *A spectral sequence associated with a nonlinear differential equation and algebra-geometric foundations of Lagrangian field theory with constraints*, Sov. Math. Dokl. **19** (1978), 144.

[15] A. M. Vinogradov, *The C-spectral sequence, Lagrangian formalism, and conservation laws. I. The linear theory*, J. Math. Anal. Appl. **100** (1984), no. 1, 1–40, DOI 10.1016/0022-247X(84)90071-4. MR739951

[16] Joseph Krasil'shchik and Alexander Verbovetsky, *Geometry of jet spaces and integrable systems*, J. Geom. Phys. **61** (2011), no. 9, 1633–1674, DOI 10.1016/j.geomphys.2010.10.012. MR2813504

[17] I. S. Krasil′shchik, V. V. Lychagin, and A. M. Vinogradov, *Geometry of jet spaces and nonlinear partial differential equations*, Advanced Studies in Contemporary Mathematics, vol. 1, Gordon and Breach Science Publishers, New York, 1986. Translated from the Russian by A. B. Sosinskiĭ. MR861121

[18] Simon L. Lyakhovich and Alexey A. Sharapov, *BRST theory without Hamiltonian and Lagrangian*, J. High Energy Phys. **3** (2005), 011, 22, DOI 10.1088/1126-6708/2005/03/011. MR2151981

[19] Maxim Grigoriev and Alexei Kotov, *Gauge PDE and AKSZ-type sigma models*, Fortschr. Phys. **67** (2019), no. 8-9, Special issue: Proceedings of the LMS/EPSRC Durham Symposium on Higher Structures in M-Theory, 1910007, 13, DOI 10.1002/prop.201910007. MR4016013

[20] M. A. Vasiliev, *Equations of motion of interacting massless fields of all spins as a free differential algebra*, Phys. Lett. B **209** (1988), no. 4, 491–497, DOI 10.1016/0370-2693(88)91179-3. MR959289

[21] M. A. Vasiliev, *Actions, charges and off-shell fields in the unfolded dynamics approach*, Int. J. Geom. Methods Mod. Phys. **3** (2006), no. 1, 37–80, DOI 10.1142/S0219887806001016. MR2209525

[22] Dennis Sullivan, *Infinitesimal computations in topology*, Inst. Hautes Études Sci. Publ. Math. **47** (1977), 269–331 (1978). MR646078

[23] R. D'Auria and P. Fré, *Geometric supergravity in $D = 11$ and its hidden supergroup*, Nuclear Phys. B **201** (1982), no. 1, 101–140, DOI 10.1016/0550-3213(82)90376-5. MR667482

[24] Jerzy Kijowski and Włodzimierz M. Tulczyjew, *A symplectic framework for field theories*, Lecture Notes in Physics, vol. 107, Springer-Verlag, Berlin-New York, 1979. MR549772

[25] Čedomir Crnković and Edward Witten, *Covariant description of canonical formalism in geometrical theories*, Three hundred years of gravitation, Cambridge Univ. Press, Cambridge, 1987, pp. 676–684. MR920461

[26] Gregg J. Zuckerman, *Action principles and global geometry*, Mathematical aspects of string theory (San Diego, Calif., 1986), Adv. Ser. Math. Phys., vol. 1, World Sci. Publishing, Singapore, 1987, pp. 259–284. MR915825

[27] I. Anderson, "The variational bicomplex," tech. rep., Formal Geometry and Mathematical Physics, Department of Mathematics, Utah State University, 1989.

[28] Igor Khavkine, *Presymplectic current and the inverse problem of the calculus of variations*, J. Math. Phys. **54** (2013), no. 11, 111502, 11, DOI 10.1063/1.4828666. MR3137018

[29] A. A. Sharapov, "On presymplectic structures for massless higher-spin fields," *Eur. Phys. J.* **C76** no. 6, (2016) 305, `arXiv:1602.06393` [hep-th].
[30] K. P. Druzhkov, *Extendable symplectic structures and the inverse problem of the calculus of variations for systems of equations written in generalized Kovalevskaya form*, J. Geom. Phys. **161** (2021), Paper No. 104013, 11, DOI 10.1016/j.geomphys.2020.104013. MR4184665
[31] M. Grigoriev, "Presymplectic structures and intrinsic Lagrangians," `arXiv:1606.07532` [hep-th].
[32] Maxim Grigoriev and Vyacheslav Gritzaenko, *Presymplectic structures and intrinsic Lagrangians for massive fields*, Nuclear Phys. B **975** (2022), Paper No. 115686, 26, DOI 10.1016/j.nuclphysb.2022.115686. MR4374899
[33] K. B. Alkalaev and M. Grigoriev, "Frame-like Lagrangians and presymplectic AKSZ-type sigma models," *Int. J. Mod. Phys.* **A29** no. 18, (2014) 1450103, `arXiv:1312.5296` [hep-th].
[34] M. Alexandrov, A. Schwarz, O. Zaboronsky, and M. Kontsevich, *The geometry of the master equation and topological quantum field theory*, Internat. J. Modern Phys. A **12** (1997), no. 7, 1405–1429, DOI 10.1142/S0217751X97001031. MR1432574
[35] Alberto S. Cattaneo and Giovanni Felder, *A path integral approach to the Kontsevich quantization formula*, Comm. Math. Phys. **212** (2000), no. 3, 591–611, DOI 10.1007/s002200000229. MR1779159
[36] M. A. Grigoriev and P. H. Damgaard, "Superfield BRST charge and the master action," *Phys. Lett.* **B474** (2000) 323–330, `arXiv:hep-th/9911092` [hep-th].
[37] Igor Batalin and Robert Marnelius, *Superfield algorithms for topological field theories*, Multiple facets of quantization and supersymmetry, World Sci. Publ., River Edge, NJ, 2002, pp. 233–251, DOI 10.1142/9789812777065_0021. MR1964906
[38] Alberto S. Cattaneo and Giovanni Felder, *On the AKSZ formulation of the Poisson sigma model*, Lett. Math. Phys. **56** (2001), no. 2, 163–179, DOI 10.1023/A:1010963926853. EuroConférence Moshé Flato 2000, Part II (Dijon). MR1854134
[39] Dmitry Roytenberg, *On the structure of graded symplectic supermanifolds and Courant algebroids*, Quantization, Poisson brackets and beyond (Manchester, 2001), Contemp. Math., vol. 315, Amer. Math. Soc., Providence, RI, 2002, pp. 169–185, DOI 10.1090/conm/315/05479. MR1958835
[40] Francesco Bonechi, Pavel Mnëv, and Maxim Zabzine, *Finite-dimensional AKSZ-BV theories*, Lett. Math. Phys. **94** (2010), no. 2, 197–228, DOI 10.1007/s11005-010-0423-3. MR2733561
[41] Glenn Barnich and Maxim Grigoriev, *A Poincaré lemma for sigma models of AKSZ type*, J. Geom. Phys. **61** (2011), no. 3, 663–674, DOI 10.1016/j.geomphys.2010.11.014. MR2763627
[42] Giuseppe Bonavolontà and Alexei Kotov, *Local BRST cohomology for AKSZ field theories: a global approach*, Mathematical aspects of quantum field theories, Math. Phys. Stud., Springer, Cham, 2015, pp. 325–341. MR3330246
[43] N. Ikeda, "Lectures on AKSZ Sigma Models for Physicists," in *Workshop on Strings, Membranes and Topological Field Theory*, pp. 79–169. WSPC, 2017. `arXiv:1204.3714` [hep-th].
[44] Maxim Grigoriev and Alexei Kotov, *Presymplectic AKSZ formulation of Einstein gravity*, J. High Energy Phys. **9** (2021), Paper No. 181, 23, DOI 10.1007/jhep09(2021)181. MR4327135
[45] I. Dneprov and M. Grigoriev, "Presymplectic BV-AKSZ formulation of conformal gravity," *Eur. Phys. J. C* **83** no. 1, (2023) 6, `arXiv:2208.02933` [hep-th].
[46] Alexey Sharapov and Evgeny Skvortsov, *Higher spin gravities and presymplectic AKSZ models*, Nuclear Phys. B **972** (2021), Paper No. 115551, 56, DOI 10.1016/j.nuclphysb.2021.115551. MR4318854
[47] Albert Schwarz, *Semiclassical approximation in Batalin-Vilkovisky formalism*, Comm. Math. Phys. **158** (1993), no. 2, 373–396. MR1249600
[48] Albert Schwarz, *Geometry of Batalin-Vilkovisky quantization*, Comm. Math. Phys. **155** (1993), no. 2, 249–260. MR1230027
[49] Alexei Kotov and Thomas Strobl, *Characteristic classes associated to Q-bundles*, Int. J. Geom. Methods Mod. Phys. **12** (2015), no. 1, 1550006, 26, DOI 10.1142/S0219887815500061. MR3293862
[50] Friedemann Brandt, *Local BRST cohomology and covariance*, Comm. Math. Phys. **190** (1997), no. 2, 459–489, DOI 10.1007/s002200050248. MR1489579
[51] Marc Henneaux, *Elimination of the auxiliary fields in the antifield formalism*, Phys. Lett. B **238** (1990), no. 2-4, 299–304, DOI 10.1016/0370-2693(90)91739-X. MR1050736

[52] Alain Dresse, Philippe Grégoire, and Marc Henneaux, *Path-integral equivalence between the extended and nonextended Hamiltonian formalisms*, Phys. Lett. B **245** (1990), no. 2, 192–196, DOI 10.1016/0370-2693(90)90132-P. MR1068704

[53] Maxim Grigoriev, *Parent formulation at the Lagrangian level*, J. High Energy Phys. **7** (2011), 061, 24, DOI 10.1007/JHEP07(2011)061. MR2875964

[54] Maxim Grigoriev, *Parent formulations, frame-like Lagrangians, and generalized auxiliary fields*, J. High Energy Phys. **12** (2012), 048, front matter + 32, DOI 10.1007/JHEP12(2012)048. MR3045284

[55] Ian M. Anderson, *Introduction to the variational bicomplex*, Mathematical aspects of classical field theory (Seattle, WA, 1991), Contemp. Math., vol. 132, Amer. Math. Soc., Providence, RI, 1992, pp. 51–73, DOI 10.1090/conm/132/1188434. MR1188434

[56] L. Dickey, "Soliton equations and hamiltonian systems (second edition)," 2003. `https://books.google.de/books?id=A73UCgAAQBAJ`.

[57] Glenn Barnich, *Brackets in the jet-bundle approach to field theory*, Secondary calculus and cohomological physics (Moscow, 1997), Contemp. Math., vol. 219, Amer. Math. Soc., Providence, RI, 1998, pp. 17–27, DOI 10.1090/conm/219/03064. MR1640441

[58] Alberto S. Cattaneo, Pavel Mnev, and Nicolai Reshetikhin, *Classical BV theories on manifolds with boundary*, Comm. Math. Phys. **332** (2014), no. 2, 535–603, DOI 10.1007/s00220-014-2145-3. MR3257656

[59] A. A. Sharapov, *Variational tricomplex of a local gauge system, Lagrange structure and weak Poisson bracket*, Internat. J. Modern Phys. A **30** (2015), no. 25, 1550152, 32, DOI 10.1142/S0217751X15501523. MR3394764

[60] Raymond Stora, *Algebraic structure and topological origin of anomalies*, Progress in gauge field theory (Cargèse, 1983), NATO Adv. Sci. Inst. Ser. B: Phys., vol. 115, Plenum, New York, 1984, pp. 543–562. MR782343

[61] F. M. C. Simão, A. S. Cattaneo, and M. Schiavina, "BV equivalence with boundary," `arXiv:2109.05268` [math-ph].

[62] Pavel Mnev, Michele Schiavina, and Konstantin Wernli, *Towards holography in the BV-BFV setting*, Ann. Henri Poincaré **21** (2020), no. 3, 993–1044, DOI 10.1007/s00023-019-00862-8. MR4062466

[63] Hovhannes M. Khudaverdian and Theodore Voronov, *On complexes related with calculus of variations*, J. Geom. Phys. **44** (2002), no. 2-3, 221–250, DOI 10.1016/S0393-0440(02)00075-X. MR1969783

[64] Glenn Barnich and Maxim Grigoriev, *Hamiltonian BRST and Batalin-Vilkovisky formalisms for second quantization of gauge theories*, Comm. Math. Phys. **254** (2005), no. 3, 581–601, DOI 10.1007/s00220-004-1275-4. MR2126484

[65] Xavier Bekaert and Maxim Grigoriev, *Notes on the ambient approach to boundary values of AdS gauge fields*, J. Phys. A **46** (2013), no. 21, 214008, 23, DOI 10.1088/1751-8113/46/21/214008. MR3062134

[66] Xavier Bekaert and Maxim Grigoriev, *Higher-order singletons, partially massless fields, and their boundary values in the ambient approach*, Nuclear Phys. B **876** (2013), no. 2, 667–714, DOI 10.1016/j.nuclphysb.2013.08.015. MR3143456

[67] A. Chekmenev and M. Grigoriev, "Boundary values of mixed-symmetry massless fields in AdS space," *Nucl. Phys.* **B913** (2016) 769–791, `arXiv:1512.06443` [hep-th].

[68] X. Bekaert, M. Grigoriev, and E. D. Skvortsov, "Higher Spin Extension of Fefferman-Graham Construction," *Universe* **4** no. 2, (2018) 17, `arXiv:1710.11463` [hep-th].

[69] Maxim Grigoriev and Evgeny Skvortsov, *Type-B formal higher spin gravity*, J. High Energy Phys. **5** (2018), 138, front matter+37, DOI 10.1007/jhep05(2018)138. MR3815049

[70] Alberto S. Cattaneo, Pavel Mnev, and Nicolai Reshetikhin, *Perturbative quantum gauge theories on manifolds with boundary*, Comm. Math. Phys. **357** (2018), no. 2, 631–730, DOI 10.1007/s00220-017-3031-6. MR3767705

[71] G. Canepa, A. S. Cattaneo, and M. Schiavina, *Boundary structure of general relativity in tetrad variables*, Adv. Theor. Math. Phys. **25** (2021), no. 2, 327–377, DOI 10.4310/ATMP.2021.v25.n2.a3. MR4388714

LEBEDEV PHYSICAL INSTITUTE, LENINSKY AVE. 53, 119991 MOSCOW, RUSSIA; AND INSTITUTE FOR THEORETICAL AND MATHEMATICAL PHYSICS, LOMONOSOV MOSCOW STATE UNIVERSITY, 119991 MOSCOW, RUSSIA

Email address: grig@lpi.ru

Contemporary Mathematics
Volume **788**, 2023
https://doi.org/10.1090/conm/788/15823

Minimal realizations of the KP hierarchy, its strict version and their reductions

G. F. Helminck and E. A. Panasenko

Abstract. We discuss here realizations of the KP hierarchy, its strict version and their reductions, the n-KdV hierarchy and the strict n-KdV hierarchy, that have a minimal dependence between the different coefficients of a solution of them, the so-called minimal realizations. Thereto we discuss all these deformations from a wider perspective and consider them in a presetting instead of a setting. In this more general set-up we will present a number of $\mathbb{C}$-subalgebras of R that are stable under the basic $\mathbb{C}$-linear derivations of R and such that these derivations commute on these $\mathbb{C}$-subalgebras. This we apply at the introduction of the minimal realizations of all deformations, we show how these realizations relate to solutions in different settings and use them to show that all hierarchies under consideration possess invariant scaling transformations.

1. Introduction

The KP hierarchy and its strict version are the evolution equations for deformations inside the algebra Psd of pseudo differential operators of respectively the subalgebra $\mathbb{C}[\partial]$ of constant differential operators in the derivation ∂ of the coefficient algebra R of Psd and the Lie subalgebra $\mathbb{C}[\partial]\partial$. The n-KdV hierarchy corresponds to the evolution equations of deformations of the commutative Lie subalgebra $\mathbb{C}[\partial^n], n \geqslant 2$, in the Lie algebra $R[\partial]$ and the strict n-KdV hierarchy consists of the evolution equations of deformations of the Lie subalgebra $\mathbb{C}[\partial^n]\partial^n, n \geqslant 2$, inside the Lie algebra $R[\partial]\partial$. The introduction of their minimal realizations requires to treat all of these deformations from a wider perspective and to drop the commutativity assumption on the basic derivations $\{\partial_i\}$ and to work from a presetting instead of a setting. Nevertheless we can show for all hierarchies that the derivations $\{\partial_i\}$ commute on the ∂-stable subalgebra generated by the coefficients of a solution and also on the polynomial subalgebra generated by the ∂-stable subalgebras corresponding to different solutions in the presetting $(R, \partial, \{\partial_i\})$. In particular, as far as solutions are concerned, it is sufficient to work with standard settings. Further we show how the minimal realizations relate to solutions in different settings. As

2020 *Mathematics Subject Classification.* 35Q53, 35Sxx, 37K10, 58B25 .
The first author is the orresponding author.
The second author was supported by the Russian Science Foundation (project no. 22-21-00772, https://rscf.ru/project/22-21-00772/).

Dedicated to the memory of Alexandre Mikhailovich Vinogradov.

©2023 American Mathematical Society

an application of the minimal realizations we present the scaling transformations that map solutions of both hierarchies from one setting to another one.

The contents of the various sections is as follows: Section 2 describes the more general starting point for all hierarchies mentioned and discusses the properties that survive. The next section gives the reduction to standard settings. In section 4 we treat the minimal realizations of all hierarchies and the last section presents their scaling invariance.

2. The hierarchies in a wider perspective

Both KP and strict KP correspond to deformations inside the algebra of pseudo differential operators Psd of a commutative Lie subalgebra. For KP it is $\mathbb{C}[\partial]$, where ∂ is a $\mathbb{C}$-linear derivation of a commutative $\mathbb{C}$-algebra R, whose elements serve as coefficients for the pseudo differential operators and for strict KP it is the Lie algebra $\mathbb{C}[\partial]\partial$. Recall that Psd is built up as follows: given R and ∂, one forms the algebra $R[\partial]$ of differential operators in ∂ with coefficients from R. It consists of $\mathbb{C}$-linear endomorphisms of R of the form $\sum_{i=0}^{n} a_i\partial^i, a_i \in R$. As we want an equality in $R[\partial]$ to imply the equality of the corresponding coefficients from R on both sides, we assume that the powers of ∂ are R-linear independent. Then one can extend the algebra $R[\partial]$ by adding the inverses of all the powers of ∂ and by allowing infinite sums of these negative powers. Thus one arrives at the algebra Psd$=R[\partial,\partial^{-1})$ of all formal series

$$p = \sum_{j=-\infty}^{N} p_j\partial^j, p_j \in R. \tag{2.1}$$

A p in Psd of the form (2.1) is said to have degree N if $p_N \neq 0$ and the zero element has degree $-\infty$. The product of two series $a = \sum_j a_j\partial^j$ and $b = \sum_i b_i\partial^i$ in Psd is given by the Leibnitz rule

$$ab := \sum_j \sum_i \sum_{s=0}^{\infty} \binom{j}{s} a_j\partial^s(b_i)\partial^{i+j-s}.$$

Using this formula, one shows recursively the following two properties:

LEMMA 2.1. *Let R^* denote the invertible elements in R. Inside Psd there holds:*

(a) *Any $P = \sum_{j=-\infty}^{m} p_j\partial^j$ with $p_m \in R^*$ has an inverse in Psd of degree $-m$.*

(b) *Each element P in Psd of the form $P = \partial^m + \sum_{j<m} p_j\partial^j$ has a unique m-th root $P^{\frac{1}{m}}$ in Psd, i.e. an element $P^{\frac{1}{m}} = \partial + \sum_{j\leqslant 0} q_j\partial^j$ satisfying $(P^{\frac{1}{m}})^m = P$, where each q_j is a polynomial expression in the $\partial^l(p_k)$, $k \geqslant j$, $l \geqslant 0$.*

The algebra Psd possesses various decompositions. For $s \in \mathbb{Z}$, any pseudo differential operator $P = \sum_j p_j\partial^j \in$ Psd can be split as

$$P = P_{>s} + P_{\leqslant s}, P_{>s} = P_{\geqslant s+1} = \sum_{j>s} p_j\partial^j \text{ and } P_{\leqslant s} = P_{<s+1} = \sum_{j\leqslant s} p_j\partial^j. \tag{2.2}$$

For $s = -1$, this corresponds to writing P as the sum of its differential operator part $P_{\geqslant 0}$ and its strict integral operator part $P_{<0}$ and for $s = 0$ one writes P as the sum of its pure differential operator part $P_{>0}$ and its integral operator part $P_{\leqslant 0}$.

The algebra Psd is w.r.t. the commutator a complex Lie algebra. From the multiplication rules in Psd follows that for $s = -1$ resp. $s = 0$ the decomposition (2.2) splits the Lie algebra Psd into the direct sum of two Lie subalgebras, namely

$$\{P \in \mathrm{Psd}, P = P_{<0}\} \oplus \{P \in \mathrm{Psd}, P = P_{\geqslant 0}\} := \mathrm{Psd}_{<0} \oplus \mathrm{Psd}_{\geqslant 0},$$

resp.

$$\mathrm{Psd} = \{P \in \mathrm{Psd}, P = P_{\leqslant 0}\} \oplus \{P \in \mathrm{Psd}, P = P_{>0}\} := \mathrm{Psd}_{\leqslant 0} \oplus \mathrm{Psd}_{>0},$$

We write $\pi_{\geqslant 0}$ for the projection from Psd on $\mathrm{Psd}_{\geqslant 0}$ consisting of taking the differential operator part of an element in Psd. Similarly, one defines the projections of Psd on respectively $\mathrm{Psd}_{\leqslant 0}$, $\mathrm{Psd}_{>0}$ and $\mathrm{Psd}_{<0}$, by respectively $\pi_{\leqslant 0}$, $\pi_{>0}$ and $\pi_{<0}$. For any $\alpha \in \mathbb{C}^*$, let ∂_α be the derivation $\alpha\partial$. Since $R[\partial, \partial^{-1}) = R[\partial_\alpha, \partial_\alpha^{-1})$, one could split the elements of Psd in a way similar to (2.2), w.r.t. the powers of ∂_α . This would yield the same four projections $\pi_{\geqslant 0}$, $\pi_{\leqslant 0}$, $\pi_{>0}$ and $\pi_{<0}$ of Psd as above.

Thanks to Lemma 2.1 we can associate to both Lie algebras $\mathrm{Psd}_{\leqslant 0}$ and $\mathrm{Psd}_{<0}$ a group in Psd, namely

$$D(0) = \{p_0 + \sum_{j<0} p_j \partial^j \mid p_0 \in R^*\} \text{ resp. } D(1) = \{p \in D(0) \mid p_0 = 1\}.$$

Now we discuss the deformations of $\mathbb{C}[\partial]$ and $\mathbb{C}[\partial]\partial$. The first deformation maps each $\sum_{i=0}^N a_i \partial^i \in \mathbb{C}[\partial]$ to $\sum_{i=0}^N a_i L^i$, where L is a pseudo differential operator of degree one of the form

$$L = \partial + \sum_{j=1}^{\infty} \ell_{j+1} \partial^{-j}. \tag{2.3}$$

Elements of the form (2.3) are prototypes of conjugating or *dressing* with an element from $D(1)$. Under a mild condition, any L of the form (2.3) is obtained in this way

LEMMA 2.2. *If $\partial : R \to R$ is surjective, then any L of the form* (2.3) *is a $D(1)$-deformation of ∂, i.e. $L = d\partial d^{-1}$ with $d \in D(1)$.*

At the n-KdV hierarchy one considers monic differential operators $\mathbf{L}_n \in R[\partial]$ of order n without a term of order $n-1$

$$\mathbf{L}_n = \partial^n + \sum_{j=0}^{n-2} \mathbf{l}_{n-j} \partial^j, \tag{2.4}$$

which are seen as deformations of ∂^n inside $R[\partial]$. Under the conditions of Lemma 2.2 these deformations are obtained by conjugation with a suitable element of $D(1)$. The deformation linked with the strict KP hierarchy sends each $\sum_{i=1}^N a_i \partial^i \in \mathbb{C}[\partial]\partial$ into $\sum_{i=1}^N a_i M^i$ with M a pseudo differential operator of degree one of a slightly more general form

$$M = \partial + \sum_{j=0}^{\infty} m_{j+1} \partial^{-j}. \tag{2.5}$$

One verifies directly that $D(0)$-deformations of ∂, i.e. elements $d\partial d^{-1}$ with $d \in D(0)$, have the form (2.5). Here one needs a somewhat stronger condition on R and ∂ such that any M of the form (2.5) is obtained like that

LEMMA 2.3. *Let ∂ be surjective and let for each $r \in R$ there be an $r_0 \in R^*$ such that $\partial(r_0) = r r_0$. Then any M of the form* (2.5) *is a $D(0)$-deformation of ∂.*

At the strict n-KdV hierarchy one considers monic differential operator $\mathbf{M}_n \in R[\partial]$ of order n without a constant term, i.e.

$$\mathbf{M}_n = \partial^n + \sum_{j=1}^{n-1} \mathbf{m}_{n-j}\partial^j, \tag{2.6}$$

which are seen as deformations of ∂^n inside the Lie subalgebra $R[\partial]\partial$ of $R[\partial]$. If the conditions of Lemma 2.3 hold, these deformations are obtained by conjugation with a suitable element of $D(0)$.

Before discussing the evolution equations that the deformations L, M, $\mathbf{L}_n$ and $\mathbf{M}_n$ have to satisfy, we note that any $\mathbb{C}$-linear derivation δ of R that commutes with ∂, extends to a $\mathbb{C}$-linear derivation of Psd by defining

$$\delta(P) := \sum_{j=-\infty}^{N} \delta(p_j)\partial^j, \text{ for any } P = \sum_{j=-\infty}^{N} p_j\partial^j \in \text{Psd}.$$

This we assume without further mentioning. We start in all four cases with an algebra R that is also equipped with a collection $\{\partial_i \mid i \geqslant 1\}$ of $\mathbb{C}$-linear derivations of R that all commute with ∂. Each ∂_i should be seen as an algebraic substitute for differentiation w.r.t. to the flow parameter of the flow corresponding to the Lie algebra element ∂^i. The set of data $(R, \partial, \{\partial_i\})$ is called a *presetting* for each hierarchy. A presetting is called *standard* if $\partial_1 = \partial$ and we denote such a presetting then as $(R,\{\partial_i\})$. If all the $\{\partial_i\}$ mutually commute, then we speak of a *setting.*

We start with the evolution equations for an L of the form (2.3). Let $B_i, i \geqslant 1$, be the short hand notation for $\pi_{\geqslant 0}(L^i)$. Then we want L to satisfy the equations

$$\partial_i(L) = [B_i, L] = [L, \pi_{<0}(L^i)], i \geqslant 1. \tag{2.7}$$

The equations (2.7) for an operator L in Psd of the form (2.3), are called the *Lax equations of the KP hierarchy* and L is a *solution* of the hierarchy in the presetting $(R, \partial, \{\partial_i\})$. The *trivial* solution of the hierarchy is $L = \partial$. Next we discuss the evolution equations for an M of the form (2.5). Let $C_i, i \geqslant 1$, be the short hand notation for $\pi_{>0}(M^i)$. Then we want M to satisfy the equations

$$\partial_i(M) = [C_i, M] = [M, \pi_{\leqslant 0}(M^i)], i \geqslant 1. \tag{2.8}$$

The equations (2.8) for an operator M in Psd of the form (2.5), are called the *Lax equations of the strict KP hierarchy* and M is a *solution* of the hierarchy in the presetting $(R, \partial, \{\partial_i\})$. Also here we have the *trivial* solution $M = \partial$.

For differential operators $\mathbf{L}_n$ resp. $\mathbf{M}_n$ of the form (2.4) resp. (2.6) we look for Lax equations that look like

$$\partial_P(\mathbf{L}_n) = [P, \mathbf{L}_n] \text{ resp. } \partial_Q(\mathbf{M}_n) = [Q, \mathbf{M}_n], \tag{2.9}$$

where P and Q are in $R[\partial]$, Q has no constant term and ∂_P and ∂_Q are derivations of R commuting with ∂. According to Lemma 2.1 both $\mathbf{L}_n$ and $\mathbf{M}_n$ possess unique n-th roots $L(n) = (\mathbf{L}_n)^{\frac{1}{n}}$ resp. $M(n) = (\mathbf{M}_n)^{\frac{1}{n}}$ that are deformations of the form (2.3) resp. (2.5). It was shown in [**3**] that the differential operators P and Q for which the equations (2.9) make sense are all $P = \sum_{i=0}^{N} d_i\pi_{\geqslant 0}(L(n)^i)$, with each $d_i \in \text{Ker}(\partial)$, and all $Q = \sum_{i=1}^{N} c_i\pi_{>0}(M(n)^i)$, with each $c_i \in \text{Ker}(\partial)$. This led for $\mathbf{L}_n$ to the system of Lax equations in $R[\partial]$

$$\partial_i(\mathbf{L}_n) = [\pi_{\geqslant 0}(L(n)^i), \mathbf{L}_n], \text{ for all } i \geqslant 1. \tag{2.10}$$

This system is called the n-KdV hierarchy as the equation (2.10) for $n = 2$ and $i = 3$ is the KdV-equation. Note that the equations (2.10) show that each $\partial_i(\mathbf{l}_{n-j})$ is a polynomial expression in the $\{\partial^s(\mathbf{l}_{n-k})\}$, where $0 \leqslant k \leqslant n-2$ and $s \geqslant 0$. The differential operator $\mathbf{L}_n$ is called a *solution* of the n-KdV hierarchy in the presetting $(R, \partial, \{\partial_i\})$. The trivial solution is $\mathbf{L}_n = \partial^n$. For each $K \in \mathbb{C}[\mathbf{L}_n]$ hold the same Lax equations as for the generator $\mathbf{L}_n$. Similarly the form of all possible Q led to another system of Lax equations in $R[\partial]$

$$\partial_i(\mathbf{M}_n) = [\pi_{>0}(M(n)^i), \mathbf{M}_n], \text{ for all } i \geqslant 1. \tag{2.11}$$

This system is called the strict n-KdV hierarchy. Again we see that the equations (2.11) show that each $\partial_i(\mathbf{m}_{n-j})$ is a polynomial expression in the $\{\partial^s(\mathbf{m}_{n-k})\}$, where $1 \leqslant k \leqslant n-1$ and $s \geqslant 0$. The differential operator $\mathbf{M}_n$ is then called a *solution* of the strict n-KdV hierarchy in the presetting $(R, \partial, \{\partial_i\})$ and also here we have the trivial solution $\mathbf{M}_n = \partial^n$. All $N \in \mathbb{C}[\mathbf{M}_n]$ satisfy the same Lax equations as $\mathbf{M}_n$.

The reason to call the n-KdV hierarchy a reduction of the KP hierarchy and the strict n-KdV hierarchy a reduction of the strict KP hierarchy is explained in the following result, whose proof can be found in [**3**].

PROPOSITION 2.1. *Given deformations $\mathbf{L}_n$ resp. $\mathbf{M}_n$ of the form (2.4) resp. (2.6) with their n-th roots $L(n)$ resp. $M(n)$. Then there holds:*

(a) *$\mathbf{L}_n$ is a solution of the n-KdV hierarchy if and only if $L(n)$ is a solution of the KP hierarchy.*
(b) *$\mathbf{M}_n$ is a solution of the strict n-KdV hierarchy if and only if $M(n)$ is a solution of the strict KP hierarchy.*

All four hierarchies have an alternative form that we need later on. It is equivalent with their Lax form. We start with the KP hierarchy and its strict version. The Lax equations (2.7) and (2.8) imply the following useful relations

THEOREM 2.4. *Both sets of Lax equations (2.7) and (2.8) are so-called compatible systems, i.e. the projections $\{B_i := \pi_{\geqslant 0}(L^i) \mid i \geqslant 1\}$ satisfy the zero curvature relations*

$$\partial_{i_1}(B_{i_2}) - \partial_{i_2}(B_{i_1}) - [B_{i_1}, B_{i_2}] = 0 \tag{2.12}$$

and the projections $\{C_i := \pi_{>0}(M^i) \mid i \geqslant 1\}$ satisfy the zero curvature relations

$$\partial_{i_1}(C_{i_2}) - \partial_{i_2}(C_{i_1}) - [C_{i_1}, C_{i_2}] = 0. \tag{2.13}$$

PROOF. The idea of the proof is to show that the left hand side of the identities (2.12) resp. (2.13) belongs to

$$\pi_{\geqslant 0}(\mathrm{Psd}) \cap \pi_{<0}(\mathrm{Psd}) \text{ resp. } \pi_{>0}(\mathrm{Psd}) \cap \pi_{\leqslant 0}(\mathrm{Psd})$$

and thus has to be zero. We give the proof for the $\{C_i\}$, that for the $\{B_i\}$ is similar and is left to the reader. The inclusion in $\pi_{>0}(\mathrm{Psd})$ is clear as both C_i and $\partial_n(C_i)$ belong to the Lie subalgebra $\pi_{>0}(\mathrm{Psd})$. To show the other one, we use the Lax equations (2.8). Recall that similar Lax equations hold for all the $\{M^N \mid N \geqslant 1\}$

$$\partial_i(M^N) = [\pi_{>0}(M^i), M^N].$$

Now we substitute $C_{i_k} = M^{i_k} - \pi_{\leqslant 0}(M^{i_k}), k = 1, 2$, and get for

$$\begin{aligned}\partial_{i_1}(C_{i_2}) - \partial_{i_2}(C_{i_1}) &= \partial_{i_1}(M^{i_2}) - \partial_{i_1}(\pi_{\leqslant 0}(M^{i_2})) - \partial_{i_2}(M^{i_1}) + \partial_{i_2}(\pi_{\leqslant 0}(M^{i_1})) \\ &= [C_{i_1}, M^{i_2}] - [C_{i_2}, M^{i_1}] - \partial_{i_1}(\pi_{\leqslant 0}(M^{i_2})) + \partial_{i_2}(\pi_{\leqslant 0}(M^{i_1}))\end{aligned}$$

and for

$$\begin{aligned}[C_{i_1}, C_{i_2}] &= [M^{i_1} - \pi_{\leqslant 0}(M^{i_1}), M^{i_2} - \pi_{\leqslant 0}(M^{i_2})] \\ &= -[\pi_{\leqslant 0}(M^{i_1}), M^{i_2}] + [\pi_{\leqslant 0}(M^{i_2}), M^{i_1}] + [\pi_{\leqslant 0}(M^{i_1}), \pi_{\leqslant 0}(M^{i_2})].\end{aligned}$$

Taking into account the second identity in (2.8), we see that the left hand side of (2.13) is equal to

$$-\partial_{i_1}(\pi_{\leqslant 0}(M^{i_2})) + \partial_{i_2}(\pi_{\leqslant 0}(M^{i_1})) - [\pi_{\leqslant 0}(M^{i_1}), \pi_{\leqslant 0}(M^{i_2})].$$

This element belongs to the Lie subalgebra $\pi_{\leqslant 0}(\text{Psd})$ and that proves the claim. □

On the other hand, there holds

THEOREM 2.5. *Given a presetting $(R, \partial, \{\partial_i\})$ in which we want to realize KP and strict KP. Suppose we have in Psd a deformation L of the form* (2.3) *and a deformation M of the form* (2.5). *Then there holds:*

(a) *Assume that the projections $\{B_i := \pi_{\geqslant 0}(L^i) \mid i \geqslant 1\}$ satisfy the zero curvature relations* (2.12). *Then L is a solution of the KP hierarchy.*
(b) *Similarly, if the projections $\{C_i := \pi_{>0}(\mathcal{M}^i) \mid i \geqslant 1\}$ satisfy the zero curvature relations* (2.13), *then M is a solution of the strict KP hierarchy.*

PROOF. Again we prove the statement for M, that for L is shown in a similar way. As a first step we show that the zero curvature conditions imply that for each fixed $i \geqslant 1$ and all $N \geqslant 1$ the degree in ∂ of $\partial_i(M^N) - [C_i, M^N]$ has an upper bound i. Substitute namely $M^N = C_N + \pi_{\leqslant 0}(M^N)$ in this expression and use the zero curvature relation for $i_1 = i$ and $i_2 = N$. Then we get

$$\begin{aligned}&\partial_i(C_N) + \partial_i(\pi_{\leqslant 0}(M^N)) - [C_i, C_N] - [C_i, \pi_{\leqslant 0}(M^N)] = \\ &\partial_N(C_i) + \partial_i(\pi_{\leqslant 0}(M^N)) - [C_i, \pi_{\leqslant 0}(M^N)]\end{aligned}$$

and this last expression has degree i or less in ∂. So, assume that there is one Lax equation (2.8) that does not hold. Then there is an $i_0 \geqslant 1$ such that

$$\partial_{i_0}(M) - [C_{i_0}, M] = \sum_{i \leqslant k(i_0)} \alpha_i \partial^i, \text{ with } \alpha_{k(i_0)} \neq 0 \text{ and all } \alpha_i \in R.$$

Since both $\partial_{i_0}(M)$ and $[C_{i_0}, M]$ are of degree smaller than or equal to zero in ∂, we know that $k(i_0) \leqslant 0$. Further, we can say that for all $N \geqslant 1$

$$\begin{aligned}\partial_{i_0}(M^N) - [C_{i_0}, M^N] &= \sum_{i=0}^{N-1} M^i(\partial_{i_0}(M) - [C_{i_0}, M])M^{N-i-1} \\ &= \left(\sum_{i=0}^{N-1} M^i(\partial_{i_0}(M) - [C_{i_0}, M])M^{-i}\right) M^{N-1}\end{aligned}$$

Then each term of the summation in the right hand side has a leading term in ∂ equal to $\alpha_{k(i_0)}\partial^{k(i_0)}$. So, the leading term of the total expression becomes

$$N\alpha_{k(i_0)}\partial^{N-1+k(i_0)}$$

for all $N \geqslant 1$ and thus $\partial_{i_0}(M^N) - [C_{i_0}, M^N]$ can only have a bounded degree in ∂ if $\alpha_{k(i_0)} = 0$. This contradicts our assumption that one of the Lax equations (2.8) would not hold and proves the claim in the proposition for M. □

Because of the equivalence between the Lax equations (2.7) for L and the zero curvature relations (2.12) for the $\{B_i \mid i \geqslant 1\}$, we call this last set of equations also the *zero curvature form* of the KP hierarchy. Similarly, the zero curvature relations (2.13) for the $\{C_i \mid i \geqslant 1\}$ are called the *zero curvature form* of the strict KP hierarchy. Also the remaining two reductions possess a zero curvature form. By combining Proposition 2.1 and Theorems 2.4 and 2.5 one obtains

PROPOSITION 2.2. *Let* $\mathbf{L}_n$ *resp.* $\mathbf{M}_n$ *be differential operators of the form* (2.4) *resp.* (2.6) *with* n*-th roots* $L(n)$ *resp.* $M(n)$. *Then there holds:*

(a) $\mathbf{L}_n$ *satisfies the Lax equations of the* n*-KdV hierarchy if and only if the zero curvature relations* (2.12) *hold for the* $\{\mathbf{B}_i := \pi_{\geqslant 0}(L(n)^i)\}$.

(b) $\mathbf{M}_n$ *satisfies the Lax equations of the strict* n*-KdV hierarchy if and only if the zero curvature relations* (2.13) *hold for the* $\{\mathbf{C}_i := \pi_{>0}(M(n)^i)\}$

3. Reduction to a standard setting

Let $(R,\, \partial,\, \{\partial_i\})$ be a given presetting for any of the hierarchies. The goal of this section is to present for each hierarchy two $\mathbb{C}$-subalgebras R_1 and R_2 of R, $R_1 \subset R_2$, such that $\partial|R_j, j = 1,2$, and all the $\{\partial_i|R_j\}$ are derivations of R_j and each set of data $(R_j, \partial|R_j, \{\partial_i|R_j\})$ is a standard setting for the hierarchy under consideration. Moreover, the coefficients of any solution of the relevant hierarchy in $(R,\, \partial,\, \{\partial_i\})$ belong to R_2. To do that we need the zero curvature form of each hierarchy as described in the last section.

First we have a look at the subset R_c of R, where all the $\{\partial_i\}$ mutually commute, i.e.

$$R_c = \{r \in R \mid \partial_{i_1}\partial_{i_2}(r) = \partial_{i_2}\partial_{i_1}(r) \text{ for all } i_1, i_2 \geqslant 1\}.$$

This set has the property

LEMMA 3.1. R_c *is a* $\mathbb{C}$*-subalgebra of* R.

PROOF. As the $\{\partial_i\}$ are $\mathbb{C}$-linear maps, it suffices to show that R_c is closed under multiplication. For each $r_1, r_2 \in R$ we have

$$\partial_{i_1}\partial_{i_2}(r_1r_2) = \partial_{i_1}\partial_{i_2}(r_1)r_2 + \partial_{i_2}(r_1)\partial_{i_1}(r_2) + \partial_{i_1}(r_1)\partial_{i_2}(r_2) + r_1\partial_{i_1}\partial_{i_2}(r_2),$$
$$\partial_{i_2}\partial_{i_1}(r_1r_2) = \partial_{i_2}\partial_{i_1}(r_1)r_2 + \partial_{i_1}(r_1)\partial_{i_2}(r_2) + \partial_{i_2}(r_1)\partial_{i_1}(r_2) + r_1\partial_{i_2}\partial_{i_1}(r_2).$$

Comparing both expressions shows that, if both r_1 and r_2 are in the kernel of the commutator of ∂_{i_1} and ∂_{i_2}, then their product also belongs to this kernel. This proves the claim. □

Now we present the smaller subalgebra R_1 for all four hierarchies. Each R_1 is formed around a single solution of the relevant hierarchy in the presetting $(R,\, \partial,\, \{\partial_i\})$. So, let L denote a single solution of the KP hierarchy, M one of the strict KP hierarchy, $\mathbf{L}_n$ one of the n-KdV hierarchy and $\mathbf{M}_n$ one of the strict n-KdV hierarchy. Let K denote one of these four solutions and let $C(K)$ be the collection of coefficients of K not equal to 1. Then we denote by $R(K)$ the $\mathbb{C}$-subalgebra of R of all polynomial expressions in the $\{\partial^s(k) \mid k \in C(K), s \geqslant 0\}$, which is clearly mapped by ∂ into $R(K)$. We denote the restriction of ∂ to $R(K)$ by ∂_K. The Lax equations of each hierarchy show that every ∂_i maps $R(K)$ into itself and we write $\partial_{i,K}$ for the restriction of ∂_i to $R(K)$. Each $R(K)$ is the announced R_1, for there holds

THEOREM 3.2. *All presettings $(R(K), \partial_K, \{\partial_{i,K}\})$, where $K = L, M, \mathbf{L}_n$ or $\mathbf{M}_n$, are standard settings.*

PROOF. First we show that all four presettings are standard. Since $B_1 = C_1 = \mathbf{B}_1 = \mathbf{C}_1 = \partial$, the Lax equation for $i = 1$ becomes $\partial_1(K) = [\partial, K] = \partial(K)$. Moreover, the derivation ∂_1 commutes with ∂, so that we have $\partial_1(\partial^s(k)) = \partial^{s+1}(k)$ for all $k \in C(K)$ and all $s \geqslant 0$. Hence ∂_K is equal to $\partial_{1,K}$ on $R(K)$. Now we merely have to show that the derivations $\{\partial_{i,K}\}$ commute.

Since the proof in all four cases follows exactly the same lines, we only give the proof for the $\{\partial_{i,M}\}$. The algebra $R(M)$ is the polynomial algebra generated by the $\{\partial^s(m_{j+1}) \mid j \geqslant 1, s \geqslant 0\}$ so by Lemma 3.1 it suffices to show that their actions on these elements commute. Further, all the $\{\partial_{i,M}\}$ commute with ∂_M and that reduces the problem to demonstrating for all $i_1 \geqslant 1$ and $i_2 \geqslant 1$ that

$$\partial_{i_1}\partial_{i_2}(M) - \partial_{i_2}\partial_{i_1}(M) = 0. \tag{3.1}$$

Using the Lax equations for M we get

$$\partial_{i_1}\partial_{i_2}(M) = \partial_{i_1}([C_{i_2}, M]) = [\partial_{i_1}(C_{i_2}), M] + [C_{i_2}, [C_{i_1}, M]]$$

and likewise

$$\partial_{i_2}\partial_{i_1}(M) = [\partial_{i_2}(C_{i_1}), M] + [C_{i_1}, [C_{i_2}, M]].$$

Since $\mathrm{ad}([C_{i_1}, C_{i_2}]) = [\mathrm{ad}(C_{i_1}), \mathrm{ad}(C_{i_2})]$ we see that the left hand side of equation (3.1) is equal to

$$[\partial_{i_1}(C_{i_2}) - \partial_{i_2}(C_{i_1}) - [C_{i_1}, C_{i_2}], M]$$

and, because M is a solution of the strict KP hierarchy, the left component of this commutator is zero by Theorem 2.4. This proves the statement in the theorem. □

Next we present the larger algebras R_2. For each hierarchy, we let K vary over all solutions in the presetting $(R, \partial, \{\partial_i\})$ and we consider in all four cases the complex subalgebra of R of all polynomial expressions in the

$$\bigcup_{\text{solutions } K} \{\partial^s(k) \mid k \in C(K), s \geqslant 0\}.$$

This assigns a so-called R_2-subalgebra in R to each hierarchy: $R_2(I)$ to KP, $R_2(II)$ to strict KP, $R_2(III)$ to n-KdV and $R_2(IV)$ to strict n-KdV. Note that each R_2-subalgebra contains the coefficients of all solutions of the hierarchy it concerns, in the presetting we started with. Since each $\partial_K = \partial_{1,K}$, the restrictions of ∂ and ∂_1 to each of these 4 subalgebras are equal. Because each $\partial_{i,K}$ maps $R(K)$ to $R(K)$, every ∂_i maps each of these 4 R_2-subalgebras to itself. Finally, Theorem 3.2 and the same argument as in the proof of Lemma 3.1 yield that all the $\{\partial_i\}$ mutually commute on the R_2-subalgebra of each hierarchy. We resume the result in a theorem

THEOREM 3.3. *The four R_2-subalgebras of R together with the restrictions to each of them of the derivations ∂ and the $\{\partial_i\}$ form a standard setting for the hierarchy they are linked with.*

Hence, by restricting R and its derivations ∂ and the $\{\partial_i\}$ to the R_2-subalgebra of the hierarchy one does not loose relevance for the solutions of the hierarchy and one ends up with a standard setting.

4. The minimal realizations

In this section we want to discuss minimal realizations of respectively the equations (2.7), (2.8), (2.10) and (2.11) in the sense that there are a minimal number of relations between the coefficients of the potential solutions L, M, $\mathbf{L}_n$ and $\mathbf{M}_n$ and those of their derivatives w.r.t. ∂. Thereto we extend ideas, used in [**4**] for single Lax equations for differential operators related to decomposition (2.2) for $s = -1$, to the present context. We start with four proper complex coefficient algebras $\tilde{R}$, $\tilde{R}_s$, $\tilde{R}(n)$ and $\tilde{R}_s(n)$ that each possess a privileged $\mathbb{C}$-linear derivation $\tilde{\partial}$. Keep in mind that any $\mathbb{C}$-linear derivation Δ of a polynomial ring $\mathbb{C}[x_s, s \in S]$ in any number of variables S, is determined uniquely by prescribing the images $\Delta(x_s)$ of all the $\{x_s\}$ thanks to the derivation property

$$\Delta(fg) = \Delta(f)g + f\Delta(g), \text{ for all } f \text{ and } g \in \mathbb{C}[x_s].$$

Moreover, one can choose the $\Delta(x_s)$ arbitrarily. Consider thereto the following four sets of unknowns

$$\begin{aligned}&\{\tilde{\ell}_{j+1}^{(k)} \mid j \geqslant 1, k \geqslant 0\}, \{\tilde{m}_{j+1}^{(k)} \mid j \geqslant 0, k \geqslant 0\},\\&\{\tilde{\mathbf{l}}_{n-j}^{(k)} \mid 0 \leqslant j \leqslant n-2, k \geqslant 0\} \text{ and } \{\tilde{\mathbf{m}}_{n-j}^{(k)} \mid 1 \leqslant j \leqslant n-1, k \geqslant 0\}\end{aligned}$$

and denote the index set for the (j,k) respectively by $I, I_s, I(n)$ and $I_s(n)$. To these sets we associate the *minimal* coefficient algebras

$$\begin{aligned}&\tilde{R} := \mathbb{C}[\tilde{\ell}_{j+1}^{(k)} \mid (j,k) \in I], \tilde{R}_s := \mathbb{C}[\tilde{m}_{j+1}^{(k)} \mid (j,k) \in I_s],\\&\tilde{R}(n) := \mathbb{C}[\tilde{\mathbf{l}}_{n-j}^{(k)} \mid (j,k) \in I(n)] \text{ and } \tilde{R}_s(n) := \mathbb{C}[\tilde{\mathbf{m}}_{n-j}^{(k)} \mid (j,k) \in I_s(n)].\end{aligned}$$

On each of them we have a $\mathbb{C}$-linear derivation $\tilde{\partial}$ that is defined for all indices (j,k) by respectively

$$\tilde{\partial}(\tilde{\ell}_{j+1}^{(k)}) := \tilde{\ell}_{j+1}^{(k+1)}, \tilde{\partial}(\tilde{m}_{j+1}^{(k)}) := \tilde{m}_{j+1}^{(k+1)}, \tilde{\partial}(\tilde{\mathbf{l}}_{n-j}^{(k)}) := \tilde{\mathbf{l}}_{n-j}^{(k+1)} \text{ and } \tilde{\partial}(\tilde{\mathbf{m}}_{n-j}^{(k)}) := \tilde{\mathbf{m}}_{n-j}^{(k+1)}.$$

It is convenient to put a multiplicative grading on the monomials in the unknown of all four $\mathbb{C}$-algebras, according to the following prescription on their respective building blocks

$$\begin{aligned}&\deg(\tilde{\ell}_{j+1}^{(k)}) = j+1+k, \deg(\tilde{m}_{j+1}^{(k)}) = j+1+k,\\&\deg(\tilde{\mathbf{l}}_{n-j}^{(k)}) = n-j+k \text{ and } \deg(\tilde{\mathbf{m}}_{n-j}^{(k)}) = n-j+k\end{aligned}$$

and, if $\tilde{\mathbf{R}}$ denotes one of these four coefficient algebras, we have a decomposition

$$\tilde{\mathbf{R}} = \oplus_{p \geqslant 0} \tilde{\mathbf{R}}^{(p)}, \text{ with } \tilde{\mathbf{R}}^{(p)} \text{ the span of the monomials of degree } p \text{ in } \tilde{\mathbf{R}}$$

Then the action of $\tilde{\partial}$ on each $\tilde{\mathbf{R}}$ is a $\mathbb{C}$-linear map of order one w.r.t. the grading and thus any $r\tilde{\partial}^n, n \geqslant 0$, with r a homogeneous element of degree p in $\tilde{\mathbf{R}}$ determines a $\mathbb{C}$-linear map from $\tilde{\mathbf{R}}^{(m)}$ to $\tilde{\mathbf{R}}^{(m+p+n)}$. Using this property, it is a straightforward verification that each pair $(\tilde{\mathbf{R}}, \tilde{\partial})$ is a proper starting point in the sense that

LEMMA 4.1. *For each of the four coefficient algebras $\tilde{\mathbf{R}}$ the powers of $\tilde{\partial}$ are $\tilde{\mathbf{R}}$-linear independent in $\tilde{\mathbf{R}}[\tilde{\partial}]$.*

The grading on $\tilde{\mathbf{R}}$ extends to a grading on $\tilde{\mathbf{R}}[\tilde{\partial}]$ by assigning the order $m+p$ to the $\mathbb{C}$-linear maps $r\tilde{\partial}^m, r \in \tilde{\mathbf{R}}^{(p)}$. We will use two homogeneous elements of degree

n in $\tilde{R}(n)[\tilde{\partial}]$ resp. $\tilde{R}_s(n)[\tilde{\partial}]$

$$\tilde{\mathbf{L}}_n = \tilde{\partial}^n + \sum_{j=0}^{n-2} \tilde{\mathbf{l}}_{n-j}^{(0)} \tilde{\partial}^j \text{ resp. } \tilde{\mathbf{M}}_n = \tilde{\partial}^n + \sum_{j=1}^{n-1} \tilde{\mathbf{m}}_{n-j}^{(0)} \tilde{\partial}^j.$$

Likewise, we can call an element P of $\tilde{\mathbf{R}}[\tilde{\partial}, \tilde{\partial}^{-1})$ *homogeneous of degree* m, if P can be written as

$$P = \sum_{i \leqslant N} p_{m-i} \tilde{\partial}^i, \text{ with all } p_{m-i} \in \tilde{\mathbf{R}}^{(m-i)}$$

Examples of homogeneous elements of degree one in $\tilde{R}[\tilde{\partial}, \tilde{\partial}^{-1})$ resp. $\tilde{R}_s[\tilde{\partial}, \tilde{\partial}^{-1})$ are

$$\tilde{L} = \tilde{\partial} + \sum_{j=1}^{\infty} \tilde{\ell}_{j+1}^{(0)} \tilde{\partial}^{-j} \text{ resp. } \tilde{M} = \tilde{\partial} + \sum_{j=0}^{\infty} \tilde{m}_{j+1}^{(0)} \tilde{\partial}^{-j}.$$

The multiplication rules in $\tilde{\mathbf{R}}[\tilde{\partial}, \tilde{\partial}^{-1})$ are such that the product of two homogeneous elements of degrees m_1 resp. m_2 yields a homogeneous element of degree $m_1 + m_2$. A careful inspection of the construction of the n-th root of a homogeneous monic element of degree n in $\tilde{\mathbf{R}}[\tilde{\partial}, \tilde{\partial}^{-1})$ shows that this n-th root is homogeneous of degree one. Hence $\tilde{L}(n) := (\tilde{\mathbf{L}}_n)^{\frac{1}{n}}$ and $\tilde{M}(n) := (\tilde{\mathbf{M}}_n)^{\frac{1}{n}}$ are homogeneous of degree one. All these properties imply that for each $i \geqslant 1$ the projections $\tilde{B}_i := \pi_{\geqslant 0}(\tilde{L}^i)$, $\tilde{\mathbf{B}}_i := \pi_{\geqslant 0}(\tilde{L}(n)^i)$, $\tilde{C}_i := \pi_{>0}(\tilde{M}^i)$ and $\tilde{\mathbf{C}}_i := \pi_{>0}(\tilde{M}(n)^i)$ are homogeneous of degree i.

Next we introduce for all four algebras a set of $\mathbb{C}$-linear derivations $\{\tilde{\partial}_i \mid i \geqslant 1\}$ of $\tilde{\mathbf{R}}$ that all commute with $\tilde{\partial}$ and such that in the presetting $(\tilde{\mathbf{R}}, \tilde{\partial}, \{\tilde{\partial}_i\})$ that concerns them, the elements $\tilde{L}$, $\tilde{\mathbf{L}}_n$, $\tilde{M}$ and $\tilde{\mathbf{M}}_n$ are a solution of respectively the Lax equations (2.7), (2.10), (2.8) and (2.11). Since in each case the set $\{\tilde{\partial}_i\}$ has to commute with $\tilde{\partial}$, there has to hold in each $\tilde{\mathbf{R}}$ for all $i \geqslant 1$, all $k \geqslant 0$ and all relevant j that

$$\tilde{\partial}_i(\tilde{\ell}_{j+1}^{(k)}) = \tilde{\partial}^k(\tilde{\partial}_i(\tilde{\ell}_{j+1}^{(0)})), \tilde{\partial}_i(\tilde{m}_{j+1}^{(k)}) = \tilde{\partial}^k(\tilde{\partial}_i(\tilde{m}_{j+1}^{(0)})),$$
$$\tilde{\partial}_i(\tilde{\mathbf{l}}_{n-j}^{(k)}) = \tilde{\partial}^k(\tilde{\partial}_i(\tilde{\mathbf{l}}_{n-j}^{(0)})) \text{ and } \tilde{\partial}_i(\tilde{\mathbf{m}}_{n-j}^{(k)}) = \tilde{\partial}^k(\tilde{\partial}_i(\tilde{\mathbf{m}}_{n-j}^{(0)})).$$

From these relations you see that in $\tilde{R}$ one merely has to prescribe the action of the derivations $\{\tilde{\partial}_i\}$ on the set of coefficients of the (pseudo) differential operator concerned, i.e. $\tilde{L}$, $\tilde{M}$, $\tilde{\mathbf{L}}_n$ or $\tilde{\mathbf{M}}_n$, and that choice is determined by the Lax equations they have to satisfy, as we will see. Note, first of all, that all four differential operators $\tilde{B}_1$, $\tilde{C}_1$, $\tilde{\mathbf{B}}_1$ and $\tilde{C}_1$ are equal to $\tilde{\partial}$. Hence, the Lax equation for $i = 1$ in (2.7), (2.8), (2.10) and (2.11) becomes respectively

$$\tilde{\partial}_1(\tilde{L}) = \tilde{\partial}(\tilde{L}), \tilde{\partial}_1(\tilde{M}) = \tilde{\partial}(\tilde{M}), \tilde{\partial}_1(\tilde{\mathbf{L}}_n) = \tilde{\partial}(\tilde{\mathbf{L}}_n), \text{ and } \tilde{\partial}_1(\tilde{\mathbf{M}}_n) = \tilde{\partial}(\tilde{\mathbf{M}}_n).$$

Therefore we choose in all four cases $\tilde{\partial}_1 := \tilde{\partial}$. We handle each case now separately. For $\tilde{L}$, the operator $\tilde{B}_i - \tilde{L}^i$ is a pseudo differential operator of strict negative degree in $\tilde{\partial}$ and its commutator with $\tilde{L}$ yields also an operator of strict negative degree in $\tilde{\partial}$. Therefore we can use this commutator to define each $\tilde{\partial}_i$ by

$$\tilde{\partial}_i(\tilde{L}) := [\tilde{B}_i - \tilde{L}^i, \tilde{L}] = [\tilde{B}_i, \tilde{L}] \tag{4.1}$$

In this way the operator $\tilde{L}$ becomes, practically by definition, a solution of the KP hierarchy in the presetting $(\tilde{R}, \tilde{\partial}, \{\tilde{\partial}_i\})$, which is in fact a standard setting by

Theorem 3.2, since $\tilde{R} = \tilde{R}(\tilde{L})$. The operator $\tilde{L}$ together with the setting $(\tilde{R}, \{\tilde{\partial}_i\})$ we call a *minimal realization of the KP hierarchy.*

In the case of $\tilde{\mathbf{L}}_n$, we use the operator $\tilde{\mathbf{B}}_i - \tilde{L}(n)^i$ that is again a pseudo differential operator of strict negative degree in $\tilde{\partial}$ and its commutator with $\tilde{\mathbf{L}}_n$ is of degree $n-2$ or lower in $\tilde{\partial}$. Moreover, this commutator belongs to $\mathrm{Psd}_{\geqslant 0}$ as it equals $[\tilde{\mathbf{B}}_i, \tilde{\mathbf{L}}_n]$. So, for the n-KdV hierarchy this is the proper candidate to define each $\tilde{\partial}_i$ by

$$\tilde{\partial}_i(\tilde{\mathbf{L}}_n) := [\tilde{\mathbf{B}}_i - \tilde{L}(n)^i, \tilde{\mathbf{L}}_n] = [\tilde{\mathbf{B}}_i, \tilde{\mathbf{L}}_n] \tag{4.2}$$

In particular, the differential operator $\tilde{\mathbf{L}}_n$ is a solution of the n-KdV hierarchy in the presetting $(\tilde{R}(n), \tilde{\partial}, \{\tilde{\partial}_i\})$, which is also a standard setting by the same argument as before. The operator $\tilde{\mathbf{L}}_n$ together with the setting $(\tilde{R}(n), \{\tilde{\partial}_i\})$ we call a *minimal realization of the n-KdV hierarchy.*

At the operator $\tilde{M}$ we consider the pseudo differential operator $\tilde{C}_i - \tilde{M}^i$, which has degree zero or less in $\tilde{\partial}$. Its commutator with $\tilde{M}$ yields also an operator in $\mathrm{Psd}_{\leqslant 0}$ and that agrees with the range of possible degrees in any $\tilde{\partial}_i(\tilde{M})$. Therefore we use this commutator to define each $\tilde{\partial}_i$ by

$$\tilde{\partial}_i(\tilde{M}) := [\tilde{C}_i - \tilde{M}^i, \tilde{M}] = [\tilde{C}_i, \tilde{M}]. \tag{4.3}$$

Then the operator $\tilde{M}$ is a solution of the strict KP hierarchy in the presetting $(\tilde{R}_s, \tilde{\partial}, \{\tilde{\partial}_i\})$, which is even a standard setting thanks to Theorem 3.2. The operator $\tilde{M}$ together with the setting $(\tilde{R}_s, \{\tilde{\partial}_i\})$ we call a *minimal realization of the strict KP hierarchy.*

Finally, we treat the case $\tilde{\mathbf{M}}_n$. Here we use the operator $\tilde{\mathbf{C}}_i - \tilde{M}(n)^i$ that is a pseudo differential operator of degree zero or less in $\tilde{\partial}$ and its commutator with $\tilde{\mathbf{M}}_n$ is of degree $n-1$ or lower in $\tilde{\partial}$. Moreover, this commutator belongs to $\mathrm{Psd}_{>0}$ as it equals $[\tilde{\mathbf{C}}_i, \tilde{\mathbf{M}}_n]$. So, this is a good candidate to define for the strict n-KdV hierarchy all $\tilde{\partial}_i$ by

$$\tilde{\partial}_i(\tilde{\mathbf{M}}_n) := [\tilde{\mathbf{C}}_i - \tilde{M}(n)^i, \tilde{\mathbf{M}}_n] = [\tilde{\mathbf{C}}_i, \tilde{\mathbf{M}}_n]. \tag{4.4}$$

Hence the differential operator $\tilde{\mathbf{M}}_n$ is a solution of the strict n-KdV hierarchy in the presetting $(\tilde{R}_s(n), \tilde{\partial}, \{\tilde{\partial}_i\})$, which is also a standard setting by the same argument as before. The operator $\tilde{\mathbf{M}}_n$ together with the setting $(\tilde{R}_s(n), \{\tilde{\partial}_i\})$ we call a *minimal realization of the strict n-KdV hierarchy.*

Next we describe in an algebraic way how other realizations of solutions of the four hierarchies are related to these minimal realizations. Consider any standard setting $(R, \{\partial_i\})$ for each of this hierarchies and in $R[\partial, \partial^{-1})$ a set $\mathcal{K} = \{L, M, \mathbf{L}_n, \mathbf{M}_n\}$ of potential solutions of each hierarchy, i.e. elements of respectively the form (2.3), (2.5), (2.4) or (2.6). For each $K \in \mathcal{K}$, let $\tilde{K}$ denote the corresponding element in the set $\tilde{\mathcal{K}} = \{\tilde{L}, \tilde{M}, \tilde{\mathbf{L}}_n, \tilde{\mathbf{M}}_n\}$. Now there is a $\mathbb{C}$-algebra morphism i_K from the minimal coefficient algebra of $\tilde{K}$ to R. These four morphisms are determined uniquely by the following prescriptions for all relevant j and k:

$$i_L(\tilde{\ell}_{j+1}^{(k)}) := \partial^k(\ell_{j+1}), i_M(\tilde{m}_{j+1}^{(k)}) := \partial^k(m_{j+1}),$$
$$i_{\mathbf{L}_n}(\tilde{\mathbf{l}}_{n-j}^{(k)}) := \partial^k(\mathbf{l}_{n-j}), i_{\mathbf{M}_n}(\tilde{\mathbf{m}}_{n-j}^{(k)}) := \partial^k(\mathbf{m}_{n-j})$$

By definition all these i_K satisfy

$$i_K \circ \tilde{\partial} = \partial \circ i_K. \tag{4.5}$$

Let $\tilde{\mathbf{R}}$ be the minimal coefficient algebra corresponding to $\tilde{K}$. Define the extension of i_K to $\tilde{\mathbf{R}}[\tilde{\partial}, \tilde{\partial}^{-1})$ by

$$i_K(\sum_{i=-\infty}^{N} \mathbf{r}_i \tilde{\partial}^i) = \sum_{i=-\infty}^{N} i_K(\mathbf{r}_i)\partial^i.$$

By property (4.5) the map i_K is a $\mathbb{C}$-algebra morphism from the pseudo differential operators $\tilde{\mathbf{R}}[\tilde{\partial}, \tilde{\partial}^{-1})$ to $R[\partial, \partial^{-1})$. In particular, these maps satisfy

$$i_L(\tilde{L}) = L, i_L(\tilde{B}_i) = \pi_{\geqslant 0}(L^i) = B_i; i_M(\tilde{M}) = M, i_M(\tilde{C}_i) = \pi_{>0}(M^i) = C_i;$$

$$i_{\mathbf{L}_n}(\tilde{L}(n)) = \mathbf{L}_n^{\frac{1}{n}}, i_{\mathbf{L}_n}(\tilde{\mathbf{B}}_i) = \pi_{\geqslant 0}(\mathbf{L}_n^{\frac{i}{n}}); i_{\mathbf{M}_n}(\tilde{M}(n)) = \mathbf{M}_n^{\frac{1}{n}}, i_{\mathbf{M}_n}(\tilde{\mathbf{C}}_i) = \pi_{>0}(\mathbf{M}_n^{\frac{i}{n}}).$$

Assume now for example that L is a solution of the Lax equations of the KP hierarchy, then we have for all $i \geqslant 1$ that

$$\begin{aligned}\partial_i(L) = \partial_i \circ i_L(\tilde{L}) = [B_i, L] = [i_L(\tilde{B}_i), i_L(\tilde{L})] = \\ = i_L([\tilde{B}_i, \tilde{L}]) = i_L \circ \tilde{\partial}_i(\tilde{L})\end{aligned}$$

Thus the $\mathbb{C}$-linear maps $\partial_i \circ i_L$ and $i_L \circ \tilde{\partial}_i$ are equal on the coefficients of $\tilde{L}$, but, because of relation (4.5) and the fact that the derivations $\{\partial_i\}$ commute with ∂, we get on $\tilde{R}[\tilde{\partial}, \tilde{\partial}^{-1})$ the compatibilities

$$\partial_i \circ i_L = i_L \circ \tilde{\partial}_i, \text{ for all } i \geqslant 1.$$

A similar reasoning applies, if one assumes M is a solution of strict KP, $\mathbf{L}_n$ of n-KdV or $\mathbf{M}_n$ of strict n-KdV and thus we get for all $K \in \mathcal{K}$ that, as soon as K is a solution of the same hierarchy as $\tilde{K}$, then the following compatibilities hold:

$$\partial_i \circ i_K = i_K \circ \tilde{\partial}_i, \text{ for all } i \geqslant 1. \tag{4.6}$$

On the other hand, if the compatibilities (4.6) hold for the map i_K, then one applies these identities to $\tilde{K}$ and, as i_K is a $\mathbb{C}$-algebra morphism, we obtain the required Lax equations for K. So we may conclude

PROPOSITION 4.1. *The relations* (4.6) *for the map i_K are equivalent to K being a solution in the setting $(R, \{\partial_i\})$ of the hierarchy for which $\tilde{K}$ is a minimal realization.*

5. Scaling invariance

Transformations, where the dependent and independent variables change by a constant nonzero factor, are called *scaling transformations* and the four hierarchies discussed here possess such transformations that map solutions into new solutions. These transformations are generalizations of the scaling invariance of the Khadomtsev-Petviashvilii (KP) equation from plasma physics ([**5**]). This nonlinear equation for a function $u = u(x, y, t)$ reads

$$\frac{3}{4}u_{yy} = (u_t - \frac{1}{4}u_{xxx} - \frac{3}{2}uu_x)_x.$$

Consider now the following transformation of the dependent and independent variables

$$\hat{x} = \alpha x, \hat{y} = \alpha^2 y, \hat{t} = \alpha^3 t, u = \alpha^2 \hat{u}, \text{ with } \alpha \neq 0.$$

A direct computation shows that the KP equation in these new coordinates becomes

$$\frac{3}{4}\alpha^2(\alpha^2)^2\hat{u}_{\hat{y}\hat{y}} = \alpha(\alpha^2\alpha^3\hat{u}_{\hat{t}} - \frac{1}{4}\alpha^3\alpha^2\hat{u}_{\hat{x}\hat{x}\hat{x}} - \frac{3}{2}\alpha(\alpha^2)^2\hat{u}\hat{u}_{\hat{x}})_{\hat{x}}$$

and this shows that $\hat{u}$ satisfies the KP equation w.r.t. $(\hat{x}, \hat{y}, \hat{t})$. In this section we will show that the Lax equations of the four hierarchies possess a similar scaling invariance.

The independent and dependent variables relevant for the four hierarchies considered in this paper are the flow parameters of the derivations $\{\partial_i\}$ from the standard setting $(R, \{\partial_i\})$ and the nontrivial coefficients of a solution of the hierarchy in question. Nonzero scalar multiplication of the flow parameters corresponds to nonzero scalar multiplication of the derivations $\{\partial_i\}$ of the setting, but one can not expect that nonzero scalar multiplication of the coefficients of a solution corresponds to a transformation of the whole coefficient algebra R. Therefore we go over to the minimal realizations of the four hierarchies and look at these transformations in that context. For, they do possess that property. We start with the minimal realizations of KP and strict KP. So, for the minimal realization of the KP hierarchy we consider general scaling transformations of the form

$$\tilde{\partial} = \alpha_1\hat{\tilde{\partial}}, \tilde{\partial}_i = \alpha_i\hat{\tilde{\partial}}_i, \tilde{\ell}^{(0)}_{j+1} = \beta_j\hat{\tilde{\ell}}^{(0)}_{j+1}, \tag{5.1}$$

with $\{\alpha_i \in \mathbb{C}^* \mid i \geqslant 1\}$ and $\{\beta_j \in \mathbb{C}^* \mid j \geqslant 1\}$. We proceed similarly with the minimal realization of the strict KP hierarchy, taking the same scalar multiplication of the derivations, and we look at the scaling transformations

$$\tilde{\partial} = \alpha_1\hat{\tilde{\partial}}, \tilde{\partial}_i = \alpha_i\hat{\tilde{\partial}}_i, \tilde{m}^{(0)}_{j+1} = \gamma_j\hat{\tilde{m}}^{(0)}_{j+1}, \tag{5.2}$$

with $\{\gamma_j \in \mathbb{C}^* \mid j \geqslant 0\}$. Note that the data $(\tilde{R}, \{\hat{\tilde{\partial}}_i\})$ and $(\tilde{R}_s, \{\hat{\tilde{\partial}}_i\})$ form again standard settings. In both cases we introduce a notation for the action of the new derivation $\hat{\tilde{\partial}}$ on the new dependent variables

$$\hat{\tilde{\ell}}^{(k)}_{j+1} := (\hat{\tilde{\partial}})^k(\hat{\tilde{\ell}}^{(0)}_{j+1}) \text{ and } \hat{\tilde{m}}^{(k)}_{j+1} := (\hat{\tilde{\partial}})^k(\hat{\tilde{m}}^{(0)}_{j+1})$$

Then the new dependent variables relate to the old ones by

$$\tilde{\ell}^{(k)}_{j+1} = \alpha_1^k\beta_j\hat{\tilde{\ell}}^{(k)}_{j+1} \text{ resp. } \tilde{m}^{(k)}_{j+1} = \alpha_1^k\gamma_j\hat{\tilde{m}}^{(k)}_{j+1}$$

and these expressions define identifications between respectively the algebras $\tilde{R}$ and $\mathbb{C}[\hat{\tilde{\ell}}^{(k)}_{j+1}]$ and the algebras $\tilde{R}_s$ and $\mathbb{C}[\hat{\tilde{m}}^{(k)}_{j+1}]$. Combining both isomorphisms with the substitution $\tilde{\partial} = \alpha_1\hat{\tilde{\partial}}$ from (5.1) and (5.2) yields isomorphisms between respectively $\tilde{R}[\tilde{\partial}, \tilde{\partial}^{-1})$ and $\mathbb{C}[\hat{\tilde{\ell}}^{(k)}_{j+1}][\hat{\tilde{\partial}}, \hat{\tilde{\partial}}^{-1})$ and the algebras $\tilde{R}_s[\tilde{\partial}, \tilde{\partial}^{-1})$ and $\mathbb{C}[\hat{\tilde{m}}^{(k)}_{j+1}][\hat{\tilde{\partial}}, \hat{\tilde{\partial}}^{-1})$. The element $\tilde{L}$ expresses as follows in $\tilde{R}[\hat{\tilde{\partial}}, \hat{\tilde{\partial}}^{-1})$

$$\tilde{L} = \alpha_1(\hat{\tilde{\partial}}) + \sum_{j=1}^{\infty}\beta_j\alpha_1^{-j}\hat{\tilde{\ell}}_{j+1}(\hat{\tilde{\partial}})^{-j}.$$

Hence, if we take from now on $\beta_j = \alpha_1^{j+1}$ for all $j \geqslant 1$, then $\tilde{L} = \alpha_1\hat{\tilde{L}}$ with $\hat{\tilde{L}} = (\hat{\tilde{\partial}}) + \sum_{j=1}^{\infty}\hat{\tilde{\ell}}_{j+1}(\hat{\tilde{\partial}})^{j}$ of the form (2.3). So, we get $\tilde{B}_i = \alpha_1^i\pi_{\geqslant 0}(\hat{\tilde{L}}^i) =: \alpha_1^i\hat{\tilde{B}}_i$ for all $i \geqslant 1$. Now, under the present scaling transformation, the Lax equations for $\tilde{L}$ become

$$\tilde{\partial}_i(\tilde{L}) = \alpha_i\alpha_1\hat{\tilde{\partial}}_i(\hat{\tilde{L}}) = [\tilde{B}_i, \tilde{L}] = \alpha_1^{1+i}[\hat{\tilde{B}}_i, \hat{\tilde{L}}]. \tag{5.3}$$

If we choose, moreover, all $\alpha_i = \alpha_1^i$, then the equations (5.3) show that $\hat{\tilde{L}}$ is a solution of the KP hierarchy in the standard setting $(\tilde{R}, \hat{\tilde{\partial}}_i)$. Similarly, we express the element $\tilde{M}$ in $\tilde{R}_s[\hat{\tilde{\partial}}, \hat{\tilde{\partial}}^{-1})$

$$\tilde{M} = \alpha_1(\hat{\tilde{\partial}}) + \sum_{j=0}^{\infty} \gamma_j\alpha_1^{-j}\hat{\tilde{m}}_{j+1}(\hat{\tilde{\partial}})^{-j}.$$

Hence, if we take from now on $\gamma_j = \alpha_1^{j+1}$ for all $j \geqslant 0$, then $\tilde{M} = \alpha_1\hat{\tilde{M}}$ with $\hat{\tilde{M}} = (\hat{\tilde{\partial}}) + \sum_{j=0}^{\infty}\hat{\tilde{m}}_{j+1}(\hat{\tilde{\partial}})^j$ in $\tilde{R}_s[\hat{\tilde{\partial}}, \hat{\tilde{\partial}}^{-1})$ of the form (2.5). In particular, there holds $\tilde{C}_i = \alpha_1^i\pi_{>0}(\hat{\tilde{M}}^i) =: \alpha_1^i\hat{\tilde{C}}_i$ for all $i \geqslant 1$. Thus the present scaling transformation transforms the Lax equations for $\tilde{M}$ into

$$\tilde{\partial}_i(\tilde{M}) = \alpha_i\alpha_1\hat{\tilde{\partial}}_i(\hat{\tilde{M}}) = [\tilde{C}_i, \tilde{M}] = \alpha_1^{1+i}[\hat{\tilde{C}}_i, \hat{\tilde{M}}]. \tag{5.4}$$

If we choose, again, all $\alpha_i = \alpha_1^i$, then the equations (5.4) show that $\hat{\tilde{M}}$ is a solution of the strict KP hierarchy in the standard setting $(\tilde{R}_s, \hat{\tilde{\partial}}_i)$.

The results found in (5.3) and (5.4) directly translate into relations in the pseudo differential operators for more general standard settings by applying respectively the maps i_L and i_M from Section 4. Here L is a solution of the KP hierarchy and M is a solution of the strict KP hierarchy, both in a standard setting $(R, \{\partial_i\})$. So, for L we consider now the scaling transformations

$$\partial = \alpha_1\hat{\partial}, \partial_i = \alpha_1^i\hat{\partial}_i, \ell_{j+1} = \alpha_1^{j+1}\hat{\ell}_{j+1}, j \geqslant 1, \tag{5.5}$$

and for M

$$\partial = \alpha_1\hat{\partial}, \partial_i = \alpha_1^i\hat{\partial}_i, m_{j+1} = \alpha_1^{j+1}\hat{m}_{j+1}, j \geqslant 0. \tag{5.6}$$

Applying the maps i_L and i_M to the identities in (5.3) and (5.4) results in the following scaling invariance for solutions of the KP hierarchy and its strict version:

THEOREM 5.1. *Let L resp. M be solutions in the standard setting $(R, \{\partial_i\})$ of the KP hierarchy resp. its strict version and let α_1 be any element in $\mathbb{C}^*$. Then there holds:*

(a) *Consider the scaling transformation (5.5) for L. Substitution of this transformation into L yields an $\hat{L} = \alpha_1^{-1}L$ in $R[\hat{\partial}, \hat{\partial}^{-1})$ that is a solution of the KP hierarchy in the standard setting $(R, \{\hat{\partial}_i\})$.*

(b) *Consider the scaling transformation (5.6) for M. The substitution of this transformation into M yields an $\hat{M} = \alpha_1^{-1}M$ in $R[\hat{\partial}, \hat{\partial}^{-1})$ that is a solution of the strict KP hierarchy in the standard setting $(R, \{\hat{\partial}_i\})$.*

Next we consider the minimal realizations of the n-KdV hierarchy and its strict version. At the first hierarchy we consider general scaling transformations of the form

$$\tilde{\partial} = \alpha_1\hat{\tilde{\partial}}, \tilde{\partial}_i = \alpha_i\hat{\tilde{\partial}}_i, \tilde{\mathbf{1}}_{n-j}^{(0)} = \eta_j\hat{\tilde{\mathbf{1}}}_{n-j}^{(0)}, \tag{5.7}$$

with $\{\alpha_i \in \mathbb{C}^* \mid i \geqslant 1\}$ and $\{\eta_j \in \mathbb{C}^* \mid 0 \leqslant j \leqslant n-2\}$. At the strict version we take the scaling transformations of the form

$$\tilde{\partial} = \alpha_1 \hat{\tilde{\partial}}, \tilde{\partial}_i = \alpha_i \hat{\tilde{\partial}}_i, \tilde{\mathbf{m}}_{n-j}^{(0)} = \varepsilon_j \hat{\tilde{\mathbf{m}}}_{n-j}^{(0)}, \tag{5.8}$$

with the $\{\alpha_i\}$ as in (5.7) and the $\{\varepsilon_j \mid 1 \leqslant j \leqslant n-1\}$ all in $\mathbb{C}^*$. The data $(\tilde{R}(n), \{\hat{\tilde{\partial}}_i\})$ resp. $(\tilde{R}_s(n), \{\hat{\tilde{\partial}}_i\})$ form again a standard setting for the n-KdV hierarchy resp. its strict version. Again we introduce for both hierarchies a convenient notation for the action of the derivation $\hat{\tilde{\partial}}$ on the new dependent variables

$$\hat{\tilde{\mathbf{l}}}_{n-j}^{(k)} := (\hat{\tilde{\partial}})^k (\hat{\tilde{\mathbf{l}}}_{n-j}^{(0)}) \text{ and } \hat{\tilde{\mathbf{m}}}_{n-j}^{(k)} := (\hat{\tilde{\partial}})^k (\hat{\tilde{\mathbf{m}}}_{n-j}^{(0)}).$$

The relation between the new dependent variables and the old ones is given by

$$\tilde{\mathbf{l}}_{n-j}^{(k)} = \alpha_1^k \eta_j \hat{\tilde{\mathbf{l}}}_{n-j}^{(k)} \text{ resp. } \tilde{\mathbf{m}}_{n-j}^{(k)} = \alpha_1^k \varepsilon_j \hat{\tilde{\mathbf{m}}}_{n-j}^{(k)},$$

which determine identifications of respectively the algebras $\tilde{R}(n)$ and $\mathbb{C}[\hat{\tilde{\mathbf{l}}}_{n-j}^{(k)}]$ and the algebras $\tilde{R}_s(n)$ and $\mathbb{C}[\hat{\tilde{\mathbf{m}}}_{n-j}^{(k)}]$. Next we combine these isomorphisms with the substitution $\tilde{\partial} = \alpha_1 \hat{\tilde{\partial}}$ from (5.7) resp. (5.8) to get isomorphisms respectively between the algebras $\tilde{R}(n)[\tilde{\partial}]$ and $\tilde{R}(n)[\hat{\tilde{\partial}}]$ and the algebras $\tilde{R}_s(n)[\tilde{\partial}]$ and $\tilde{R}_s(n)[\hat{\tilde{\partial}}]$. The element $\tilde{\mathbf{L}}_n$ expresses as follows in $\tilde{R}(n)[\hat{\tilde{\partial}}]$

$$\tilde{\mathbf{L}}_n = \alpha_1^n \hat{\tilde{\partial}}^n + \sum_{j=0}^{n-2} \eta_j \alpha_1^j \hat{\tilde{\mathbf{l}}}_{n-j}^{(0)} (\hat{\tilde{\partial}})^j.$$

Hence, if we take from now on $\eta_j = \alpha_1^{n-j}$ for all $j \geqslant 1$, then $\tilde{\mathbf{L}}_n = \alpha_1^n \hat{\tilde{\mathbf{L}}}_n$ with the differential operator $\hat{\tilde{\mathbf{L}}}_n = (\hat{\tilde{\partial}})^n + \sum_{j=0}^{n-2} \hat{\tilde{\mathbf{l}}}_{n-j}^{(0)} (\hat{\tilde{\partial}})^j$ of the form (2.3). So, we get $\tilde{\mathbf{B}}_i = \alpha_1^i \hat{\tilde{\mathbf{B}}}_i = \alpha_1^i \pi_{\geqslant 0}(\hat{\tilde{L}}(n)^i)$ for all $i \geqslant 1$. Now, under the present scaling transformation, the Lax equations for $\tilde{\mathbf{L}}_n$ become

$$\tilde{\partial}_i(\tilde{\mathbf{L}}_n) = \alpha_i \alpha_1^n \hat{\tilde{\partial}}_i(\hat{\tilde{\mathbf{L}}}_n) = [\tilde{B}_i, \tilde{\mathbf{L}}_n] = \alpha_1^{i+n} [\hat{\tilde{B}}_i, \hat{\tilde{\mathbf{L}}}_n]. \tag{5.9}$$

If we choose, moreover, all $\alpha_i = \alpha_1^i$, then the equations (5.9) show that $\hat{\tilde{\mathbf{L}}}_n$ is a solution of the n-KdV hierarchy in the standard setting $(\tilde{R}(n), \hat{\tilde{\partial}}_i)$. In the same way, we get for $\tilde{\mathbf{M}}_n$ in $\tilde{R}_s(n)[\hat{\tilde{\partial}}])$ the expression

$$\tilde{\mathbf{M}}_n = \alpha_1^n (\hat{\tilde{\partial}})^n + \sum_{j=1}^{n-1} \varepsilon_j \alpha_1^j \hat{\tilde{\mathbf{m}}}_{n-j} (\hat{\tilde{\partial}})^j.$$

Hence, if we take from now on $\varepsilon_j = \alpha_1^{n-j}$ for all $j, 1 \leqslant j \leqslant n-1$, then $\tilde{\mathbf{M}}_n = \alpha_1^n \hat{\tilde{\mathbf{M}}}_n$ with $\hat{\tilde{\mathbf{M}}}_n = (\hat{\tilde{\partial}})^n + \sum_{j=1}^{n-1} \hat{\tilde{\mathbf{m}}}_{n-j} (\hat{\tilde{\partial}})^j$ of the form (2.6). Since both $\tilde{M}(n)$ and $\hat{\tilde{M}}(n)$ are homogeneous of degree one, it follows that $\tilde{M}(n) = \alpha_1 \hat{\tilde{M}}(n) = \alpha_1 (\hat{\tilde{\mathbf{M}}}_n)^{\frac{1}{n}}$ and $\tilde{C}_i = \alpha_1^i \pi_{>0}(\hat{\tilde{M}}(n)^i) =: \alpha_1^i \hat{\tilde{C}}_i$ for all $i \geqslant 1$. Hence, under the present scaling transformation the Lax equations for $\tilde{\mathbf{M}}_n$ become

$$\tilde{\partial}_i(\tilde{\mathbf{M}}_n) = \alpha_i \alpha_1^n \hat{\tilde{\partial}}_i(\hat{\tilde{\mathbf{M}}}_n) = [\tilde{C}_i, \tilde{\mathbf{M}}_n] = \alpha_1^{i+n} [\hat{\tilde{C}}_i, \hat{\tilde{\mathbf{M}}}_n]. \tag{5.10}$$

If we choose again all $\alpha_i = \alpha_1^i$, then the equations (5.10) show that $\hat{\tilde{\mathbf{M}}}_n$ is a solution of the strict n-KdV hierarchy in the setting $(\tilde{R}_s(n), \hat{\tilde{\partial}}_i)$.

The results found in (5.9) and (5.10) translate again into relations for differential operators in more general standard settings by applying respectively the maps $i_{\mathbf{L}_n}$ and $i_{\mathbf{M}_n}$ from Section 4. Here $\mathbf{L}_n$ is a solution of the n-KdV hierarchy and $\mathbf{M}_n$ one of the strict n-KdV hierarchy, both in the standard setting $(R, \{\partial_i\})$. So, for $\mathbf{L}_n$ we consider now the scaling transformations

$$\partial = \alpha_1 \hat{\partial}, \partial_i = \alpha_1^i \hat{\partial}_i, \mathbf{l}_{n-j} = \alpha_1^{n-j} \hat{\mathbf{l}}_{n-j}, 0 \leqslant j \leqslant n-2 \tag{5.11}$$

and for $\mathbf{M}_n$

$$\partial = \alpha_1 \hat{\partial}, \partial_i = \alpha_1^i \hat{\partial}_i, \mathbf{m}_{n-j} = \alpha_1^{n-j} \hat{\mathbf{m}}_{n-j}, 1 \leqslant j \leqslant n-1. \tag{5.12}$$

Applying the maps $i_{\mathbf{L}_n}$ and $i_{\mathbf{M}_n}$ to the relations found in (5.9) and (5.10) yields the following scaling invariance for solutions of the n-KdV hierarchy and its strict version:

THEOREM 5.2. *Let $\mathbf{L}_n$ resp. $\mathbf{M}_n$ be solutions of the n-KdV hierarchy resp. its strict version in the standard setting $(R, \{\partial_i\})$ and let α_1 be any element in $\mathbb{C}^*$. Then there holds:*

(a) *Consider the scaling transformation (5.11) for the differential operator $\mathbf{L}_n$. Then substitution of this transformation into $\mathbf{L}_n$ yields an $\hat{\mathbf{L}}_n = \alpha_1^{-n}\mathbf{L}_n$ in $R[\hat{\partial}]$ that is a solution of the n-KdV hierarchy in the setting $(R, \{\hat{\partial}_i\})$.*

(b) *Consider the scaling transformation (5.12) for the differential operator $\mathbf{M}_n$. Then substitution of this transformation into $\mathbf{M}_n$ yields an $\hat{\mathbf{M}}_n = \alpha_1^{-n}\mathbf{M}_n$ in $R[\hat{\partial}]\hat{\partial}$ that is a solution of the strict n-KdV hierarchy in the setting $(R, \{\hat{\partial}_i\})$.*

References

[1] Graeme Segal and George Wilson, *Loop groups and equations of KdV type*, Inst. Hautes Études Sci. Publ. Math. **61** (1985), 5–65. MR783348

[2] G. F. Helminck, A. G. Helminck, and E. A. Panasenko, *Integrable deformations in the algebra of pseudodifferential operators from a Lie algebraic perspective*, Theoret. and Math. Phys. **174** (2013), no. 1, 134–153, DOI 10.1007/s11232-013-0011-7. Russian version appears in Teoret. Mat. Fiz. **174** (2013), no. 1, 154–176. MR3172958

[3] G. F. Helminck and E. A. Panasenko, *Reductions of the strict KP hierarchy* (Russian, with Russian summary), Teoret. Mat. Fiz. **205** (2020), no. 2, 190–207, DOI 10.4213/tmf9916; English transl., Theoret. and Math. Phys. **205** (2020), no. 2, 1411–1425. MR4169736

[4] I. M. Gel′fand and L. A. Dikiĭ, *Fractional powers of operators, and Hamiltonian systems* (Russian), Funkcional. Anal. i Priložen. **10** (1976), no. 4, 13–29. MR0433508

[5] B.B. Khadomtsev, V.I. Petviashvilii, *On the stability of solitary waves in weakly dispersing media.* Sov. Phys. Doklady **15**, 539–541.

KORTEWEG-DE VRIES INSTITUTE FOR MATHEMATICS, UNIVERSITY OF AMSTERDAM, P.O. BOX 94248, 1090 GE AMSTERDAM, THE NETHERLANDS
Email address: g.f.helminck@uva.nl

DERZHAVIN TAMBOV STATE UNIVERSITY, INTERNATSIONALNAYA 33, 392622 TAMBOV, RUSSIA
Email address: panlena_t@mail.ru

Contemporary Mathematics
Volume **788**, 2023
https://doi.org/10.1090/conm/788/15824

Minimal coupling of gravitational and electromagnetic fields in General Relativity

Josef Janyška and Marco Modugno

Devoted to Alexandre Vinogradov

ABSTRACT. In a general relativistic spacetime $\boldsymbol{E}$, we consider the *gravitational field* and the *electromagnetic field*, which are represented, respectively, by the Levi–Civita connection and the scaled 2–form

$$K^{\natural} : T\boldsymbol{E} \to T^*\boldsymbol{E} \otimes TT\boldsymbol{E} \qquad \text{and} \qquad F : \boldsymbol{E} \to (\mathbb{L}^{1/2} \otimes \mathbb{M}^{1/2}) \otimes \Lambda^2 T^*\boldsymbol{E}.$$

Then, with reference to a *charged particle*, with mass and charge (m, q), we obtain a *minimal coupling* of the gravitational connection $K^{\natural}$ with the electromagnetic field F, so yielding the *joined (non linear) connection*

$$K := K^{\natural} + K^{\mathfrak{e}} : \mathbb{T}^* \otimes T\boldsymbol{E} \to T^*\boldsymbol{E} \otimes T(\mathbb{T}^* \otimes T\boldsymbol{E}),$$

where we have set

$$K^{\mathfrak{e}} := -\frac{q}{m} \mathrm{T} \widehat{F} : \mathbb{T}^* \otimes T\boldsymbol{E} \to T^*\boldsymbol{E} \otimes (\mathbb{T}^* \otimes T\boldsymbol{E}),$$
$$\widehat{F} := g^{\sharp 2}(F) : \boldsymbol{E} \to (\mathbb{L}^{-3/2} \otimes \mathbb{M}^{1/2}) \otimes T^*\boldsymbol{E} \otimes T\boldsymbol{E},$$
$$\mathrm{T} : \mathbb{T}^* \otimes T\boldsymbol{E} \to \mathbb{R} : v \to -\frac{1}{c^2} g(v, v).$$

Actually, the standard *Lorentz law of motion*

$$\nabla^{\natural}{}_{ds} ds = -\frac{q}{m} g^{\sharp}(ds \lrcorner F)$$

turns out to be equivalent, in terms of the joined connection K, to the *law*

$$\nabla_{ds} ds = 0.$$

Then, chosen a general observer o, we rephrase the above joined objects in terms of the observed electric and magnetic fields $\vec{E}[o]$ and $\vec{B}[o]$.

The above results extend to an einsteinian general relativistic framework a minimal coupling of gravitational and electromagnetic fields, which has been found for classical and quantum mechanics in the galileian framework.

Introduction

A minimal coupling of gravitational and electromagnetic fields has been a possible goal from the very beginning of General Relativity. Such a research thread is often referred to as "unified field theory". Several authors proposed different approaches in different directions and established interesting results (see, for instance,

---

2020 *Mathematics Subject Classification.* Primary 83-02; Secondary 83-06.

Key words and phrases. General relativity, gravitational field, electromagnetic field, minimal coupling.

The second author is the corresponding author.

©2023 American Mathematical Society

[**4–6**, **8**, **14**, **19**]). A valuable detailed history of this research thread can be found in [**7**], along with an extended and commented bibliography.

In our covariant formulation of quantum mechanics in a galileian framework we found a natural minimal coupling of electromagnetic field with the gravitational connection [**11**]. Such a coupling works for classical field theory in several directions and also for quantum mechanics.

In the present paper, we some how rephrase, in an einsteinian general relativistic framework, the coupling achieved in the covariant galileian framework, limiting our attention to the Lorentz law of motion for a charged particle.

Namely, in the einsteinian general relativistic framework, we couple in a covariant way the electromagnetic field $F : \boldsymbol{E} \to (\mathbb{L}^{1/2} \otimes \mathbb{M}^{1/2}) \otimes \Lambda^2 T^*\boldsymbol{E}$ with the gravitational connection $K^\natural : \mathbb{T}^* \otimes T\boldsymbol{E} \to T^*\boldsymbol{E} \otimes T(\mathbb{T}^* \otimes T\boldsymbol{E})$, obtaining a "joined connection" $K = K^\natural + K^\mathfrak{e}$, where the additional electromagnetic term $K^\mathfrak{e}$ is non linear.

Then, we can easily show that the standard Lorentz law for the motion $s : \boldsymbol{T} \to \boldsymbol{E}$ of a charged particle can be expressed in a compact way through the above joined connection K, by writing $\nabla_{ds} ds = 0$. Thus, in a sense, we can say that the charged particle "sees" just the joined connection K, instead of "seeing" the gravitational and the electromagnetic fields $K^\natural$ and F, separately.

We first propose a fully straightforward intrinsic approach to this matter; then, we consider a general observer o and provide an observed description of the above results in terms of the observed electric and magnetic fields.

This approach might suggest a hierarchical viewpoint concerning classical fields. Namely, first we have the gravitational connection $K^\natural$, which is a section of an affine bundle; then, the metric field g, which is a "constrained" section of a vector bundle and, further, the electromagnetic field F, which is a section of a vector bundle. In our joined formulation, the metric field g yields an interaction between the gravitational field $K^\natural$ and the electromagnetic field F into the joined connection K.

We speculate that perhaps such a viewpoint might be extended also to other aspects of classical field theory and possibly to quantum field theory.

We stress that our formulation of the joined expression of the Lorentz law of motion has been easily achieved via an intrinsic mathematical language; the same result might have been hardly guessed directly via the standard coordinate language.

In order to achieve a formulation explicitly covariant, not only with respect to coordinates, but also with respect to units of measurement, we deal with the language of *physical scales* [**11**, **12**] and refer to the following *scale spaces*:

- the positive space of *time intervals* $\mathbb{T}$,
- the positive space of *lengths* $\mathbb{L}$,
- the positive space of *masses* $\mathbb{M}$.

We are also dealing with *time unit of measurements* and their *duals*

$$u_0 \in \mathbb{T} \qquad \text{and} \qquad u^0 \in \mathbb{T}^*.$$

The *light velocity*, the *mass* and the *charge* of a particle are represented by the following scale factors:

$$c = c_0 u^0 \in \mathbb{T}^* \otimes \mathbb{L}, \qquad c_0 \in \mathbb{L}, \qquad m \in \mathbb{M}, \qquad q \in \mathbb{T}^{-1} \otimes \mathbb{L}^{3/2} \otimes \mathbb{M}^{1/2}.$$

Accordingly, the *metric field* and the *electromagnetic field* are represented by a scaled tensor and a scaled 2–form

$$g : \boldsymbol{E} \to \mathbb{L}^2 \otimes (T^*\boldsymbol{E} \otimes T^*\boldsymbol{E}) \qquad \text{and} \qquad F : \boldsymbol{E} \to (\mathbb{L}^{l/2} \otimes \mathbb{M}^{1/2}) \otimes \Lambda^2 T^*\boldsymbol{E}.$$

The scales of the charge and of the electromagnetic field are determined by the Lorentz law of motion and the 2nd Maxwell equation [**11**].

Besides the standard tensor product $\otimes$ of real vector spaces, we deal also with the tensor product $\otimes$ of a "positive space" with a real vector space and with the tensor product $\otimes$ of two "positive spaces" [**12**]. By abuse of language, we use the same notation $\otimes$ in the three cases above; the context clarifies, without ambiguity, which kind of tensor product we are dealing with.

In the galileian theory [**11**], the "universal" space $\boldsymbol{T}$ allows us to find a general link between the time coordinate x^0 and the time unit u_0.

In the einsteinian case, this circumstance does not occur. Therefore, we must distinguish the two roles of the index 0. Thus, the index 0 deals sometimes with the coordinate x^0 and sometimes with the time unit u_0; indeed, in some cases this double role of the index 0 occurs in the same formula. The context clarifies, without ambiguity, which kind of role we are dealing with.

1. Spacetime

1.1. Spacetime and lorentzian metric. We represent the general relativistic *spacetime* by a four dimensional manifold $\boldsymbol{E}$ equipped with a *scaled lorentzian metric* [**11, 12, 17, 18**]

$$g : \boldsymbol{E} \to \mathbb{L}^2 \otimes (T^*\boldsymbol{E} \otimes T^*\boldsymbol{E})$$

of signature $(-,+,+,+)$. We suppose spacetime to be oriented and time oriented.

We deal with local *unscaled spacetime charts* [**11, 12**]

$$(x^\lambda) = (x^0, x^i) : \boldsymbol{E} \to \mathbb{R}^4 = \mathbb{R} \times \mathbb{R}^3$$

adapted to the lorentzian structure; thus, by definition, the vector field ∂_0 is timelike and future oriented.

We denote the *contravariant metric* and the *metric isomorphisms* by

$$\bar{g} : \boldsymbol{E} \to \mathbb{L}^{-2} \otimes (T\boldsymbol{E} \otimes T\boldsymbol{E}),$$
$$g^\flat : T\boldsymbol{E} \to \mathbb{L}^2 \otimes T^*\boldsymbol{E} \qquad \text{and} \qquad g^\sharp : T^*\boldsymbol{E} \to \mathbb{L}^{-2} \otimes T\boldsymbol{E}.$$

We have the coordinate expressions

$$g = g_{\lambda\mu} d^\lambda \otimes d^\mu, \qquad \bar{g} = g^{\lambda\mu} \partial_\lambda \otimes \partial_\mu,$$
$$g_{\lambda\mu} : \boldsymbol{E} \to \mathbb{L}^2 \otimes \mathbb{R}, \qquad g^{\lambda\mu} : \boldsymbol{E} \to \mathbb{L}^{-2} \otimes \mathbb{R}.$$

We deal with the scaled *spacetime volume form* and the *spacetime volume vector* induced by the metric and the orientation of spacetime

$$\upsilon : \boldsymbol{E} \to \mathbb{L}^4 \otimes \Lambda^4 T^*\boldsymbol{E} \qquad \text{and} \qquad \bar{\upsilon} : \boldsymbol{E} \to \mathbb{L}^{-4} \otimes \Lambda^4 T\boldsymbol{E}.$$

We have the coordinate expressions

$$\upsilon = \sqrt{|\det g|}\, d^0 \wedge d^1 \wedge d^2 \wedge d^3 \quad \text{and} \quad \bar{\upsilon} = \sqrt{|\det \bar{g}|}\, \partial_0 \wedge \partial_1 \wedge \partial_2 \wedge \partial_3.$$

By definition, we have

$$(\Lambda^4 \bar{g})(\upsilon, \upsilon) = -1, \qquad i_{\bar{\upsilon}} \upsilon = 1, \qquad (\Lambda^4 g)(\bar{\upsilon}, \bar{\upsilon}) = -1.$$

1.2. Tangent and cotangent bundles. We deal with the *tangent bundle* and *cotangent bundle* of spacetime and the associated *scaled tangent bundle* and *scaled cotangent bundle*

$$TE \to E, \qquad T^*E \to E \qquad \text{and} \qquad \mathbb{T}^* \otimes TE \to E, \qquad \mathbb{T} \otimes T^*E \to E.$$

Their local fibred charts naturally induced by a spacetime chart and a time unit of measurement are denoted by

$$(x^\lambda, \dot{x}^\lambda), \qquad (x^\lambda, \dot{x}_\lambda) \qquad \text{and} \qquad (x^\lambda, \dot{x}^\lambda_0), \qquad (x^\lambda, \dot{x}^0_\lambda).$$

We define the *positive timelike tangent bundle* and the *positive timelike cotangent bundle* of spacetime to be the subbundles over E

$$T^-E := \{v \in TE \mid g(v,v) < 0,\ \mathrm{f}(v) = +\} \subset TE,$$
$$T^*_-E := \{\phi \in T^*E \mid \bar{g}(\phi,\phi) < 0,\ \mathrm{f}\big(g^\sharp(\phi)\big) = +\} \subset T^*E,$$

where f denotes *positive* or *negative time orientation* of timelike vectors.

Moreover, we consider the *scaled bundles* over E

$$\mathbb{T}^* \otimes T^-E \to E \qquad \text{and} \qquad \mathbb{T} \otimes T^*_-E \to E.$$

The above bundles T^-E, T^*_-E, $\mathbb{T}^* \otimes T^-E$, $\mathbb{T} \otimes T^*_-E$ turn out to be "positive bundles" over $\mathbb{R}^+$ in the sense discussed in [**12**].

Further, we define the *timelike shell tangent bundle* and the *timelike shell cotangent bundle* to be the $(4+3)$–dimensional subbundles over E

$$\widehat{J}_1E := \{v \in \mathbb{T}^* \otimes TE \mid g(v,v) = -c^2,\ \mathrm{f}(v) = +\} \subset \mathbb{T}^* \otimes T^-E \subset \mathbb{T}^* \otimes TE,$$
$$\widehat{J}^*_1E := \{\phi \in \mathbb{T} \otimes T^*E \mid \bar{g}(\phi,\phi) = -1/c^2,\ \mathrm{f}\big(g^\sharp(\phi)\big) = +\} \subset \mathbb{T} \otimes T^*_-E \subset \mathbb{T} \otimes T^*E.$$

The coordinate expressions of the above constraints are

$$g_{\lambda\mu}\dot{x}^\lambda_0\dot{x}^\mu_0 = -(c_0)^2 \qquad \text{and} \qquad g^{\lambda\mu}\dot{x}^0_\lambda\dot{x}^0_\mu = -1/(c_0)^2.$$

1.3. Motions. We define a *motion* [**9, 11, 17**] to be the inclusion map of a timelike 1–dimensional submanifold T

$$s : T \to E.$$

Every (local) spacetime chart (x^λ) yields the following (local) chart of T

$$\breve{x}^0 := x^0|_T : T \to \mathbb{R}.$$

Given a motion $s : T \to E$, the spacetime scaled metric

$$\frac{1}{c^2}g : E \to \mathbb{T}^2 \otimes (T^*E \otimes T^*E)$$

naturally yields on the 1–dimensional submanifold $T \subset E$ a (local) metric, hence a (local) affine structure associated with the vector space $\bar{\mathbb{T}} := \mathbb{T} \otimes \mathbb{R}$.

Namely, we define the *proper time map* of the motion s to be the (local) bijection

$$\sigma : T \to \bar{\mathbb{T}} : t \mapsto \frac{1}{c}\int_{[t_0,t]} \|\frac{ds}{d\breve{x}^0}\| d\breve{x}^0, \tag{1.1}$$

with reference to a "time origin" $t_0 \in T$.

Indeed, this (local) affine structure does not depend on the choice of the initial time t_0. We say also that T is the *proper time* of the motion s.

We stress that every motion $s : \boldsymbol{T} \to \boldsymbol{E}$ has its own affine submanifold $\boldsymbol{T} \subset \boldsymbol{E}$, while the same vector space $\bar{\mathbb{T}}$ is associated with all motions; this fact reflects the circumstance that "good clocks" make sense in General Relativity.

We define the *velocity* of a motion s to be the *1st differential* calculated with respect to the proper time σ, i.e. with respect to the above affine structure of $\boldsymbol{T}$,

$$ds := \frac{ds}{d\sigma} : \boldsymbol{T} \to \mathbb{T}^* \otimes T\boldsymbol{E}. \tag{1.2}$$

PROPOSITION 1.1. We have the coordinate expression

$$ds = c\alpha^0[s]\big((\partial_0 \circ s) + \partial_0 s^i (\partial_i \circ s)\big), \tag{1.3}$$

where

$$\alpha^0[s] := 1/\sqrt{|(g_{00} \circ s) + 2(g_{0j} \circ s)\partial_0 s^j + (g_{ij} \circ s)\partial_0 s^i \partial_0 s^j|} : \boldsymbol{T} \to \mathbb{L}^* \otimes \mathbb{R}^+.$$

PROOF. By definition, we have $\partial_0 s^0 = 1$, hence

$$\frac{ds}{dx^0} = \partial_0 \circ s + \partial_0 s^i (\partial_i \circ s).$$

Moreover, equation (1.1) yields the equality

$$\frac{d\sigma}{d\breve{x}^0} = \frac{1}{c}\|\frac{ds}{d\breve{x}^0}\| = \frac{1}{c}\sqrt{|(g_{00} \circ s) + 2(g_{0j} \circ s)\partial_0 s^j + (g_{ij} \circ s)\partial_0 s^i \partial_0 s^j|},$$

hence

$$\frac{d\breve{x}^0}{d\sigma} = c/\|\frac{ds}{d\breve{x}^0}\| = c/\sqrt{|(g_{00} \circ s) + 2(g_{0j} \circ s)\partial_0 s^j + (g_{ij} \circ s)\partial_0 s^i \partial_0 s^j|}.$$

Therefore, we obtain the coordinate expression

$$\begin{aligned} ds := \frac{ds}{d\sigma} &= \frac{ds^\lambda}{d\sigma}(\partial_\lambda \circ s) = \frac{ds^\lambda}{d\breve{x}^0}\frac{d\breve{x}^0}{d\sigma}(\partial_\lambda \circ s) = \\ &= \frac{c\big((\partial_0 \circ s) + \partial_0 s^i(\partial_i \circ s)\big)}{\sqrt{|(g_{00} \circ s) + 2(g_{0j} \circ s)\partial_0 s^j + (g_{ij} \circ s)\partial_0 s^i \partial_0 s^j|}}, \end{aligned}$$

where the partial derivatives $\partial_0 s^i$ and the vector fields ∂_λ are computed with reference to a spacetime chart x^λ. □

Thus, the factor $c\alpha^0[s]$ transforms derivatives with respect to the time coordinate $\breve{x}^0$ into derivatives with respect to the proper time σ.

PROPOSITION 1.2. The definition of σ yields the equality

$$g(ds, ds) = -c^2 \in \mathbb{T}^{-2} \otimes \mathbb{L}^2.$$

Hence, for each motion s, we can write

$$ds : \boldsymbol{T} \to \widehat{J}_1\boldsymbol{E} \subset \mathbb{T}^* \otimes T^-\boldsymbol{E} \subset \mathbb{T}^* \otimes T\boldsymbol{E}. \quad \square$$

1.4. Phase space. We define the *phase space* to be the bundle of 1st–order jets of motions [**9**, **13**]

$$J_1\boldsymbol{E} \to \boldsymbol{E}.$$

For each motion $s : \boldsymbol{T} \to \boldsymbol{E}$, its 1st jet prolongation is an injective map

$$j_1 s : \boldsymbol{T} \to J_1\boldsymbol{E}.$$

Every spacetime chart (x^λ) naturally yields a fibred chart (x^λ, x^i_0) of $J_1\boldsymbol{E}$, according to the coordinate expression [**10**]

$$x^i_0 \circ j_1 s = \breve{\partial}_0 s^i.$$

1.5. Contact structure of the phase space. The lorentzian metric g yields the following maps.

PROPOSITION 1.3. We have a unique fibred isomorphism over $\boldsymbol{E}$, called *contact map*, [**9, 13**]

$$\text{д} : J_1\boldsymbol{E} \to \widehat{J}_1\boldsymbol{E} \subset \mathbb{T}^* \otimes T^-\boldsymbol{E} \subset \mathbb{T}^* \otimes T\boldsymbol{E}, \tag{1.4}$$

such that, for each motion $s : \boldsymbol{T} \to \boldsymbol{E}$,

$$\text{д} \circ j_1 s = ds.$$

Then, we define the *time form* to be the fibred morphism over $\boldsymbol{E}$

$$\tau := -\frac{1}{c^2} g^\flat(\text{д}) : J_1\boldsymbol{E} \to \mathbb{T} \otimes T^*\boldsymbol{E}. \tag{1.5}$$

We have the identities

$$g(\text{д}, \text{д}) = -c^2, \qquad \bar{g}(\tau, \tau) = -1/c^2, \qquad \tau(\text{д}) = 1, \qquad g^\sharp(\tau) = -\frac{1}{c^2}\text{д},$$

and the coordinate expressions

$$\text{д} = c\alpha^0(\partial_0 + x_0^i \partial_i) \qquad \text{and} \qquad \tau = -\frac{1}{c}\alpha^0(g_{0\mu} + g_{i\mu} x_0^i) d^\mu,$$

where we have set

$$\alpha^0 := 1/\sqrt{|g_{00} + 2g_{0j} x_0^j + g_{ij} x_0^i x_0^j|} : J_1\boldsymbol{E} \to \mathbb{L}^* \otimes \mathbb{R}^+. \quad \square$$

In a sense, the above einsteinian time form τ based on $J_1\boldsymbol{E}$ is the analogous of the galileian time form dt based on $\boldsymbol{E}$ [**11**]. We stress that dt is exact, while τ needs not to be closed.

Moreover, we observe that the form τ defines a scaled *contact structure* on $J_1\boldsymbol{E}$ [**10**].

1.6. Orthogonal splitting of the tangent space. The contact structure of the phase space yields a natural *orthogonal splitting* of the tangent and cotangent spaces over the phase space, according to the following procedure.

PROPOSITION 1.4. We consider the fibred product bundles

$$T^\uparrow\boldsymbol{E} := J_1\boldsymbol{E} \underset{\boldsymbol{E}}{\times} T\boldsymbol{E} \to J_1\boldsymbol{E} \qquad \text{and} \qquad T^{*\uparrow}\boldsymbol{E} := J_1\boldsymbol{E} \underset{\boldsymbol{E}}{\times} T^*\boldsymbol{E} \to J_1\boldsymbol{E},$$

and define the vector subbundles over $J_1\boldsymbol{E}$

$$T^\uparrow{}_\parallel\boldsymbol{E} := \{(j, v) \in T^\uparrow\boldsymbol{E} \mid v = \lambda\text{д}(j); \lambda \in \mathbb{T} \otimes \mathbb{R}\} \subset T^\uparrow\boldsymbol{E},$$
$$T^{*\uparrow}{}_\parallel\boldsymbol{E} := \{(j, \omega) \in T^{*\uparrow}\boldsymbol{E} \mid g^\sharp(\omega) = \lambda\tau(j); \lambda \in \mathbb{T} \otimes \mathbb{R}\} \subset T^{*\uparrow}\boldsymbol{E},$$

and

$$T^\uparrow{}_\perp\boldsymbol{E} := \{(j, v) \in T^\uparrow\boldsymbol{E} \mid g\big(\text{д}(j), v\big) = 0\} \subset T^\uparrow\boldsymbol{E},$$
$$T^{*\uparrow}{}_\perp\boldsymbol{E} := \{(j, \omega) \in T^{*\uparrow}\boldsymbol{E} \mid \bar{g}\big(\tau(j), \omega\big) = 0\} \subset T^{*\uparrow}\boldsymbol{E}.$$

Then, we obtain the projection fibred morphisms over $J_1\boldsymbol{E}$

$$\theta := \mathrm{Id} - \text{д} \otimes \tau : T^\uparrow\boldsymbol{E} \to T^\uparrow{}_\perp\boldsymbol{E} \quad : (j, v) \mapsto \Big(j,\ v - \langle\tau(j), v\rangle\text{д}(j)\Big),$$
$$\theta^* := \mathrm{Id} - \tau \otimes \text{д} : T^{*\uparrow}\boldsymbol{E} \to T^{*\uparrow}{}_\perp\boldsymbol{E} : (j, \omega) \mapsto \Big(j,\ \omega - \langle\omega, \text{д}(j)\rangle\tau(j)\Big). \quad \square$$

1.7. Quadratic form of the tangent space. In view of the joined spacetime connection 2.3, we introduce the following object.

DEFINITION 1.5. We define the *quadratic form of the scaled tangent space* to be the map

$$\text{(1.6)}\qquad \mathrm{T} : \mathbb{T}^* \otimes T\boldsymbol{E} \to \mathbb{R} : v \to -\frac{1}{c^2} g(v, v),$$

with coordinate expression

$$\mathrm{T} = -\frac{1}{(c_0)^2}(g_{00}\dot{x}_0^0\dot{x}_0^0 + 2g_{0j}\dot{x}_0^0\dot{x}_0^j + g_{ij}\dot{x}_0^i\dot{x}_0^j). \quad \square$$

We emphasise that, for each motion $s : \boldsymbol{T} \to \boldsymbol{E}$, we have

$$\mathrm{T} \circ ds = 1.$$

The above identity is analogous to the identity $ds \lrcorner dt = 1$, which holds in the galileian framework [**11**].

2. Gravitational and electromagnetic fields

We define a *spacetime connection* to be a linear symmetric connection of spacetime

$$K : T\boldsymbol{E} \to T^*\boldsymbol{E} \otimes TT\boldsymbol{E},$$

with coordinate expression

$$K = d^\lambda \otimes (\partial_\lambda + K_\lambda{}^\mu{}_\nu \dot{x}^\nu \dot{\partial}_\mu), \qquad \text{where} \qquad K_\lambda{}^\mu{}_\nu = K_\nu{}^\mu{}_\lambda \in \mathrm{map}(\boldsymbol{E}, \mathbb{R}).$$

Moreover, the above connection naturally yields a connection on the scaled bundle $\mathbb{T}^* \otimes T\boldsymbol{E} \to \boldsymbol{E}$ (denoted by the same symbol)

$$K : \mathbb{T}^* \otimes T\boldsymbol{E} \to T^*\boldsymbol{E} \otimes T(\mathbb{T}^* \otimes T\boldsymbol{E}),$$

with coordinate expression

$$K = d^\lambda \otimes (\partial_\lambda + K_\lambda{}^\mu{}_\nu \dot{x}_0^\nu \dot{\partial}_\mu^0), \quad \text{where} \quad K_\lambda{}^\mu{}_\nu = K_\mu{}^\mu{}_\lambda : \boldsymbol{E} \to \mathbb{R}, \quad \dot{x}_0^\nu : \mathbb{T}^* \otimes T\boldsymbol{E} \to \mathbb{R}.$$

2.1. Gravitational connection. As usual, we define the *gravitational spacetime connection* to be the Levi–Civita connection of spacetime induced by the metric g [**17, 18**]

$$K^\natural : T\boldsymbol{E} \to T^*\boldsymbol{E} \otimes TT\boldsymbol{E},$$

whose coordinate expression is

$$K^\natural = d^\lambda \otimes (\partial_\lambda + K^\natural{}_\lambda{}^\nu{}_\mu \dot{x}^\mu \dot{\partial}_\nu),$$

where

$$K^\natural{}_\lambda{}^\nu{}_\mu = -\frac{1}{2} g^{\nu\alpha}(\partial_\lambda g_{\alpha\mu} + \partial_\mu g_{\alpha\lambda} - \partial_\alpha g_{\lambda\mu}) : \boldsymbol{E} \to \mathbb{R}.$$

2.2. Electromagnetic field. The *electromagnetic field* is represented by a scaled 2–form [**11, 12, 17**]

$$F : \boldsymbol{E} \to (\mathbb{L}^{1/2} \otimes \mathbb{M}^{1/2}) \otimes \Lambda^2 T^* \boldsymbol{E}.$$

Further, we consider the *electromagnetic mixed tensor*

$$\widehat{F} := g^{\sharp 2}(F) : \boldsymbol{E} \to (\mathbb{L}^{-3/2} \otimes \mathbb{M}^{1/2}) \otimes T^* \boldsymbol{E} \otimes T\boldsymbol{E}.$$

We have the coordinate expressions

$$F = F_{\lambda\mu} d^\lambda \wedge d^\mu, \qquad \widehat{F} = F_\lambda{}^\nu d^\lambda \otimes \partial_\nu = g^{\mu\nu} F_{\lambda\mu} d^\lambda \otimes \partial_\nu,$$

with

$$F_{\lambda\mu} : \boldsymbol{E} \to (\mathbb{L}^{1/2} \otimes \mathbb{M}^{1/2}) \otimes \mathbb{R}, \qquad F_\lambda{}^\nu : \boldsymbol{E} \to (\mathbb{L}^{-3/2} \otimes \mathbb{M}^{1/2}) \otimes \mathbb{R}.$$

2.3. Joined spacetime connection. Now, let us consider a charged particle of charge and mass

$$m \in \mathbb{M} \qquad \text{and} \qquad q \in \mathbb{T}^{-1} \otimes \mathbb{L}^{3/2} \otimes \mathbb{M}^{1/2}.$$

In view of coupling the electromagnetic field with the gravitational field, we define the following object.

DEFINITION 2.1. We define the *electromagnetic coupling term* to be the quadratic fibred morphism over $\boldsymbol{E}$

$$K^{\mathfrak{e}} := -\frac{q}{m} \mathrm{T}\widehat{F} : \mathbb{T}^* \otimes T\boldsymbol{E} \to T^*\boldsymbol{E} \otimes (\mathbb{T}^* \otimes T\boldsymbol{E}) \simeq T^*\boldsymbol{E} \otimes V_{\boldsymbol{E}}(\mathbb{T}^* \otimes T\boldsymbol{E}), \tag{2.1}$$

with coordinate expression

$$K^{\mathfrak{e}} = \frac{q}{m} \frac{1}{(c_0)^2} (g_{\alpha\beta} \dot{x}_0^\alpha \dot{x}_0^\beta) g^{\mu\nu} F_{\lambda\mu} d^\lambda \otimes \partial_\nu. \quad \square$$

DEFINITION 2.2. We define the (non linear) *joined connection* to be the connection of the bundle $\mathbb{T}^* \otimes T\boldsymbol{E} \to \boldsymbol{E}$

$$K := K^\natural + K^{\mathfrak{e}} = K^\natural - \frac{q}{m} \mathrm{T}\widehat{F} : \mathbb{T}^* \otimes T\boldsymbol{E} \to T^*\boldsymbol{E} \otimes T(\mathbb{T}^* \otimes T\boldsymbol{E}). \quad \square \tag{2.2}$$

2.4. Joined Lorentz law of motion. Let us consider the standard Lorentz *law of motion* of a charged particle effected by the gravitational and electromagnetic fields [**17, 18**]

$$\nabla^\natural{}_{ds} ds = -\frac{q}{m} g^\sharp (ds \lrcorner F), \tag{2.3}$$

where $\nabla^\natural$ is the covariant differential associated with the gravitational spacetime connection

$$K^\natural : \mathbb{T}^* \otimes T\boldsymbol{E} \to T^*\boldsymbol{E} \otimes T(\mathbb{T}^* \otimes T\boldsymbol{E})$$

and

$$-\frac{q}{m} g^\sharp (ds \lrcorner F) : \boldsymbol{T} \to \mathbb{T}^{-2} \otimes T\boldsymbol{E}$$

is the Lorentz force.

LEMMA 2.3. *For each motion* $s : \boldsymbol{T} \to \boldsymbol{E}$, *we have*

$$ds \lrcorner (K^{\mathfrak{e}} \circ ds) = -\frac{q}{m} ds \lrcorner \widehat{F} = -\frac{q}{m} g^\sharp (ds \lrcorner F).$$

PROOF. We have

$$ds \lrcorner (K^{\mathfrak{e}} \circ ds) = -\frac{q}{m}(\tau \circ ds)(ds \lrcorner \widehat{F}) = -\frac{q}{m} g^{\sharp}(ds \lrcorner F).$$

□

THEOREM 2.4. *The Lorentz law of motion* (2.3) *can be written as*

$$\nabla_{ds} ds = 0, \tag{2.4}$$

where ∇ *is the covariant differential associated with the joined connection*

$$K := K^{\natural} + K^{\mathfrak{e}} : \mathbb{T}^* \otimes T\boldsymbol{E} \to T^*\boldsymbol{E} \otimes T(\mathbb{T}^* \otimes T\boldsymbol{E})$$

on the bundle $\mathbb{T}^* \otimes T\boldsymbol{E} \to \boldsymbol{E}$.

PROOF. The theorem follows from the above Lemma and from the equality

$$\nabla_{ds} ds = \nabla^{\natural}{}_{ds} ds - ds \lrcorner (K^{\mathfrak{e}} \circ ds).$$

□

3. Observed expressions

Now, we choose a generic observer o and express the joined spacetime connection in terms of the observed electric and magnetic fields [**1**–**3**, **16**].

3.1. Observers.

DEFINITION 3.1. We define an *observer* to be a (local) continuum motion, i.e. a (local) 1–parameter group of motions

$$o : \mathbb{T} \times \boldsymbol{E} \to \boldsymbol{E}.$$

Therefore, by definition, we have (locally) the maps

$$o_0 = \mathrm{Id} : \boldsymbol{E} \to \boldsymbol{E}, \qquad o_{u+v} = o_v \circ o_u : \boldsymbol{E} \to \boldsymbol{E}, \quad \text{for each} \quad u, v \in \mathbb{T},$$

and the timelike 1–dimensional submanifolds

$$C_e : \boldsymbol{T}_e := o_e(\mathbb{T}) \to \boldsymbol{E}, \quad \text{for every} \quad e \in \boldsymbol{E}.$$

In coordinates, we write (locally)

$$o^{\lambda} := x^{\lambda} \circ o : \mathbb{T} \times \boldsymbol{E} \to \mathbb{R}.$$

The partial 1–jet of o, with respect to the parameter, is the section [**13**]

$$j_1 o : \boldsymbol{E} \to J_1\boldsymbol{E},$$

with coordinate expression

$$o_0^{\lambda} \equiv x_0^{\lambda} \circ j_1 o := \partial_0 o^{\lambda} : \boldsymbol{E} \to \mathbb{R},$$

where ∂ denotes the partial derivative with respect to the parameter, evaluated at $0 \in \mathbb{T}$. □

PROPOSITION 3.2. We define the *observed scaled vector field* (*velocity* of o) and the *observed scaled time 1-form*

$$д[o] := д \circ j_1 o : \boldsymbol{E} \to \widehat{J}_1\boldsymbol{E} \subset \mathbb{T}^* \otimes T^{-}\boldsymbol{E} \subset \mathbb{T}^* \otimes T\boldsymbol{E},$$

$$\tau[o] := -\frac{1}{c^2} g^{\flat}(д[o]) : \boldsymbol{E} \to \widehat{J}_1^*\boldsymbol{E} \subset \mathbb{T} \otimes T_{-}^*\boldsymbol{E} \subset \mathbb{T} \otimes T^*\boldsymbol{E}.$$

We have the identities

$$g(\text{д}[o], \text{д}[o]) = -c^2, \quad \bar{g}(\tau[o], \tau[o]) = -1/c^2, \quad \tau[o](\text{д}[o]) = 1, \quad g^\sharp(\tau[o]) = -\frac{1}{c^2}\text{д}[o],$$

and the coordinate expressions

$$\text{д}[o] = c\alpha^0[o](\partial_0 + o_0^i \partial_i) \qquad \text{and} \qquad \tau[o] = -\frac{1}{c}\alpha^0[o](g_{0\mu} + g_{j\mu} o_0^j) d^\mu,$$

where we have set

$$\alpha^0[o] := 1/\sqrt{|g_{00} + 2g_{0j} o_0^j + g_{ij} o_0^i o_0^j|} : \boldsymbol{E} \to \mathbb{L}^* \otimes \mathbb{R}^+. \qquad \square$$

In a sense, we can regard $\tau[o]$ as the einsteinian observer dependent analogous of the galileian observer independent scaled time form dt [**11**].

We stress that, in general, the scaled form $\tau[o]$ is not closed.

In the particular case when $\tau[o]$ is closed, it is locally exact and yields an observed local timelike foliation of spacetime, which locally replaces the galileian global time fibring $t : \boldsymbol{E} \to \boldsymbol{T}$.

Next, let us consider an observer o and analyse induced distinguished splittings.

3.2. Observed splittings of the tangent and cotangent bundles. With reference to an observer o, we have the observed splitting of the tangent and cotangent bundles of spacetime.

PROPOSITION 3.3. We split $T\boldsymbol{E}$ into its subbundles *parallel* and *orthogonal* to $\text{д}[o]$

$$T\boldsymbol{E} = T_\| \boldsymbol{E}[o] \underset{\boldsymbol{E}}{\oplus} T_\perp \boldsymbol{E}[o], \qquad \text{where} \qquad T_\| \boldsymbol{E}[o] \subset T\boldsymbol{E}, \quad T_\perp \boldsymbol{E}[o] \subset T\boldsymbol{E},$$

via the fibred projections over $\boldsymbol{E}$

$$\pi_\|[o] : T\boldsymbol{E} \to T_\| \boldsymbol{E}[o] \qquad \text{and} \qquad \pi_\perp[o] : T\boldsymbol{E} \to T_\perp \boldsymbol{E}[o].$$

Namely, for each vector field $X : \boldsymbol{E} \to T\boldsymbol{E}$, we obtain the splitting

$$X = X_\|[o] + X_\perp[o],$$

where

$$X_\|[o] = i_X \tau[o] \text{д}[o] = -\frac{1}{c^2} g(X, \text{д}[o]) \text{д}[o],$$

$$X_\perp[o] = X - i_X \tau[o] \text{д}[o] = X + \frac{1}{c^2} g(X, \text{д}[o]) \text{д}[o]. \quad \square$$

PROPOSITION 3.4. We split $T^*\boldsymbol{E}$ into its subbundles *parallel* and *orthogonal* to $\tau[o]$

$$T^*\boldsymbol{E} = T_\|^* \boldsymbol{E}[o] \underset{\boldsymbol{E}}{\oplus} T_\perp^* \boldsymbol{E}[o],$$

via the fibred projections over $\boldsymbol{E}$

$$\pi_\|[o] : T^*\boldsymbol{E} \to T_\|^* \boldsymbol{E}[o] \qquad \text{and} \qquad \pi_\perp[o] : T^*\boldsymbol{E} \to T_\perp^* \boldsymbol{E}[o].$$

Namely, for each 1–form $\omega : \boldsymbol{E} \to T^*\boldsymbol{E}$, we obtain the splitting

$$\omega = \omega_\|[o] + \omega_\perp[o],$$

where

$$\omega_\|[o] = i_{\text{д}[o]} \omega \tau[o] \qquad \text{and} \qquad \omega_\perp[o] = \omega - i_{\text{д}[o]} \omega \tau[o]. \quad \square$$

COROLLARY 3.5. The mutually inverse linear fibred isomorphisms over $\boldsymbol{E}$

$$g^\flat : T\boldsymbol{E} \to \mathbb{L}^2 \otimes T^*\boldsymbol{E} \qquad \text{and} \qquad g^\sharp : T^*\boldsymbol{E} \to \mathbb{L}^{-2} \otimes T\boldsymbol{E}$$

restrict to mutually inverse fibred linear isomorphisms

$$\begin{aligned} g^\flat{}_{\|}[o] : T_{\|}\boldsymbol{E}[o] \to \mathbb{L}^2 \otimes T^*_{\|}\boldsymbol{E}[o] \qquad &\text{and} \qquad g^\sharp{}_{\|}[o]\ : T^*_{\|}\boldsymbol{E}[o] \to \mathbb{L}^{-2} \otimes T_{\|}\boldsymbol{E}[o],\\ g^\flat{}_{\perp}[o] : T_{\perp}\boldsymbol{E}[o] \to \mathbb{L}^2 \otimes T^*_{\perp}\boldsymbol{E}[o] \qquad &\text{and} \qquad g^\sharp{}_{\perp}[o] : T^*_{\perp}\boldsymbol{E}[o] \to \mathbb{L}^{-2} \otimes T_{\perp}\boldsymbol{E}[o]. \quad \square \end{aligned}$$

NOTE 3.6. We notice that the metric g naturally makes

$$(T_{\|}\boldsymbol{E}[o])^* = T^*_{\|}\boldsymbol{E}[o] \qquad \text{and} \qquad (T_{\perp}\boldsymbol{E}[o])^* = T^*_{\perp}\boldsymbol{E}[o]$$

via the equalities

$$\omega_{\|}(X_{\|}) := g\big(g^\sharp(\omega_{\|}), X_{\|}\big) \qquad \text{and} \qquad \omega_{\perp}(X_{\perp}) := g\big(g^\sharp(\omega_{\perp}), X_{\perp}\big),$$

for each

$$\omega_{\|} \in T^*_{\|}\boldsymbol{E}[o], \qquad \omega_{\perp} \in T^*_{\perp}\boldsymbol{E}[o], \qquad X_{\|} \in T_{\|}\boldsymbol{E}[o], \qquad X_{\perp} \in T_{\perp}\boldsymbol{E}[o]. \quad \square$$

The orthogonal bundles are naturally equipped with a scaled volume form and a scaled volume vector.

PROPOSITION 3.7. We denote by

$$\eta[o] : \boldsymbol{E} \to \mathbb{L}^3 \otimes \Lambda^3 T^*_{\perp}\boldsymbol{E}[o] \qquad \text{and} \qquad \bar{\eta}[o] : \boldsymbol{E} \to \mathbb{L}^{-3} \otimes \Lambda^3 T_{\perp}\boldsymbol{E}[o] \tag{3.1}$$

the observer dependent scaled *volume form* and scaled *volume vector* induced on the tangent space orthogonal to д$[o]$ by the metric g, the orientation of spacetime and the time orientation.

By definition, we have

$$\Lambda^3\bar{g}\big(\eta[o], \eta[o]\big) = 1, \qquad i_{\bar{\eta}[o]}\eta[o] = 1, \qquad \Lambda^3 g\big(\bar{\eta}[o], \bar{\eta}[o]\big) = 1.$$

$$\upsilon = c\tau[o] \wedge \eta[o], \qquad \bar{\upsilon} = \frac{1}{c}\text{д}[o] \wedge \bar{\eta}[o],$$

$$\eta[o] = \frac{1}{c} i_{\text{д}[o]}\upsilon, \qquad \bar{\eta}[o] = c\, i_{\tau[o]}\bar{\upsilon}. \quad \square$$

3.3. Observed splitting of the velocity of a motion. With reference to an observer o, we have the observed splitting of the velocity of a motion.

PROPOSITION 3.8. For each motion $s : \boldsymbol{T} \to \boldsymbol{E}$, according to the observed splitting of $T\boldsymbol{E}$, we have the observed splitting

$$ds = (\alpha[o] \circ s)(\text{д}[o] \circ s + \vec{ds}[o]),$$

where

$$\vec{ds}[o] := (\pi_{\perp} \circ ds)/(\alpha[o] \circ s) : \boldsymbol{T} \to \mathbb{T}^* \otimes T_{\perp}\boldsymbol{E}[o],$$

$$\alpha[o] \circ s := -\frac{1}{c^2} g(ds, \text{д}[o] \circ s) = 1/\sqrt{1 - \|\vec{ds}[o]\|^2/c^2} : \boldsymbol{T} \to \mathbb{R}^+. \quad \square$$

3.4. Observed splitting of the electromagnetic field. With reference to an observer o, we have the observed splitting of the electromagnetic field in terms of the observed electric and magnetic fields.

PROPOSITION 3.9. According to the observed splitting

$$\Lambda^2 T^* \boldsymbol{E} = \big(T^*_{\|}\boldsymbol{E}[o] \wedge T^*_{\perp}\boldsymbol{E}[o]\big) \underset{\boldsymbol{E}}{\oplus} \Lambda^2 T^*_{\perp}\boldsymbol{E}[o],$$

we can uniquely split the electromagnetic field F as

$$F = -2\tau[o] \wedge E[o] + F_{\perp}[o],$$

where

$$E[o] := -\frac{1}{2} i_{д[o]} F : \boldsymbol{E} \to (\mathbb{T}^{-1} \otimes \mathbb{L}^{1/2} \otimes \mathbb{M}^{1/2}) \otimes T^*_{\perp}\boldsymbol{E}[o],$$

$$F_{\perp}[o] := F + 2\tau[o] \wedge E[o] : \boldsymbol{E} \to (\mathbb{L}^{1/2} \otimes \mathbb{M}^{1/2}) \otimes \Lambda^2 T^*_{\perp}\boldsymbol{E}[o]. \quad \square$$

DEFINITION 3.10. We define the *observed electric field* and the *observed magnetic field* to be the scaled vector fields

$$\vec{E}[o] := g^{\sharp}(E[o]) \quad : \boldsymbol{E} \to (\mathbb{T}^{-1} \otimes \mathbb{L}^{-3/2} \otimes \mathbb{M}^{1/2}) \otimes T_{\perp}\boldsymbol{E}[o],$$

$$\vec{B}[o] := \frac{c}{2} i_{F_{\perp}[o]} \bar{\eta}[o] : \boldsymbol{E} \to (\mathbb{T}^{-1} \otimes \mathbb{L}^{-3/2} \otimes \mathbb{M}^{1/2}) \otimes T_{\perp}\boldsymbol{E}[o]. \quad \square$$

COROLLARY 3.11. We have the equality

$$F_{\perp}[o] = \frac{2}{c} i_{\vec{B}[o]} \eta[o],$$

hence we obtain

$$F = 2(-\tau[o] \wedge E[o] + \frac{1}{c} i_{\vec{B}[o]} \eta[o]).$$

$$\widehat{F} = -\tau[o] \otimes \vec{E}[o] - \frac{1}{c^2} E[o] \otimes д[o] + \frac{2}{c} g^{\sharp 2}(i_{\vec{B}[o]} \eta[o]). \quad \square$$

3.5. Observed splitting of the Lorentz force. With reference to an observer o, we have the observed splitting of the Lorentz force [**17**].

PROPOSITION 3.12. We have the observed splitting

$$g^{\sharp}(ds \lrcorner F) = ds \lrcorner \widehat{F} =$$

$$= -(\alpha[o] \circ s)\Big(\frac{1}{c^2} g(\vec{ds}[o], \vec{E}[o]) д[o] + (\vec{E}[o] + \frac{1}{c} \vec{ds}[o] \times \vec{B}[o])\Big),$$

where the cross product $\times$ is computed in the vector bundle $T_{\perp}\boldsymbol{E}[o]$.

PROOF. We have

$$g^{\sharp}(ds \lrcorner F) = ds \lrcorner \widehat{F} =$$

$$= (\alpha[o] \circ s)\big(д[o] \circ s + \vec{ds}[o]\big) \lrcorner \big(- \tau[o] \otimes \vec{E}[o] - \frac{1}{c^2} E[o] \otimes д[o] + \frac{2}{c} g^{\sharp 2}(i_{\vec{B}[o]} \eta[o])\big)$$

$$= (\alpha[o] \circ s)\big(- \vec{E}[o] - \frac{1}{c^2} (\vec{ds}[o] \lrcorner E[o]) д[o] + \frac{2}{c} \vec{ds}[o] \lrcorner g^{\sharp 2}(i_{\vec{B}[o]} \eta[o])\big)$$

$$= -(\alpha[o] \circ s)\big(\frac{1}{c^2} g(\vec{ds}[o], \vec{E}[o]) д[o] + (\vec{E}[o] + \frac{1}{c} \vec{ds}[o] \times \vec{B}[o])\big).$$

$\square$

3.6. Observed splitting of the joined connection. With reference to an observer o, we have the observed splitting of the joined connection.

PROPOSITION 3.13. We have the observed splitting (cfr. [**17**])

$$ds \lrcorner (K^{\mathfrak{e}} \circ ds) =$$

$$= \frac{q}{m}\sqrt{1 - \|\vec{ds}[o]\|^2/c^2}\Big(\frac{1}{c^2} g(\vec{ds}[o], \vec{E}[o])(д[o] \circ s) + (\vec{E}[o] \circ s) + \frac{1}{c}\vec{ds}[o] \times \vec{B}[o]\Big).$$

PROOF. The proof follows from Proposition 3.12 and the equalities

$$ds \lrcorner (K^{\mathfrak{e}} \circ ds) = -\frac{q}{m}(ds \lrcorner \widehat{F}) \qquad \text{and} \qquad \alpha[o] \circ s = 1/\sqrt{1 - \|\vec{ds}[o]\|^2/c^2}.$$

□

References

[1] Donato Bini, *Observers, observables and measurements in general relativity*, General relativity, cosmology and astrophysics, Fundam. Theor. Phys., vol. 177, Springer, Cham, 2014, pp. 67–90. MR3329390

[2] C. Cattaneo, *General relativity: relative standard mass, momentum, energy and gravitational field in a general system of reference* (English, with Italian summary), Nuovo Cimento (10) **10** (1958), 318–337. MR103105

[3] Fernando de Felice and Donato Bini, *Classical measurements in curved space-times*, Cambridge Monographs on Mathematical Physics, Cambridge University Press, Cambridge, 2010, DOI 10.1017/CBO9780511777059. MR2760869

[4] Albert Einstein, *A generalized theory of gravitation*, Rev. Modern Physics **20** (1948), 35–39, DOI 10.1103/revmodphys.20.35. MR0026460

[5] Albert Einstein, *The meaning of relativity. Third edition, revised, including The Generalized Theory of Gravitation*, Princeton University Press, Princeton, N. J., 1950. MR0048945

[6] Albert Einstein, *The meaning of relativity*, Princeton University Press, Princeton, N. J., 1955. 5th ed. MR0076496

[7] Hubert F. M. Goenner, *On the history of unified field theories*, Living Rev. Relativ. **7** (2004), 2004-2, 152, DOI 10.12942/lrr-2004-2. MR2037618

[8] F. W. Hehl and Y. N. Obukhov, *How does the electromagnetic field couple to gravity, in particular to metric, nonmetricity, torsion, and curvature?*, in gyros, clocks, interferometers. . . : Testing relativistic gravity in space (C. Lämmerzahl, C.W.F. Everitt, and F.W. Hehl, eds.), Lecture Notes in Physics, Vol. 562, Springer-Verlag, Berlin, Heidelberg, 2001.

[9] Josef Janyška and Marco Modugno, *Classical particle phase space in general relativity*, Differential geometry and applications (Brno, 1995), Masaryk Univ., Brno, 1996, pp. 573–602. MR1406377

[10] Josef Janyška and Marco Modugno, *Geometric structures of the classical general relativistic phase space*, Int. J. Geom. Methods Mod. Phys. **5** (2008), no. 5, 699–754, DOI 10.1142/S021988780800303X. MR2445392

[11] Josef Janyška and Marco Modugno, *An introduction to covariant quantum mechanics*, Fundamental Theories of Physics, vol. 205, Springer, Cham, [2022] ©2022, DOI 10.1007/978-3-030-89589-1. MR4417779

[12] Josef Janyška, Marco Modugno, and Raffaele Vitolo, *An algebraic approach to physical scales*, Acta Appl. Math. **110** (2010), no. 3, 1249–1276, DOI 10.1007/s10440-009-9505-6. MR2639169

[13] I. S. Krasil′shchik, V. V. Lychagin, and A. M. Vinogradov, *Geometry of jet spaces and nonlinear partial differential equations*, Advanced Studies in Contemporary Mathematics, vol. 1, Gordon and Breach Science Publishers, New York, 1986. Translated from the Russian by A. B. Sosinskiĭ. MR861121

[14] LX Li, *A new unified theory of electromagnetic and gravitational interactions*, Front. Phys. 11, 110402 (2016). `https://doi.org/10.1007/s11467-016-0588-z`.

[15] Enrico Massa, *Reference systems in general relativity* (Italian, with English summary), A great mathematician of the nineteenth century. Papers in honor of Eugenio Beltrami (1835–1900) (Italian), Ist. Lombardo Accad. Sci. Lett. Incontr. Studio, vol. 39, LED–Ed. Univ. Lett. Econ. Diritto, Milan, 2007, pp. 211–223. MR2374678

[16] Enrico Massa and Enrico Pagani, *Jet bundle geometry, dynamical connections, and the inverse problem of Lagrangian mechanics* (English, with English and French summaries), Ann. Inst. H. Poincaré Phys. Théor. **61** (1994), no. 1, 17–62. MR1303184

[17] Charles W. Misner, Kip S. Thorne, and John Archibald Wheeler, *Gravitation*, W. H. Freeman and Co., San Francisco, Calif., 1973. MR0418833

[18] Rainer Kurt Sachs and Hung Hsi Wu, *General relativity for mathematicians*, Graduate Texts in Mathematics, Vol. 48, Springer-Verlag, New York-Heidelberg, 1977. MR0503498

[19] M. Tajmar and C. J. de Matos, *Coupling Electromagnetism with Gravitation in the Weak Field Approximation*, `arXiv:gr-qc/0003011`.

Department of Mathematics and Statistics, Masaryk University, Kotlářská 2, 611 37 Brno, Czech Republic

Email address: `janyska@math.muni.cz`

Department of Mathematics and Informatics "U. Dini", University of Florence, Via S. Marta 3, 50139 Florence, Italy

Email address: `marco.modugno@unifi.it`

Contemporary Mathematics
Volume **788**, 2023
https://doi.org/10.1090/conm/788/15825

Projective-geometric aspects of bi-Hamiltonian structures of KdV type

Paolo Lorenzoni and Raffaele Vitolo

Dedicated to the memory of Alexandre Mikhailovich Vinogradov (1938–2019)

Abstract. We introduce the problem of classification of bi-Hamiltonian structures of KdV type under projective reciprocal transformations. This problem leads naturally to studying the compatibility of a first order localizable homogeneous Hamiltonian operator with a higher order homogeneous Hamiltonian operator. We study the simplest third-order case where the orbit contains a constant operator. Computations with weakly non local Hamiltonian operators have been made by techniques developed in a previous paper.

1. Introduction

Bi-Hamiltonian structures play a fundamental role in integrability theories since their introduction by F. Magri [**22**]. In this paper we are concerned with bi-Hamiltonian structures of KdV-type. The concept was introduced in [**25**], and extended in [**21**] by a full classification of such structures in dimension 2. Let us define the main object of this paper.

We say a bi-Hamiltonian structure of KdV-type to be a pair of compatible Hamiltonian operators

$$P = P_1, \qquad Q = Q_1 + R_i \tag{1.1}$$

such that P_1, Q_1 are compatible first-order Hamiltonian operators of Dubrovin–Novikov type [**11**] and R_i is an i-th order Hamiltonian operator of Dubrovin–Novikov type [**10**] that is compatible with both P_1 and Q_1. We stress that 'compatibility' is formulated as the vanishing of the Schouten bracket (see *e.g.* [**6, 19**] for the definition of the Schouten bracket) of all pairs of operators: $[P_1, Q_1] = [P_1, R_i] = [Q_1, R_i] = 0$, and that is equivalent to a quite complicated nonlinear system of PDEs imposed on the coefficients of the operators (see Section 3).

2020 *Mathematics Subject Classification.* Primary 37K10; Secondary 37K20, 37K25.

The second author is the corresponding author.

This research has been partially supported by the Department of Mathematics and Applications of the Università di Milano-Bicocca, Department of Mathematics and Physics of the Università del Salento, GNFM of the Istituto Nazionale di Alta Matematica (INdAM), the research project Mathematical Methods in Non Linear Physics (MMNLP) by the Commissione Scientifica Nazionale – Gruppo 4 – Fisica Teorica of the Istituto Nazionale di Fisica Nucleare (INFN). P. L. is supported by funds of H2020-MSCA-RISE-2017 Project No. 778010 IPaDEGAN.

©2023 American Mathematical Society

The main example of bi-Hamiltonian structure of KdV-type is the bi-Hamiltonian pair for the Korteweg–de Vries equation:

$$P = P_1 = \partial_x, \qquad Q = Q_1 + R_3, \quad Q_1 = 2u\partial_x + u_x, \quad R_3 = \partial_x^3. \tag{1.2}$$

There are many nonlinear PDEs that have a similar bi-Hamiltonian structure: Camassa–Holm, Kaup–Broer, coupled Harry-Dym ($i = 3$), AKNS, Kaup–Broer ($i = 2$) (see **[21]** and references therein) and the two-component Camassa-Holm equation (see **[20]** and **[12]**). In **[1, 23, 24]** the one-component case was completely classified, while in **[21]** there is a two-component classification, see below.

Given the importance of integrable systems of PDEs of KdV-type, it is interesting to understand the geometric meaning of the trios of operators. This article is a first attempt in that direction.

A first observation is that the problem of classifying bi-Hamiltonian structures of KdV-type is particularly interesting for $i = 2$ and $i = 3$. Indeed, for $i > 3$ such structures and the corresponding integrable hierarchies can be reduced to their dispersionless limit by means of a Miura transformation (see **[4, 8]**).

A second observation is that tackling the problem in the most general form of the operators is probably hopeless, in view of the complexity of their coefficients, especially of the higher-order operators. Moreover, using the canonical form $P_1 = \eta^{ij}\partial_x$ (where η^{ij} is a constant non-degenerate matrix) of the (non-degenerate) first-order operators does not lead to any simplification of the higher-order operators.

On the other hand, it is known **[2, 7, 26–28]** that higher-order operators (of low orders) admit a canonical form with a reduced number of coefficients (2.3). That canonical form is defined up to affine transformations of the dependent variables. An affine classification of the higher-order operators exists only for the scalar case and the 2-component case (*i.e.*, systems of 2 PDEs in 2 independent variables). The affine classification starts to be problematic from 3-component operators in view of the huge number of equivalence classes.

In **[21]** we classified, using explicit calculations and the affine classification, the trios of compatible operators that give rise to bi-Hamiltonian systems of KdV-type, both in the scalar case and in the 2-component case.

Thus, it is desirable to

- uncover geometric properties of compatible trios of operators of the form (1.1);
- extend the classification of **[21]** using the above geometric properties.

In this paper, we will focus on the first item, leaving the classification problem to future research.

Recently, full classifications became available up to 8-component operators ($i = 2$) and up to 4-component operators ($i = 3$) when one considers the group of reciprocal projective transformations (see eq. (2.4)) **[16, 17, 31]**.

The above classifications are based on the relation between higher order Hamiltonian operators of Dubrovin–Novikov type and algebraic varieties of lines in a projective space. It is then interesting to know if compatible first-order operators share similar projective-geometric properties.

A major problem of the projective-geometric approach to the classification of trios is that, while reciprocal projective transformations preserve the canonical form of higher-order operators, they transform local first-order operators into nonlocal operators of Ferapontov type **[13]**. That implies that compatibility of P_1 (Q_1) and

R_i should take into account the fact that the operators P_1 (Q_1) can be nonlocal, even if they are of a type that is said to be *localizable* [**15**], as it is a member of the orbit of a local first-order operator under the action of the reciprocal projective transformation group.

Luckily, now there exist a theoretical framework [**6**] and computational tools [**5**] that can be used for compatibility calculations involving nonlocal operators of Ferapontov type.

The problem of the compatibility of a second-order or third-order operator in the canonical form (2.3) with a first-order nonlocal operator is rather hard, especially in the third-order case. So, in this paper we make an initial step towards the general compatibility problem, and consider the following case: calculate the compatibility conditions between

$$P_1 = g^{ij}\partial_x + \Gamma^{ij}_k u^k_x + w^i_k u^k_x \partial_x^{-1} u^j_x + u^i_x \partial_x^{-1} w^j_k u^k_x \quad \text{and} \quad R_3 = \eta^{ij}\partial_x^3, \tag{1.3}$$

where (η^{ij}) is a symmetric constant non-degenerate matrix ($\det(\eta^{ij}) \neq 0$), see Section 2 for the notation. Indeed, the operator R_3 generates the simplest orbit in every projective-geometric classification of third-order operators under the action of the projective reciprocal transformation group (2.4). Accordingly, the group action transforms a local first-order operator P_1 (2.2) into a nonlocal operator of the above form.

In spite of the apparent simplicity of the problem, the calculations are already significant and the results that we found admit a non-trivial geometric interpretation. In particular, the compatibility between P and R implies that:

- the Christoffel symbols Γ^{ij}_k define a Frobenius algebra structure on the cotangent bundle of the manifold of dependent variables (u^i);
- the operator $P = L + N$ splits into its local part L and its nonlocal part N, and they are independently Hamiltonian operators;
- for $n \geq 3$, $N = 0$, and the operator becomes purely local.

The operators that we obtain from the last result are not new, being a special case of results in [**29**] (see also [**3, 18**]). However, we obtain them without the initial assumption of locality of the first-order operators. We have preliminary results in case of a second-order operator R_2 and general third-order operator R_3 that show that the non-local part does not vanish.

2. Preliminaries

We will consider n unknown functions $u^k = u^k(t,x)$, $k = 1, \ldots, n$, of two independent variables (t,x). An evolutionary system of PDEs is said to be *Hamiltonian* with respect to a local Hamiltonian operator A, if it can be written in the form

$$u^k_t = f^k(t,x,u^j,u^j_x,u^j_{xx},\ldots) = A^{ij}\frac{\delta H}{\delta u^j}, \tag{2.1}$$

where $A^{ij} = A^{ij\sigma}\partial_\sigma$ is a matrix differential operator in total derivatives $\partial_\sigma = \partial_x \circ \cdots \circ \partial_x$ (σ times) and with coefficients $A^{ij\sigma}$ that are smooth functions of $(t,x,u^j,u^j_x,u^j_{xx},\ldots)$, and $H = \int h(t,x,u^j,u^j_x,u^j_{xx},\ldots)\,dx$ is a Hamiltonian functional. The operator is required to define a Poisson bracket between functionals; that is equivalent to A being formally skew-adjoint and its Schouten bracket $[A,A] = 0$ identically vanishing.

First-order Hamiltonian operators of Dubrovin–Novikov type have the form

$$P_1 = g^{ij}\partial_x + \Gamma^{ij}_k u^k_x, \tag{2.2}$$

where the coefficients g^{ij} and Γ^{ij}_k depend on (u^k) and, in the non-degenerate case $\det(g^{ij}) \neq 0$, P_1 is Hamiltonian if and only if g_{ij} (the inverse of g^{ij}) is a flat pseudo-Riemannian metric and $\Gamma^j_{hk} = -g_{hi}\Gamma^{ij}_k$ are the Christoffel symbols of the associated Levi-Civita connection.

Higher-order Dubrovin–Novikov operators have a much more complicated form, see [**10**] for details. However, it was proved [**2, 7, 26, 28**] that if the order is 2 or 3, they admit, respectively, the canonical forms

$$R_2 = \partial_x \circ f^{ij} \circ \partial_x, \qquad R_3 = \partial_x \circ (\ell^{ij}\partial_x + c^{ij}_k u^k_x) \circ \partial_x. \tag{2.3}$$

It was proved in [**16, 31**] that the above canonical forms are invariant with respect to *projective reciprocal transformations*, *i.e.* transformations of the type

$$d\tilde{x} = \Delta dx, \qquad \tilde{u}^k = \frac{T^k_j u^j + T^k_0}{\Delta}, \tag{2.4}$$

where $\Delta = T^0_j u^j + T^0_0$. The transformation is local with respect to the dependent variables but it is nonlocal with respect to the independent variables. The result of such a transformation on an operator of the form P_1 is an operator of the form

$$\tilde{P}_1 = \tilde{g}^{ij}\partial_{\tilde{x}} + \tilde{\Gamma}^{ij}_k \tilde{u}^k_{\tilde{x}} + w^i_k \tilde{u}^k_{\tilde{x}} \partial^{-1}_{\tilde{x}} \tilde{u}^j_{\tilde{x}} + \tilde{u}^i_{\tilde{x}} \partial^{-1}_{\tilde{x}} w^j_k \tilde{u}^k_{\tilde{x}}, \tag{2.5}$$

where $w^i_k = w^i_k(\tilde{u}^p)$. Such an operator is said to be a nonlocal, localizable, homogeneous operator of Ferapontov type [**13, 15**].

In our paper we will focus on the compatibility conditions between the operators P_1 of the form (2.5) and $R_3 = \eta^{ij}\partial^3_x$, with (η^{ij}) a constant matrix. While the operator R_3 is Hamiltonian if and only if $\eta^{ij} = \eta^{ji}$, the operator P_1 is Hamiltonian if and only if:

(1) g^{ij} is a contravariant pseudo-Riemannian metric and Γ^{ij}_k are the contravariant Christoffel symbols of its Levi-Civita connection;
(2) the following equations hold:

$$g^{is}w^j_s = g^{js}w^i_s, \tag{2.6}$$

$$\nabla_i w^j_k = \nabla_k w^j_i, \tag{2.7}$$

$$R^{ij}_{kh} = w^i_k\delta^j_h - w^j_k\delta^i_h - w^i_h\delta^j_k + w^j_h\delta^i_k, \tag{2.8}$$

where ∇ is the Levi-Civita connection of g^{ij} and $R^j_{pkh} = g_{pi}R^{ij}_{kh}$ is the Riemannian curvature tensor of g^{ij}.

We stress that *the problem is well-posed, since the trios of Hamiltonian operator that we are considering are invariant under the action of the projective reciprocal transformation group.*

3. Conditions of compatibility

In this Section we calculate the conditions that come from the compatibility of P and R, *i.e.* $[P, R] = 0$, for a pair of Hamiltonian operators, where $P = P_1$ as in (1.3) and $R = R_3 = \eta^{ij}\partial^3_x$, with (η^{ij}) a constant symmetric non-degenerate matrix. There will be interesting projective-geometric consequences that will be discussed afterwards.

We will adopt the 'distributional' notation for operators see [**5**, **6**], and assume the operators of the form

$$P^{ij}_{xy} = g^{ij}\delta'(x-y) + \Gamma^{ij}_s u^s_x \delta(x-y) + u^i_x \nu(x-y) w^j_s u^s_y + w^i_s u^s_x \nu(x-y) u^j_y \tag{3.1}$$

$$R^{ij}_{xy} = \eta^{ij}\delta'''(x-y), \tag{3.2}$$

where

$$\nu(x-y) = \begin{cases} \frac{1}{2} & \text{if } x > y \\ -\frac{1}{2} & \text{if } x < y \end{cases} \tag{3.3}$$

is the Heaviside step function.

LEMMA 3.1. *The condition $[P, R] = 0$ implies the following conditions:*

$$\Gamma^{ik}_l \eta^{lj} = \Gamma^{jk}_l \eta^{li}, \tag{3.4}$$

$$\Gamma^{kj}_l = \Gamma^{jk}_l, \tag{3.5}$$

$$\Gamma^{sj}_p \Gamma^{ir}_s - \Gamma^{sr}_p \Gamma^{ij}_s = 0, \tag{3.6}$$

$$3\frac{\partial \Gamma^{jk}_s}{\partial u^l} + \delta^j_l w^k_s + \delta^k_s w^j_l + \eta_{ls} w^j_m \eta^{mk} + \eta_{ms} w^m_l \eta^{jk} + \delta^k_l w^j_s + \delta^j_s w^k_l = 0, \tag{3.7}$$

$$w^i_l \eta^{lj} - w^j_l \eta^{li} = 0, \tag{3.8}$$

$$w^k_l \delta^j_s + w^j_s \delta^k_l - w^j_l \delta^k_s - w^k_s \delta^j_l = 0. \tag{3.9}$$

PROOF. We use Dubrovin-Zhang formula [**9**] for Schouten bracket:

$$[P,R]^{ijk}_{x,y,z} =$$
$$\frac{\partial P^{ij}_{x,y}}{\partial u^l_{(s)}(x)} \partial^s_x R^{lk}_{x,z} + \frac{\partial R^{ij}_{x,y}}{\partial u^l_{(s)}(x)} \partial^s_x P^{lk}_{x,z} + \frac{\partial P^{ij}_{x,y}}{\partial u^l_{(s)}(y)} \partial^s_y P^{lk}_{y,z} + \frac{\partial R^{ij}_{x,y}}{\partial u^l_{(s)}(y)} \partial^s_y P^{lk}_{y,z} +$$
$$+ \frac{\partial P^{ki}_{z,x}}{\partial u^l_{(s)}(z)} \partial^s_z R^{lj}_{z,y} + \frac{\partial R^{ki}_{z,x}}{\partial u^l_{(s)}(z)} \partial^s_z P^{lj}_{z,y} + \frac{\partial P^{ki}_{z,x}}{\partial u^l_{(s)}(x)} \partial^s_x R^{lj}_{x,y} + \frac{\partial R^{ki}_{z,x}}{\partial u^l_{(s)}(x)} \partial^s_x P^{lj}_{x,y} +$$
$$+ \frac{\partial P^{jk}_{y,z}}{\partial u^l_{(s)}(y)} \partial^s_y R^{li}_{y,x} + \frac{\partial R^{jk}_{y,z}}{\partial u^l_{(s)}(y)} \partial^s_y P^{li}_{y,x} + \frac{\partial P^{jk}_{y,z}}{\partial u^l_{(s)}(z)} \partial^s_z R^{li}_{z,x} + \frac{\partial R^{jk}_{y,z}}{\partial u^l_{(s)}(z)} \partial^s_z P^{li}_{z,x}.$$

In our case it reduces to

$$[P,R]^{ijk}_{x,y,z} =$$
$$\frac{\partial P^{ij}_{x,y}}{\partial u^l(x)} R^{lk}_{x,z} + \frac{\partial P^{ij}_{x,y}}{\partial u^l(y)} R^{lk}_{y,z} + \frac{\partial P^{ki}_{z,x}}{\partial u^l(z)} R^{lj}_{z,y} + \frac{\partial P^{ki}_{z,x}}{\partial u^l(x)} R^{lj}_{x,y}$$
$$+ \frac{\partial P^{jk}_{y,z}}{\partial u^l(y)} R^{li}_{y,x} + \frac{\partial P^{jk}_{y,z}}{\partial u^l(z)} R^{li}_{z,x} + \frac{\partial P^{ij}_{x,y}}{\partial u^l_x} \partial_x R^{lk}_{x,z} + \frac{\partial P^{ij}_{x,y}}{\partial u^l_y} \partial_y R^{lk}_{y,z}$$
$$+ \frac{\partial P^{ki}_{z,x}}{\partial u^l_z} \partial_z R^{lj}_{z,y} + \frac{\partial P^{ki}_{z,x}}{\partial u^l_x} \partial_x R^{lj}_{x,y} + \frac{\partial P^{jk}_{y,z}}{\partial u^l_y} \partial_y R^{li}_{y,x} + \frac{\partial P^{jk}_{y,z}}{\partial u^l_z} \partial_z R^{li}_{z,x}.$$

Using the Algorithm described in [6], we can bring the Schouten bracket $[P, R]$ in normal form. After long computations, we get the following expressions.

$$\frac{\partial P^{ij}_{x,y}}{\partial u^l(x)} R^{lk}_{x,z} = \frac{\partial g^{ij}}{\partial u^l}\delta'_{xy}\eta^{lk}\delta'''_{xz} + \frac{\partial \Gamma^{ij}_s}{\partial u^l} u^s_x \delta_{xy}\eta^{lk}\delta'''_{xz} + \frac{\partial w^i_s}{\partial u^l} u^s_x \nu_{xy} u^j_y \eta^{lk}\delta'''_{xz}.$$

$$\frac{\partial P^{ij}_{x,y}}{\partial u^l(y)} R^{lk}_{y,z} =$$
$$= u^i_x \frac{\partial w^j_s}{\partial u^l} u^s_x \eta^{lk}\delta''_{xy}\delta_{xz} + 2u^i_x \partial_x\left(\frac{\partial w^j_s}{\partial u^l} u^s_x\right)\eta^{lk}\delta'_{xy}\delta_{xz} + u^i_x\partial^2_x\left(\frac{\partial w^j_s}{\partial u^l}u^s_x\right)\eta^{lk}\delta_{xy}\delta_{xz}+$$
$$- 2u^i_x \frac{\partial w^j_s}{\partial u^l} u^s_x \eta^{lk}\delta'_{xy}\delta'_{xz} - 2u^i_x\partial_x\left(\frac{\partial w^j_s}{\partial u^l}u^s_x\right)\eta^{lk}\delta_{xy}\delta'_{xz} + u^i_x \frac{\partial w^j_s}{\partial u^l} u^s_x \eta^{lk}\delta_{xy}\delta''_{xz}+$$
$$+ u^i_x \frac{\partial w^j_s}{\partial u^l} u^s_z \eta^{lk}\nu_{zx}\delta'''_{zy} + 3u^i_x\partial_z\left(\frac{\partial w^j_s}{\partial u^l}u^s_z\right)\eta^{lk}\nu_{zx}\delta''_{zy} + 3u^i_x\partial^2_z\left(\frac{\partial w^j_s}{\partial u^l}u^s_z\right)\eta^{lk}\nu_{zx}\delta'_{zy}+$$
$$+ u^i_x\partial^3_x\left(\frac{\partial w^j_s}{\partial u^l}u^s_z\right)\eta^{lk}\nu_{zx}\delta_{zy}.$$

$$\frac{\partial P^{ki}_{z,x}}{\partial u^l(z)} R^{lj}_{z,y} =$$
$$-\frac{\partial g^{ki}}{\partial u^l}\eta^{lj}\delta''''_{xy}\delta_{xz} - \frac{\partial g^{ki}}{\partial u^l}\eta^{lj}\delta'''_{xy}\delta'_{xz} - \partial_x\left(\frac{\partial g^{ki}}{\partial u^l}\right)\eta^{lj}\delta'''_{xy}\delta_{xz} + \frac{\partial \Gamma^{ki}_s}{\partial u^l}u^s_x\eta^{lj}\delta'''_{xy}\delta_{xz}+$$
$$+ \frac{\partial w^k_s}{\partial u^l} u^s_z \nu_{zx} u^i_x \eta^{lj}\delta'''_{zy}$$

$$\frac{\partial P^{ki}_{z,x}}{\partial u^l(x)} R^{lj}_{x,y} =$$
$$- 3u^k_x \frac{\partial w^i_s}{\partial u^l} u^s_x \eta^{lj}\delta''_{xy}\delta_{xz} + 3u^k_x \frac{\partial w^i_s}{\partial u^l} u^s_x \eta^{lj}\delta'_{xy}\delta'_{xz} + 3u^k_{xx}\frac{\partial w^i_s}{\partial u^l}u^s_x\eta^{lj}\delta'_{xy}\delta_{xz}+$$
$$+ u^k_z \frac{\partial w^i_s}{\partial u^l} u^s_y \eta^{lj}\nu_{yz}\delta'''_{yx} + 3u^k_z\partial_y\left(\frac{\partial w^i_s}{\partial u^l}u^s_y\right)\eta^{lj}\nu_{yz}\delta''_{yx}+$$
$$+ 3u^k_z\partial^2_y\left(\frac{\partial w^i_s}{\partial u^l}u^s_y\right)\eta^{lj}\nu_{yz}\delta'_{yx} + u^k_z\partial^3_y\left(\frac{\partial w^i_s}{\partial u^l}u^s_y\right)\eta^{lj}\nu_{yz}\delta_{yx}.$$

$$
\begin{aligned}
&\frac{\partial P^{jk}_{y,z}}{\partial u^l(y)} R^{li}_{y,x} = \\
&-\frac{\partial g^{jk}}{\partial u^l}\eta^{li}\delta'''_{xy}\delta'_{xz} - 3\partial_x\left(\frac{\partial g^{jk}}{\partial u^l}\right)\eta^{li}\delta''_{xy}\delta'_{xz} - 3\partial_x^2\left(\frac{\partial g^{jk}}{\partial u^l}\right)\eta^{li}\delta'_{xy}\delta'_{xz} + \\
&-\partial_x^3\left(\frac{\partial g^{jk}}{\partial u^l}\right)\eta^{li}\delta_{xy}\delta'_{xz} - 3\frac{\partial g^{jk}}{\partial u^l}\eta^{li}\delta''_{xy}\delta''_{xz} - 6\partial_x\left(\frac{\partial g^{jk}}{\partial u^l}\right)\eta^{li}\delta'_{xy}\delta''_{xz} + \\
&+ -3\partial_x^2\left(\frac{\partial g^{jk}}{\partial u^l}\right)\eta^{li}\delta_{xy}\delta''_{xz} - 3\frac{\partial g^{jk}}{\partial u^l}\eta^{li}\delta'_{xy}\delta'''_{xz} - 3\partial_x\left(\frac{\partial g^{jk}}{\partial u^l}\right)\eta^{li}\delta_{xy}\delta'''_{xz} + \\
&-\frac{\partial g^{jk}}{\partial u^l}\eta^{li}\delta_{xy}\delta''''_{xz} - \frac{\partial \Gamma^{jk}_s}{\partial u^l}u^s_x\eta^{li}\delta'''_{xy}\delta_{xz} - 3\partial_x\left(\frac{\partial \Gamma^{jk}_s}{\partial u^l}u^s_x\right)\eta^{li}\delta''_{xy}\delta_{xz} + \\
&-3\partial_x^2\left(\frac{\partial \Gamma^{jk}_s}{\partial u^l}u^s_x\right)\eta^{li}\delta'_{xy}\delta_{xz} - \partial_x^3\left(\frac{\partial \Gamma^{jk}_s}{\partial u^l}u^s_x\right)\eta^{li}\delta_{xy}\delta_{xz} - 3\frac{\partial \Gamma^{jk}_s}{\partial u^l}u^s_x\eta^{li}\delta''_{xy}\delta'_{xz} + \\
&-6\partial_x\left(\frac{\partial \Gamma^{jk}_s}{\partial u^l}u^s_x\right)\eta^{li}\delta'_{xy}\delta'_{xz} - 3\partial_x^2\left(\frac{\partial \Gamma^{jk}_s}{\partial u^l}u^s_x\right)\eta^{li}\delta_{xy}\delta'_{xz} - 3\frac{\partial \Gamma^{jk}_s}{\partial u^l}u^s_x\eta^{li}\delta'_{xy}\delta''_{xz} + \\
&-3\partial_x\left(\frac{\partial \Gamma^{jk}_s}{\partial u^l}u^s_x\right)\eta^{li}\delta_{xy}\delta''_{xz} - \frac{\partial \Gamma^{jk}_s}{\partial u^l}u^s_x\eta^{li}\delta_{xy}\delta'''_{xz} + \frac{\partial w^j_s}{\partial u^l}u^s_y\eta^{li}\nu_{yz}u^k_z\delta'''_{yx}.
\end{aligned}
$$

$$
\begin{aligned}
&\frac{\partial P^{jk}_{y,z}}{\partial u^l(z)} R^{li}_{z,x} = \\
&u^j_x\frac{\partial w^k_s}{\partial u^l}u^s_x\eta^{li}\delta''_{xy}\delta_{xz} + 2u^j_{xx}\frac{\partial w^k_s}{\partial u^l}u^s_x\eta^{li}\delta'_{xy}\delta_{xz} + u^j_{xxx}\frac{\partial w^k_s}{\partial u^l}u^s_x\eta^{li}\delta_{xy}\delta_{xz} + \\
&+3u^j_x\frac{\partial w^k_s}{\partial u^l}u^s_x\eta^{li}\delta'_{xy}\delta'_{xz} + 3u^j_{xx}\frac{\partial w^k_s}{\partial u^l}u^s_x\eta^{li}\delta_{xy}\delta'_{xz} + 3u^j_x\partial_x\left(\frac{\partial w^k_s}{\partial u^l}u^s_x\right)\eta^{li}\delta'_{xy}\delta_{xz} + \\
&+3u^j_{xx}\partial_x\left(\frac{\partial w^k_s}{\partial u^l}u^s_x\right)\eta^{li}\delta_{xy}\delta_{xz} + 3u^j_x\frac{\partial w^k_s}{\partial u^l}u^s_x\eta^{li}\delta_{xy}\delta''_{xz} + \\
&+6u^j_x\partial_x\left(\frac{\partial w^k_s}{\partial u^l}u^s_x\right)\eta^{li}\delta_{xy}\delta'_{xz} + 3u^j_x\partial_x^2\left(\frac{\partial w^k_s}{\partial u^l}u^s_x\right)\eta^{li}\delta_{xy}\delta_{xz} + \\
&+u^j_y\frac{\partial w^k_s}{\partial u^l}u^s_x\eta^{li}\nu_{xy}\delta'''_{xz} + 3u^j_y\partial_x\left(\frac{\partial w^k_s}{\partial u^l}u^s_x\right)\eta^{li}\nu_{xy}\delta''_{xz} + \\
&+3u^j_y\partial_x^2\left(\frac{\partial w^k_s}{\partial u^l}u^s_x\right)\eta^{li}\nu_{xy}\delta'_{xz} + u^j_y\partial_x^3\left(\frac{\partial w^k_s}{\partial u^l}u^s_x\right)\eta^{li}\nu_{xy}\delta_{xz}.
\end{aligned}
$$

$$
\begin{aligned}
&\frac{\partial P^{ij}_{x,y}}{\partial u^l_x}\partial_x R^{lk}_{x,z} = \\
&(\Gamma^{ij}_l\delta_{xy} + \delta^i_l\nu_{xy}w^j_s u^s_y + w^i_l\nu_{xy}u^j_y)\eta^{lk}\delta''''_{xz} = \\
&\Gamma^{ij}_l\eta^{lk}\delta_{xy}\delta''''_{xz} + \nu_{xy}w^j_s u^s_y\eta^{ik}\delta''''_{xz} + w^i_l\nu_{xy}u^j_y\eta^{lk}\delta''''_{xz}.
\end{aligned}
$$

$$\frac{\partial P^{ij}_{x,y}}{\partial u^l_y}\partial_y R^{lk}_{y,z} =$$
$$u^i_x w^j_l \eta^{lk}\delta_{xy}\delta'''_{xz} - u^i_x w^j_l \eta^{lk}\delta'_{xy}\delta''_{xz} - u^i_x \partial_x(w^j_l)\eta^{lk}\delta_{xy}\delta''_{xz} + u^i_x w^j_l \eta^{lk}\delta''_{xy}\delta'_{xz} +$$
$$+ 2u^i_x\partial_x(w^j_l)\eta^{lk}\delta'_{xy}\delta'_{xz} + u^i_x\partial^2_x(w^j_l)\eta^{lk}\delta_{xy}\delta'_{xz} - u^i_x w^j_l\eta^{lk}\delta'''_{xy}\delta_{xz} +$$
$$- 3u^i_x\partial_x(w^j_l)\eta^{lk}\delta''_{xy}\delta_{xz} - 3u^i_x\partial^2_x(w^j_l)\eta^{lk}\delta'_{xy}\delta_{xz} - u^i_x\partial^3_x(w^j_l)\eta^{lk}\delta_{xy}\delta_{xz} +$$
$$- u^i_x w^j_l(z)\eta^{lk}\nu_{zx}\delta''''_{zy} - 4u^i_x\partial_z(w^j_l)\eta^{lk}\nu_{zx}\delta'''_{zy} - 6u^i_x\partial^2_z(w^j_l)\eta^{lk}\nu_{zx}\delta''_{zy} +$$
$$- 4u^i_x\partial^3_z(w^j_l)\eta^{lk}\nu_{zx}\delta'_{zy} - u^i_x\partial^4_z(w^j_l)\eta^{lk}\nu_{zx}\delta_{zy} + w^i_s u^s_x\eta^{jk}\delta_{xy}\delta'''_{xz} +$$
$$- w^i_s u^s_x\eta^{jk}\delta'_{xy}\delta''_{xz} + w^i_s u^s_x\eta^{jk}\delta''_{xy}\delta'_{xz} - w^i_s u^s_x\eta^{jk}\delta'''_{xy}\delta_{xz} +$$
$$- w^i_s u^s_x\eta^{jk}\nu_{zx}\delta''''_{zy}.$$

$$\frac{\partial P^{ki}_{z,x}}{\partial u^l_z}\partial_z R^{lj}_{z,y} =$$
$$\Gamma^{ki}_l\eta^{lj}\delta''''_{xy}\delta_{xz} + \nu_{zx}w^i_s u^s_x\eta^{kj}\delta''''_{zy} + w^k_l(z)\nu_{zx}u^i_x\eta^{lj}\delta''''_{zy}.$$

$$\frac{\partial P^{ki}_{z,x}}{\partial u^l_x}\partial_x R^{lj}_{x,y} =$$
$$u^k_x w^i_l\eta^{lj}\delta_{xy}\delta'''_{xz} + 3u^k_{xx}w^i_l\eta^{lj}\delta_{xy}\delta''_{xz} + 3u^k_{xxx}w^i_l\eta^{lj}\delta_{xy}\delta'_{xz} + u^k_{xxxx}w^i_l\eta^{lj}\delta_{xy}\delta_{xz} +$$
$$+ 4u^k_x w^i_l\eta^{lj}\delta'_{xy}\delta''_{xz} + 8u^k_{xx}w^i_l\eta^{lj}\delta'_{xy}\delta'_{xz} + 4u^k_{xxx}w^i_l\eta^{lj}\delta'_{xy}\delta_{xz} + 6u^k_x w^i_l\eta^{lj}\delta''_{xy}\delta'_{xz} +$$
$$+ 6u^k_{xx}w^i_l\eta^{lj}\delta''_{xy}\delta_{xz} + 4u^k_x w^i_l\eta^{lj}\delta'''_{xy}\delta_{xz} - u^k_z w^i_l(y)\eta^{lj}\nu_{yz}\delta''''_{yx} - 4u^k_z\partial_y(w^i_l)\eta^{lj}\nu_{yz}\delta'''_{yx} +$$
$$- 6u^k_z\partial^2_y(w^i_l)\eta^{lj}\nu_{yz}\delta''_{yx} - 4u^k_z\partial^3_y(w^i_l)\eta^{lj}\nu_{yz}\delta'_{yx} - u^k_z\partial^4_y(w^i_l)\eta^{lj}\nu_{yz}\delta_{yx} +$$
$$+ w^k_s u^s_x\eta^{ij}\delta_{xy}\delta'''_{xz} + 3\partial_x(w^k_s u^s_x)\eta^{ij}\delta_{xy}\delta''_{xz} + 3\partial^2_x(w^k_s u^s_x)\eta^{ij}\delta_{xy}\delta'_{xz} +$$
$$+ \partial^3_x(w^k_s u^s_x)\eta^{ij}\delta_{xy}\delta_{xz} + 4w^k_s u^s_x\eta^{ij}\delta'_{xy}\delta''_{xz} + 8\partial_x(w^k_s u^s_x)\eta^{ij}\delta'_{xy}\delta'_{xz} +$$
$$+ 4\partial^2_x(w^k_s u^s_x)\eta^{ij}\delta'_{xy}\delta_{xz} + 6w^k_s u^s_x\eta^{ij}\delta''_{xy}\delta'_{xz} + 6\partial_x(w^k_s u^s_x)\eta^{ij}\delta''_{xy}\delta_{xz} +$$
$$+ 4w^k_s u^s_x\eta^{ij}\delta'''_{xy}\delta_{xz} - w^k_s u^s_z\eta^{ij}\nu_{yz}\delta''''_{yx}.$$

$$\frac{\partial P^{jk}_{y,z}}{\partial u^l_y}\partial_y R^{li}_{y,x} =$$
$$\Gamma^{jk}_l\eta^{li}\delta''''_{xy}\delta_{xz} + 4\partial_x(\Gamma^{jk}_l)\eta^{li}\delta'''_{xy}\delta_{xz} + 6\partial^2_x(\Gamma^{jk}_l)\eta^{li}\delta''_{xy}\delta_{xz} + 4\partial^3_x(\Gamma^{jk}_l)\eta^{li}\delta'_{xy}\delta_{xz} +$$
$$+ \partial^4_x(\Gamma^{jk}_l)\eta^{li}\delta_{xy}\delta_{xz} + 4\Gamma^{jk}_l\eta^{li}\delta'''_{xy}\delta'_{xz} + 12\partial_x(\Gamma^{jk}_l)\eta^{li}\delta''_{xy}\delta'_{xz} + 12\partial^2_x(\Gamma^{jk}_l)\eta^{li}\delta'_{xy}\delta'_{xz} +$$
$$+ 4\partial^3_x(\Gamma^{jk}_l)\eta^{li}\delta_{xy}\delta'_{xz} + 6\Gamma^{jk}_l\eta^{li}\delta''_{xy}\delta''_{xz} + 12\partial_x(\Gamma^{jk}_l)\eta^{li}\delta'_{xy}\delta''_{xz} + 6\partial^2_x(\Gamma^{jk}_l)\eta^{li}\delta_{xy}\delta''_{xz} +$$
$$+ 4\Gamma^{jk}_l\eta^{li}\delta'_{xy}\delta'''_{xz} + 4\partial_x(\Gamma^{jk}_l)\eta^{li}\delta_{xy}\delta'''_{xz} + \Gamma^{jk}_l\eta^{li}\delta_{xy}\delta''''_{xz} + w^k_s u^s_z\eta^{ji}\nu_{yz}\delta''''_{yx} +$$
$$+ w^j_l(y)\nu_{yz}u^k_z\eta^{li}\delta''''_{yx}.$$

$$
\begin{aligned}
&\frac{\partial P^{jk}_{y,z}}{\partial u^l_z}\partial_z R^{li}_{z,x} = \\
&- u^j_x w^k_l \eta^{li}\delta'''_{xy}\delta_{xz} - 3u^j_{xx} w^k_l \eta^{li}\delta''_{xy}\delta_{xz} - 3u^j_{xxx} w^k_l \eta^{li}\delta'_{xy}\delta_{xz} - u^j_{xxxx} w^k_l \eta^{li}\delta_{xy}\delta_{xz} + \\
&- 4u^j_x w^k_l \eta^{li}\delta''_{xy}\delta'_{xz} - 8u^j_{xx} w^k_l \eta^{li}\delta'_{xy}\delta'_{xz} - 4u^j_{xxx} w^k_l \eta^{li}\delta_{xy}\delta'_{xz} - 4u^j_x \partial_x(w^k_l)\eta^{li}\delta''_{xy}\delta_{xz} + \\
&- 8u^j_{xx}\partial_x(w^k_l)\eta^{li}\delta'_{xy}\delta_{xz} - 4u^j_{xxx}\partial_x(w^k_l)\eta^{li}\delta_{xy}\delta_{xz} - 6u^j_x w^k_l \eta^{li}\delta'_{xy}\delta''_{xz} + \\
&- 6u^j_{xx} w^k_l \eta^{li}\delta_{xy}\delta''_{xz} - 12u^j_x \partial_x(w^k_l)\eta^{li}\delta'_{xy}\delta'_{xz} - 12u^j_{xx}\partial_x(w^k_l)\eta^{li}\delta_{xy}\delta'_{xz} + \\
&- 6u^j_x \partial^2_x(w^k_l)\eta^{li}\delta'_{xy}\delta_{xz} - 6u^j_{xx}\partial^2_x(w^k_l)\eta^{li}\delta_{xy}\delta_{xz} - 4u^j_x w^k_l \eta^{li}\delta_{xy}\delta'''_{xz} + \\
&- 12u^j_y \partial_x(w^k_l)\eta^{li}\delta_{xy}\delta''_{xz} - 12u^j_x \partial^2_x(w^k_l)\eta^{li}\delta_{xy}\delta'_{xz} - 4u^j_x \partial^3_x(w^k_l)\eta^{li}\delta_{xy}\delta_{xz} + \\
&- u^j_y w^k_l(x)\eta^{li}\nu_{xy}\delta''''_{xz} - 4u^j_y \partial_x(w^k_l)\eta^{li}\nu_{xy}\delta'''_{xz} - 6u^j_y \partial^2_x(w^k_l)\eta^{li}\nu_{xy}\delta''_{xz} + \\
&- 4u^j_y \partial^3_x(w^k_l)\eta^{li}\nu_{xy}\delta'_{xz} - u^j_y \partial^4_x(w^k_l)\eta^{li}\nu_{xy}\delta_{xz} - w^j_s u^s_x \eta^{ki}\delta'''_{xy}\delta_{xz} + \\
&- 3\partial_x(w^j_s u^s_x)\eta^{ki}\delta''_{xy}\delta_{xz} - 3\partial^2_x(w^j_s u^s_x)\eta^{ki}\delta'_{xy}\delta_{xz} - \partial^3_x(w^j_s u^s_x)\eta^{ki}\delta_{xy}\delta_{xz} + \\
&- 4w^j_s u^s_x \eta^{ki}\delta''_{xy}\delta'_{xz} - 8\partial_x(w^j_s u^s_x)\eta^{ki}\delta'_{xy}\delta'_{xz} - 4\partial^2_x(w^j_s u^s_x)\eta^{ki}\delta_{xy}\delta'_{xz} + \\
&- 6w^j_s u^s_x \eta^{ki}\delta'_{xy}\delta''_{xz} - 6\partial_x(w^j_s u^s_x)\eta^{ki}\delta_{xy}\delta''_{xz} - 4w^j_s u^s_x \eta^{ki}\delta_{xy}\delta'''_{xz} - w^j_s u^s_y \eta^{ki}\nu_{xy}\delta''''_{xz}.
\end{aligned}
$$

The above terms are written in canonical form; according with [**6**] the Schouten bracket vanishes if and only if all the coefficients of the above canonical form vanish. Collecting similar terms we have:

- The coefficient of $\delta_{xy}\delta''''_{xz}$ is
$$-\frac{\partial g^{jk}}{\partial u^l}\eta^{li} + \Gamma^{ij}_l\eta^{lk} + \Gamma^{jk}_l\eta^{li} = \Gamma^{ij}_l\eta^{lk} - \Gamma^{kj}_l\eta^{li}.$$
- The coefficient of $\delta'_{xy}\delta'''_{xz}$ is
$$-3\frac{\partial g^{jk}}{\partial u^l}\eta^{li} + 4\Gamma^{jk}_l\eta^{li} + \frac{\partial g^{ij}}{\partial u^l}\eta^{lk} = -3\Gamma^{kj}_l\eta^{li} + \Gamma^{jk}_l\eta^{li} + \Gamma^{ij}_l\eta^{lk} + \Gamma^{ji}_l\eta^{lk}.$$
- The coefficient of $\delta''_{xy}\delta''_{xz}$ is
$$-3\frac{\partial g^{jk}}{\partial u^l}\eta^{li} + 6\Gamma^{jk}_l\eta^{li} = -3\Gamma^{kj}_l\eta^{li} + 3\Gamma^{jk}_l\eta^{li}.$$
- The coefficient of $\delta'''_{xy}\delta'_{xz}$ is
$$-\frac{\partial g^{ki}}{\partial u^l}\eta^{lj} - \frac{\partial g^{jk}}{\partial u^l}\eta^{li} + 4\Gamma^{jk}_l\eta^{li} = -(\Gamma^{ki}_l + \Gamma^{ik}_l)\eta^{lj} - (\Gamma^{jk}_l + \Gamma^{kj}_l)\eta^{li} + 4\Gamma^{jk}_l\eta^{li}.$$
- The coefficient of $\delta''''_{xy}\delta_{xz}$ is
$$-\frac{\partial g^{ki}}{\partial u^l}\eta^{lj} + \Gamma^{ki}_l\eta^{lj} + \Gamma^{jk}_l\eta^{li} = -\Gamma^{ik}_l\eta^{lj} + \Gamma^{jk}_l\eta^{li}.$$

Summarizing, all the above terms provide the conditions (3.4) and (3.5). Indeed the vanishing of the first, third and fifth coefficients are clearly equivalent to (3.4). Using (3.4), the vanishing of the second coefficient is equivalent to

$$
\begin{aligned}
(3.10)\quad & -3\Gamma^{kj}_l\eta^{li} + \Gamma^{jk}_l\eta^{li} + \Gamma^{ij}_l\eta^{lk} + \Gamma^{ji}_l\eta^{lk} \\
&= -2\Gamma^{kj}_l\eta^{li} + \Gamma^{jk}_l\eta^{li} + \Gamma^{ji}_l\eta^{lk} \\
&= (\Gamma^{jk}_l - \Gamma^{kj}_l)\eta^{li} + (\Gamma^{ji}_l - \Gamma^{ij}_l)\eta^{lk} = 0
\end{aligned}
$$

and the vanishing of the fourth coefficient turns out to be

$$
\begin{aligned}
(3.11)\quad -(\Gamma^{ki}_l+\Gamma^{ik}_l)\eta^{lj}-(\Gamma^{jk}_l+\Gamma^{kj}_l)\eta^{li}+4\Gamma^{jk}_l\eta^{li} &= \\
-\Gamma^{ki}_l\eta^{lj}-\Gamma^{kj}_l\eta^{li}+2\Gamma^{jk}_l\eta^{li} &= \\
(\Gamma^{jk}_l-\Gamma^{kj}_l)\eta^{li}+(\Gamma^{ik}_l-\Gamma^{ki}_l)\eta^{lj} &= 0.
\end{aligned}
$$

Summing up the two conditions (3.10) and (3.11) we obtain

$$3(\Gamma^{jk}_l-\Gamma^{kj}_l)\eta^{li}=0,$$

on account of which both conditions (3.10) and (3.11) vanish. That is equivalent to condition (3.5).

Let us continue to collect like terms.

- The coefficient of $\nu_{xy}\delta''''_{xz}$ is

$$w^j_s u^s_y\eta^{ik}+w^i_l u^j_y\eta^{lk}-u^j_y w^k_l\eta^{li}-w^j_s u^s_y\eta^{ki}=u^j_y(w^i_l\eta^{lk}-w^k_l\eta^{li}).$$

- The coefficient of $\nu_{zx}\delta''''_{zy}$ is

$$-u^i_x w^j_l\eta^{lk}-w^i_s u^s_x\eta^{jk}+w^i_s u^s_x\eta^{kj}+w^k_l u^i_x\eta^{lj}=-u^i_x(w^j_l\eta^{lk}-w^k_l\eta^{lj}).$$

- The coefficient of $\nu_{yz}\delta''''_{yx}$ is

$$-u^k_z w^i_l\eta^{lj}-w^k_s u^s_z\eta^{ij}+w^k_s u^s_z\eta^{ji}+w^j_l u^k_z\eta^{li}=-u^k_z(w^i_l\eta^{lj}-w^j_l\eta^{li}).$$

All the above coefficients provide the condition (3.8).

Now, we collect all coefficients of u^s_{xxxx} inside the coefficient of $\delta_{xy}\delta_{xz}$. We obtain the condition:

$$-\Gamma^{jk}_{s,l}\eta^{li}+w^i_l\delta^k_s\eta^{lj}+w^k_s\eta^{ij}+\Gamma^{jk}_{l,s}\eta^{li}-w^k_l\delta^j_s\eta^{li}-w^j_s\eta^{ki}=0. \tag{3.12}$$

Using (3.8) we can eliminate η^{li}, and using (2.8) we obtain (3.6).

Let us consider two further coefficients.

- The coefficient of $\delta'_{xy}\delta''_{xz}$ is

$$
\begin{aligned}
-6\partial_x\left(\frac{\partial g^{jk}}{\partial u^l}\right)\eta^{li}-3\frac{\partial\Gamma^{jk}_s}{\partial u^l}u^s_x\eta^{li}-u^i_x w^j_l\eta^{lk}-w^i_s u^s_x\eta^{jk}+\\
+4u^k_x w^i_l\eta^{lj}+4w^k_s u^s_x\eta^{ij}+12\partial_x(\Gamma^{jk}_l)\eta^{li}-6u^j_x w^k_l\eta^{li}-6w^j_s u^s_x\eta^{ki}.
\end{aligned}
$$

- The coefficient of $\delta''_{xy}\delta'_{xz}$ is

$$
\begin{aligned}
-3\partial_x\left(\frac{\partial g^{jk}}{\partial u^l}\right)\eta^{li}-3\frac{\partial\Gamma^{jk}_s}{\partial u^l}u^s_x\eta^{li}+u^i_x w^j_l\eta^{lk}+w^i_s u^s_x\eta^{jk}+\\
+6u^k_x w^i_l\eta^{lj}+6w^k_s u^s_x\eta^{ij}-4u^j_x w^k_l\eta^{li}-4w^j_s u^s_x\eta^{ki}+12\partial_x(\Gamma^{jk}_l)\eta^{li}.
\end{aligned}
$$

The sum and the difference of the above two coefficients are, respectively,

$$
\begin{aligned}
&6(\Gamma^{jk}_{l,s}-\Gamma^{jk}_{s,l})\eta^{li}+10(\delta^k_s w^i_l\eta^{lj}+w^k_s\eta^{ij}-\delta^j_s w^k_l\eta^{li}-w^j_s\eta^{ki}),\\
&6\Gamma^{jk}_{l,s}\eta^{li}+2\delta^i_s w^j_l\eta^{lk}+2w^i_s\eta^{jk}+2\delta^k_s w^i_l\eta^{lj}+2w^k_s\eta^{ij}+2\delta^j_s w^k_l\eta^{li}+2w^j_s\eta^{ki}.
\end{aligned}
$$

The above expressions yield, on account of (3.12), the conditions

$$
\begin{aligned}
&3\frac{\partial\Gamma^{jk}_s}{\partial u^l}+\delta^j_l w^k_s+\delta^k_s w^j_l+\eta_{ls}w^j_m\eta^{mk}+\eta_{ms}w^m_l\eta^{jk}+\delta^k_l w^j_s+\delta^j_s w^k_l=0,\\
&w^k_l\eta^{ij}+w^j_m\eta^{mi}\delta^k_l-w^j_l\eta^{ki}-w^k_m\eta^{im}\delta^j_l=0.
\end{aligned}
$$

The first condition is condition (3.7) and the second condition is (3.9) up to lowering the indices. □

REMARK 3.2. We observe that (2.8) and (3.9) imply the closure condition

$$\Gamma^{jk}_{l,s} - \Gamma^{jk}_{s,l} = 0.$$

The conditions that we obtained in the above Lemma admit an interesting geometric interpretation. First of all, we have a Frobenius algebra structure on the cotangent bundle of the manifold of dependent variables.

THEOREM 3.3. *The condition* $[P, R] = 0$ *implies that there exists a Frobenius algebra (without unity)* $(\eta, \circ)$ *which is defined by the operation*

$$(a \circ b)_k = \Gamma^{ij}_k a_i b_j \tag{3.13}$$

PROOF. The product $\circ$ defined on the cotangent space of the manifold of dependent variables is commutative in view of (3.5), associative in view of (3.6) and compatible with the Monge metric η in view of (3.4). □

Another interesting consequence of the conditions of the Lemma is that the first order operator splits into the sum of two operators that are independently Hamiltonian.

THEOREM 3.4. *The condition* $[P, R] = 0$ *implies that the operator* P *splits into the sum* $P = L + N$*, where* L *and* N *are the two Hamiltonian operators*

$$L^{ij} = g^{ij}\partial_x + \Gamma^{ij}_k u^k_x, \qquad N^{ij} = w^i_k u^k_x \partial_x^{-1} u^j_x + u^i_x \partial_x^{-1} w^j_k u^k_x \tag{3.14}$$

PROOF. The statement is a direct consequence of (3.9). □

THEOREM 3.5. *For* $n \geq 3$ *the affinor* w^i_j *vanishes, the contravariant Christoffel symbols* Γ^{ij}_k *become constant and* P *becomes a local Hamiltonian operator.*

PROOF. Let us assume that $n \geq 3$. Using the previous conditions it is easy to prove that the first order Hamiltonian operator must be local. Indeed, for $j = s \neq k \neq l \neq j$ the condition (3.9) tell us that $w^k_l = 0$ for any $k \neq l$. For $j = s \neq k = l$ we obtain $w^k_k + w^j_j = 0$ for any $k \neq j$. This implies $w^i_i = 0$ for any i since if $i \neq j \neq k \neq i$ we should have simoultenously:

$$w^i_i + w^j_j = 0, \quad w^i_i + w^k_k = 0, \quad w^k_k + w^j_j = 0.$$

The first two conditions imply $w^k_k = w^j_j$ which in turn implies $w^j_j = w^k_k = w^i_i = 0$.

Using (3.7) it is immediate to check that the vanishing of the affinor implies that the Christoffel symbols Γ^{ij}_k are constant.

For $n \geq 3$ we do not have further conditions since all the remaining terms contain the affinor or the derivatives of the Christoffel symbols. □

REMARK 3.6. We stress that the locality of the trios for $n > 2$ is not an *a priori* assumption, but it is the result of non trivial computations and it is strictly related with the choice $R^{ij}_3 = \eta^{ij}\partial_x^3$. See **[30]** for many examples of compatible pairs of P_1 (local and nonlocal) and R_3 (in normal form, with non-constant leading coefficient matrix h^{ij}) where locality *does not hold.*

Let us consider now a pair of flat contravariant linear metrics (g,h) whose Christoffel symbols satisfy the conditions (3.4), (3.5), (3.6). These metrics define a flat pencil iff the pencil $g-\lambda h$ is flat for any λ and the contravariant Christoffel symbols of the pencil coincide with the pencil of contravariant Christoffel symbols $\Gamma^{ij}_k-\lambda b^{ij}_k$ (where Γ^{ij}_k and b^{ij}_k are the contravariant Christoffel symbols of g and h respectively).

Imposing these conditions we get

$$\Gamma^{sj}_p b^{ir}_s-\Gamma^{sr}_p b^{ij}_s+b^{sj}_p\Gamma^{ir}_s-b^{sr}_p\Gamma^{ij}_s=0, \tag{3.15}$$

$$g^{il}b^{kj}_l+h^{il}\Gamma^{kj}_l=g^{kl}b^{ij}_l-h^{kl}\Gamma^{ij}_l. \tag{3.16}$$

Notice that both conditions are automatically satisfied if we choose $h=\eta$. In this case (3.15) is automatically satisfied since $b^{ij}_k=0$ and (3.16) reduces to (3.5) which is fulfilled by hypothesis.

COROLLARY 3.7. *Any Frobenius algebra structure $(\eta,\circ)$ (without unity) on the cotangent space defines a compatible trio of Hamitonian operators.*

REMARK 3.8. There is a well-known differential-geometric interpretation of nonlocal operators of the type (1.3): the metric g_{ij} should be conformally flat and w^i_j coincides with the Schouten tensor (see [**14**] for details). However, a projective-geometric interpretation is still missing. Further projective-geometric aspects of first-order (nonlocal) operators that are compatible with local third-order operators will be considered in a future research work.

4. Conclusion

The results of the present paper go back to 2017 and are part of a general project of classification of Hamiltonian trios based on normal forms of second-order and third-order operators under the action of projective reciprocal transformations. Due to the difficulties of treating the general case we have decided to publish these preliminary results.

The local trios that we get for $n>2$ are known in the literature. They can be obtained as special cases of the results of [**29**] and also appear in [**3**]. What is new in this paper is their projective-differential geometric interpretation that requires dealing with localizable Hamiltonian operators. Further investigation is needed in order to completely describe the operators P_1 when $n=2$.

Acknowledgment

We thank E. V. Ferapontov for his interest in our work.

References

[1] A. M. Astashov and A. M. Vinogradov, *On the structure of Hamiltonian operators in field theory*, J. Geom. Phys. **3** (1986), no. 2, 263–287, DOI 10.1016/0393-0440(86)90022-7. MR879368

[2] A. V. Balandin and G. V. Potëmin, *On nondegenerate differential-geometric Poisson brackets of third order* (Russian), Uspekhi Mat. Nauk **56** (2001), no. 5(341), 177–178, DOI 10.1070/RM2001v056n05ABEH000441; English transl., Russian Math. Surveys **56** (2001), no. 5, 976–977. MR1892563

[3] Alexey V. Bolsinov, Andrey Yu. Konyaev, and Vladimir S. Matveev, *Applications of Nijenhuis geometry: non-degenerate singular points of Poisson-Nijenhuis structures*, Eur. J. Math. **8** (2022), no. 4, 1355–1376, DOI 10.1007/s40879-020-00429-6. MR4510609

[4] Guido Carlet, Reinier Kramer, and Sergey Shadrin, *Central invariants revisited* (English, with English and French summaries), J. Éc. polytech. Math. **5** (2018), 149–175, DOI 10.5802/jep.66. MR3738511

[5] M. Casati, P. Lorenzoni, D. Valeri, and R. Vitolo, *Weakly nonlocal Poisson brackets: tools, examples, computations*, Comput. Phys. Commun. **274** (2022), Paper No. 108284, 18, DOI 10.1016/j.cpc.2022.108284. MR4373337

[6] Matteo Casati, Paolo Lorenzoni, and Raffaele Vitolo, *Three computational approaches to weakly nonlocal Poisson brackets*, Stud. Appl. Math. **144** (2020), no. 4, 412–448, DOI 10.1111/sapm.12302. MR4094629

[7] Philip W. Doyle, *Differential geometric Poisson bivectors in one space variable*, J. Math. Phys. **34** (1993), no. 4, 1314–1338, DOI 10.1063/1.530213. MR1210220

[8] Boris Dubrovin, Si-Qi Liu, and Youjin Zhang, *On Hamiltonian perturbations of hyperbolic systems of conservation laws. I. Quasi-triviality of bi-Hamiltonian perturbations*, Comm. Pure Appl. Math. **59** (2006), no. 4, 559–615, DOI 10.1002/cpa.20111. MR2199786

[9] B. Dubrovin and Y. Zhang, *Normal forms of integrable PDEs, Frobenius manifolds and Gromov-Witten invariants*, arXiv: `arXiv:math/0108160`.

[10] B. A. Dubrovin and S. P. Novikov, *Poisson brackets of hydrodynamic type* (Russian), Dokl. Akad. Nauk SSSR **279** (1984), no. 2, 294–297. MR770656

[11] B. A. Dubrovin and S. P. Novikov, *Hamiltonian formalism of one-dimensional systems of the hydrodynamic type and the Bogolyubov-Whitham averaging method* (Russian), Dokl. Akad. Nauk SSSR **270** (1983), no. 4, 781–785. MR715332

[12] Gregorio Falqui, *On a Camassa-Holm type equation with two dependent variables*, J. Phys. A **39** (2006), no. 2, 327–342, DOI 10.1088/0305-4470/39/2/004. MR2198963

[13] E. V. Ferapontov, *Nonlocal Hamiltonian operators of hydrodynamic type: differential geometry and applications*, Topics in topology and mathematical physics, Amer. Math. Soc. Transl. Ser. 2, vol. 170, Amer. Math. Soc., Providence, RI, 1995, pp. 33–58, DOI 10.1090/trans2/170/03. MR1355550

[14] E. V. Ferapontov, *Conformally plane metrics, systems of hydrodynamic type and nonlocal Hamiltonian operators*, Russian Math. Surveys 50 (1995), no. 4, 811–813.

[15] E. V. Ferapontov and M. V. Pavlov, *Reciprocal transformations of Hamiltonian operators of hydrodynamic type: nonlocal Hamiltonian formalism for linearly degenerate systems*, J. Math. Phys. **44** (2003), no. 3, 1150–1172, DOI 10.1063/1.1542921. MR1958260

[16] E. V. Ferapontov, M. V. Pavlov, and R. F. Vitolo, *Projective-geometric aspects of homogeneous third-order Hamiltonian operators*, J. Geom. Phys. **85** (2014), 16–28, DOI 10.1016/j.geomphys.2014.05.027. MR3253545

[17] E. V. Ferapontov, *Towards the classification of homogeneous third-order Hamiltonian operators*, Int. Math. Res. Not. **22** (2016), 6829–6855, arXiv: `arXiv:1508.02752`.

[18] A. Yu. Konyaev, *Geometry of Inhomogeneous Poisson Brackets, Multicomponent Harry Dym Hierarchies, and Multicomponent Hunter–Saxton Equations*, Russ. J. Math. Phys. **29** (2022), no. 4, 518–541, DOI 10.1134/s1061920822040100. MR4539419

[19] Joseph Krasil'shchik, Alexander Verbovetsky, and Raffaele Vitolo, *The symbolic computation of integrability structures for partial differential equations*, Texts and Monographs in Symbolic Computation, Springer, Cham, 2017, DOI 10.1007/978-3-319-71655-8. MR3753682

[20] Si-Qi Liu and Youjin Zhang, *Deformations of semisimple bihamiltonian structures of hydrodynamic type*, J. Geom. Phys. **54** (2005), no. 4, 427–453, DOI 10.1016/j.geomphys.2004.11.003. MR2144711

[21] P. Lorenzoni, A. Savoldi, and R. Vitolo, *Bi-Hamiltonian structures of KdV type*, J. Phys. A **51** (2018), no. 4, 045202, 16, DOI 10.1088/1751-8121/aa994d. MR3742010

[22] Franco Magri, *A simple model of the integrable Hamiltonian equation*, J. Math. Phys. **19** (1978), no. 5, 1156–1162, DOI 10.1063/1.523777. MR488516

[23] O. I. Mokhov, *Local third-order Poisson brackets* (Russian), Uspekhi Mat. Nauk **40** (1985), no. 5(245), 257–258. MR810820

[24] O. I. Mokhov, *Hamiltonian differential operators and contact geometry* (Russian), Funktsional. Anal. i Prilozhen. **21** (1987), no. 3, 53–60, 96. MR911774

[25] Peter J. Olver and Philip Rosenau, *Tri-Hamiltonian duality between solitons and solitary-wave solutions having compact support*, Phys. Rev. E (3) **53** (1996), no. 2, 1900–1906, DOI 10.1103/PhysRevE.53.1900. MR1401317

[26] G. V. Potëmin, *Poisson brackets of differential-geometric type* (Russian), Dokl. Akad. Nauk SSSR **286** (1986), no. 1, 39–42. MR822095

[27] G. V. Potëmin, *Some aspects of differential geometry and algebraic geometry in the theory of solitons*, Ph.D. thesis, Moscow State University, 1991, 99 pages, in Russian.

[28] G. V. Potëmin, *On third-order differential-geometric Poisson brackets* (Russian), Uspekhi Mat. Nauk **52** (1997), no. 3(315), 173–174, DOI 10.1070/RM1997v052n03ABEH001817; English transl., Russian Math. Surveys **52** (1997), no. 3, 617–618. MR1479400

[29] Ian A. B. Strachan and Błażej M. Szablikowski, *Novikov algebras and a classification of multicomponent Camassa-Holm equations*, Stud. Appl. Math. **133** (2014), no. 1, 84–117, DOI 10.1111/sapm.12040. MR3244959

[30] J. Vašíček and R. Vitolo, *WDVV equations and invariant bi-Hamiltonian formalism*, J. High Energy Phys. **8** (2021), Paper No. 129, 28, DOI 10.1007/jhep08(2021)129. MR4317141

[31] P. Vergallo and R. Vitolo, *Projective geometry of homogeneous second order Hamiltonian operators*, arXiv: arXiv:2203.04237.

Dipartimento di Matematica Applicazioni, University of Milano-Bicocca, and Sezione INFN di Milano-Bicocca, via Roberto Cozzi 53 I-20125 Milano, Italy
Email address: paolo.lorenzoni@unimib.it

Dipartimento di Matematica e Fisica "E. De Giorgi", Università del Salento, and Sezione INFN di Lecce, via per Arnesano, 73100 Lecce, Italy
Email address: raffaele.vitolo@unisalento.it

Contemporary Mathematics
Volume **788**, 2023
https://doi.org/10.1090/conm/788/15826

Measurement of random operators, jet geometry and high-order phase transitions

Valentin V. Lychagin

In memory of my teacher Sasha Vinogradov

Abstract. In this paper, we discuss various geometrical structures appearing in study of random linear operators that act between finite-dimensional vector spaces. We show that they are closely connected to the geometry of the jet bundles over a manifold of the extreme probability measures. Thus, the high-order phase transitions appear as singularity in prolongations of the state manifolds.

1. Introduction

In this paper, we discuss a relationship between the geometry of jet spaces ([**5**],[**6**]) and information geometry ([**12**], [**1**]). To be more concrete, we restrict ourselves to the geometrical theory of measurement of random operators ([**10**], [**9**]) that constitutes a more applicable part of classical information geometry. A more attractive part of this interaction consists of the relationship between the geometry of singularities in solutions of nonlinear differential equations from the jet geometry side ([**5**], [**8**]) and phase transitions from the information geometry side ([**3**], [**7**], [**13**]).

It is worth to note that geometrical theory of singularities of solutions is based mainly on singularities of projections like Arnold's theory of singularities of projections of Lagrangian manifolds ([**2**]).

On the other hand, there are two types of phase transitions ([**3**], [**7**], [**13**]): the phase transitions of the first order related to transitions from one phase to another accompanied by "jumps" of extensive variables, and the phase transitions of the high order, where we do not observe "jumps" but only non-regular behavior in higher-order derivatives.

This gives a hint that in the geometric theory of differential equations there are also singularities other than the singularities of projections.

It turns out these are the singularities of prolongations. We illustrate these phenomena by the example of the Van der Waals equation of state and the second-order phase transitions that occur there.

2020 *Mathematics Subject Classification.* 47B80, 82B26.
Key words and phrases. Random operator, phase transition, jet geometry.
The author was partially supported by Russian Science Foundation Grant No 21-71-20034.

©2023 American Mathematical Society

The paper is organized in the following way. At first, we apply methods of paper **[10]** to the measurement of random linear operators, acting between two finite dimensional vector spaces. We use the Kullback–Leibler divergence and the principle of the minimum information gain to get the description of the extreme probability measures and, finally, to get the geometrical description of the measurement procedure in terms of Legendrian or, in general, integral manifolds in the jet bundles over the manifold of all extreme probability measures.

The further interplay between geometry and probability allows us to construct a series of important geometrical structures on the state manifold of the measurement, as well as on their prolongations.

Then, at first, we use all these geometrical structures to reduce the state manifolds by eliminating the non-applicable states. Namely, we say a state is applicable if the means of positive random tensors, evaluating in the geometrical terms are positive. For example, the second central moment defines a differential symmetric 2-form on the state manifold, and positivity of the variance requires removing the domain where this form is negative. It automatically leads to the existence of a Riemannian structure on the applicable part of the state manifold.

The second half devoted to the consideration of various types of phase transitions. We find conditions, under which these transitions occur, and illustrate them in the case of Van der Waals gas.

The final part is devoted to studies of characteristic polynomials for random operators acting on a given vector space.

2. Measurement of random operators

By a random operator, we mean a measurable map $A\colon (\Omega,\mu) \to \mathrm{Hom}\,(V,W)$, where (Ω,μ) is a probability space, V and W are finite-dimensional vector spaces over $\mathbb{R}$. Especially we are interested in the case when $W = V$, where $\mathrm{Hom}\,(V,V) = \mathrm{End}\,V$ is the associative algebra of linear operators in V.

In what follows, we will interpret probability measures as "measurement devices" and by a result of μ-measurement we understand the mean of the random operator $A\,(\omega)$, i. e. the following linear operator

$$A_\mu = \int_\Omega A(\omega)\, d\mu \in \mathrm{Hom}(V,W).$$

We assume that the initial or prior measure μ_0 and the random operator A are chosen in such a way that $A_{\mu_0} = 0$.

Moreover, all probability measures μ under consideration are assumed to be absolutely continuous regarding the chosen prior measure μ_0.

This means that there is a density function ρ that equals the Radon–Nikodym derivative $d\mu/d\mu_0$.

Therefore, instead of "measurement devices" we could consider the set of non-negative measurable functions ρ such that $\int_\Omega \rho(\omega)\, d\mu_0 = 1$, and the zero set $\{\rho = 0\} \subset \Omega$ has a zero μ_0-measure.

Thus, to get operator $a \in \mathrm{Hom}(V,W)$ as the result of measurement, we need to find the probability density function ρ such that $\int_\Omega \rho(\omega)A(\omega)\, d\mu_0 = a$.

It is clear that these conditions do not define the measurement device.

Therefore, we need more conditions.

To this end, let us consider the Kullback–Leibler divergence (or information gain) between two probability measures

$$I(\mu, \mu_0) = \int_\Omega \ln\left(\frac{d\mu}{d\mu_0}\right) d\mu.$$

It satisfies two important for us properties: $I(\mu, \mu_0) \geq 0$ and $I(\mu, \mu_0) = 0$ if and only if $\mu = \mu_0$ almost everywhere.

Thus, we convert the problem of finding the measurement device into the following variational problem: find the minimum of the functional

$$I\colon \rho \mapsto \int_\Omega \rho \ln \rho \, d\mu_0,$$

subjected to two constrains

$$\int_\Omega \rho(\omega)\, d\mu_0 = 1,$$

$$\int_\Omega \rho(\omega) A(\omega)\, d\mu_0 = a.$$

Introducing the Lagrange multipliers $\lambda \in \mathbb{R}$ and $\alpha \in \mathrm{Hom}(V,W)^* = \mathrm{Hom}(W,V)$, where operators $\alpha \in \mathrm{Hom}(W,V)$, we will identify them with linear functions on $\mathrm{Hom}(V,W)$ in the following way

$$\langle \alpha, a\rangle = \mathrm{Tr}\, \alpha a.$$

To see that this correspondence is an isomorphism, we represent $\mathrm{Hom}(V,W)$ and $\mathrm{Hom}(W,V)$ in the standard way: $\mathrm{Hom}(V,W) = W \otimes V^*$, $\mathrm{Hom}(W,V) = V \otimes W^*$.

Then,

$$\mathrm{Hom}(V,W)^* = W^* \otimes V \xleftrightarrow{\text{Twist}} V \otimes W^* = \mathrm{Hom}(W,V),$$

and the paring could be seen from the following sequence of natural operators

$$\mathrm{Hom}(V,W) \otimes \mathrm{Hom}(W,V) = W \otimes V^* \otimes V \otimes W^* \xrightarrow{\textit{composition}} V \otimes V^* \xrightarrow{\mathrm{Tr}} \mathbb{R}.$$

Thus, we use the method of Lagrange multipliers and consider the following functional

$$\mathcal{F}\colon \rho \mapsto \int_\Omega \rho \ln \rho \, d\mu_0 + \lambda \left(\int_\Omega \rho(\omega)\, d\mu_0 - 1 \right) + \mathrm{Tr}\left(\alpha \left(\int_\Omega \rho(\omega) A(\omega)\, d\mu_0 - a \right)\right).$$

The standard computations (see, for example, [**10**]), lead us to the following solution

$$\rho_\alpha(\omega) = \frac{1}{Z(\alpha)} \exp\left(\mathrm{Tr}\left(\alpha A(\omega)\right)\right), \tag{1}$$

where

$$Z(\alpha) = \int_\Omega \exp\left(\mathrm{Tr}\left(\alpha A(\omega)\right)\right) d\mu_0,$$

or

$$Z(\alpha) = \int_\Omega \det\left(\exp\left(\alpha A(\omega)\right)\right) d\mu_0 \tag{2}$$

is the partition function, and $\alpha \in \operatorname{Hom}(W, V)$.

The probability measure $\mu_\alpha = \rho_\alpha \mu_0$ is called *an extreme measure.*

Remark also that μ_α equals μ_0, when $\alpha = 0$.

Moreover,

$$A_{\mu_\alpha} = \frac{\partial H}{\partial \alpha},$$

where

$$H(\alpha) = \ln Z(\alpha).$$

Moreover, it is easy to check that the information gain $I(\alpha) = I(\mu_\alpha, \mu_0)$ equals

$$I(\mu_\alpha, \mu_0) = \int_\Omega \rho_\alpha \ln \rho_a \, d\mu_0 = \int_\Omega \rho_\alpha (-\ln Z(\alpha) + \operatorname{Tr}(\alpha A)) \, d\mu_0 =$$
$$- \ln Z(\alpha) + \operatorname{Tr}(\alpha A_{\mu_\alpha}) = \operatorname{Tr}\left(\alpha \frac{\partial H}{\partial \alpha}\right) - H(\alpha).$$

In other words, the function $I(\alpha)$ of the information gain is the Legendre transform of function $H(\alpha)$.

In thermodynamics, the last function is called the Gibbs free energy (see, for example, [**10**]) and in our more general case we will continue to call the function $H(\alpha)$ the Gibbs free energy function.

To simplify further notations, we denote by P the vector space of operators $\operatorname{Hom}(V, W)$ and by $Q = P^* = \operatorname{Hom}(W, V)$ the dual vector space.

The above relations (1), (2) allow us find the central moments of the extreme probability measures.

At first, we define random tensors

$$A^{\otimes k} \colon \Omega \to P^{\otimes k},$$
$$A^{\otimes k} \colon \omega \mapsto A(\omega)^{\otimes k} \in P^{\otimes k},$$

and their means

$$m_k(\alpha) = \int_\Omega A(\omega)^{\otimes k} \, d\mu_0,$$

which are symmetric tensors in $S^k(P) \subset P^{\otimes k}$.

Remark that elements of the space $S^k(P) = S^k(Q^*)$ are homogeneous polynomials on Q.

Denote by $\Sigma^k(Q)$ the module of symmetric differential forms on the manifold Q. Then the above interpretation shows us that $m_k \in \Sigma^k(Q)$ are symmetric differential k-forms.

We call them the *k-th moments of the random operator A.*

More precisely, we get (see [**10**] for more details) the following expressions for these moments in terms of partition function

$$m_k = \frac{k!}{Z} d_k Z \in \Sigma^k(Q).$$

Here we denoted by $d_k f$ the k-th differential of a function f.

Namely, if the function f is a function of n variables, say $f(x_1, \ldots, x_n)$, then is the following symmetric differential k-form

$$d_k f = \sum_\sigma \frac{1}{\sigma!} \frac{\partial^k f}{\partial x^\sigma} dx_1^{\sigma_1} \cdot \cdots \cdot dx_N^{\sigma_N},$$

where $\sigma = (\sigma_1, \ldots, \sigma_N)$ are multi-indices and $\cdot$ stands for the symmetric product of symmetric differential forms.

The central moments $c_k = \sum_{i=0}^{k}(-1)^{k-i}\binom{k}{i}m_i \cdot m_1^{k-i} \in \Sigma^k(Q)$ have the following expression in terms of the partition function Z, (see [**10**]):

$$c_k = \sum_{i=0}^{k} \frac{i!\, d_i Z}{Z} \left(-\frac{dZ}{Z}\right)^{k-i}.$$

Thus, for example, the second central moment in terms of the Gibbs free energy has the following form

$$c_2 = d_2 H \in \Sigma^2(Q).$$

3. Jet geometry and probability

The probability analysis of a random operator A includes, as a rule, knowledge of all (central) moments of A.

As we have seen, this means, in terms of the Gibbs free energy function H, the knowledge of all jets of the function.

To this end, we consider the trivial bundle $\pi \colon \mathbb{R} \times Q \to Q$ and its k-jet bundles $\pi_k \colon \mathbf{J}^k \to Q$.

Let $\pi_{k,l} \colon \mathbf{J}^k \to \mathbf{J}^l$, where $k > l$, be the natural projections.

The fibre $\pi_k^{-1}(\alpha)$ of the bundle π_k, consists of all possible k-jets $[f]_\alpha^k$ at the point $\alpha \in Q$ of all smooth functions on Q.

The trivial affine connection on the manifold Q defines the splitting:

$$\pi_k^{-1}(\alpha) = \oplus_{i=0}^{i=k} S^i(P),$$
$$[f]_\alpha^k = f(a) \oplus d_a f \oplus \cdots \oplus d_a^k f.$$

Remark that in the case of the Gibbs free energy H or partition function Z the k-jets $[H]_\alpha^k$ and $[Z]_\alpha^k$ define and are completely defined by all moments $m_i(\alpha)$ or $c_i(\alpha)$, $1 \le i \le k$, and values $H(\alpha)$ and $Z(\alpha)$.

Denote by $L_f^{(k)} \subset \mathbf{J}^k$ the graph of k-jet of the function $f \in C^\infty(Q)$, i. e. the submanifold of dimension $\dim Q$ formed by k-jets $[f]_\alpha^k$, $\alpha \in Q$.

Then, $\pi_{k,l}(L_f^{(k)}) = L_f^{(l)}$, and we call submanifold $L_f^{(k)}$ *$(k-l)$-prolongation of the submanifold $L_f^{(l)}$*.

It is known (see, for example, [**5**]) that the manifolds $\mathbf{J}^k$, $k \ge 1$, are endowed with the natural distributions $\mathcal{C}_k$, called the Cartan distributions, and the manifolds $L_f^{(k)}$ are integral manifolds of these distributions uniquely distinguished in the class of all $(\dim Q)$-dimensional integral manifolds L of the Cartan distributions by the only requirement, that the mapping $\pi_k \colon L \to Q$ is a diffeomorphism.

In what follows, we will call a *state manifold of order* k any $(\dim Q)$-dimensional integral manifold $L \subset \mathbf{J}^k$ of the Cartan distribution.

Thus, see for example ([**10**]), in thermodynamics the Equations Of State (EOS), are just state manifolds of order 1.

3.1. Adjusted state manifolds. Let $L \subset \mathbf{J}^k$ be a state manifold and let

$$\operatorname{Sing} L = \left\{x \in L \,\middle|\, \dim\ker\left(d_x\pi_k \colon T_x L \to T_{\pi_k(x)}Q\right) \ge 1\right\}$$

be the set of all singular points of the mapping $\pi_k \colon L \to Q$.

The set of singular values of the mapping $\pi_k\colon L \to Q$, i. e. the set $\pi_k(\operatorname{Sing} L)$, we, to follow the thermodynamic terminology, call the *coexistence submanifold.*

Then, in the complement $L_{\mathrm{reg}} = L \setminus \operatorname{Sing} L$, the mapping $\pi_k\colon L_{\mathrm{reg}} \to Q$ is a local diffeomorphism.

Therefore, for any point $x \in L_{\mathrm{reg}}$ there are neighborhoods $U_x \subset L_{\mathrm{reg}}$, $O_x = \pi_k(U_x) \subset Q$, such that $x \in U_x$ and $\pi_k\colon U_x \to O_x$ is a diffeomorphism. So, there is a smooth function H_x (called *local Gibbs function*) such that $L_{H_x}^{(k)} \subset L_{\mathrm{reg}}$.

Moreover, together with this function, we get a series of symmetric differential forms $m_i(H_x)$, $c_i(H_x)$, $i \le k$, in the neighborhood O_x. Taking their pullbacks, we get symmetric differential i-forms $m_i \in \Sigma^i(U_x)$, $m_i \in \Sigma^i(U_x)$.

Summing up these forms, we get symmetric differential i-forms, $1 \le i \le k$,

$$c_i \in \Sigma^i(L_{\mathrm{reg}}),\ m_i \in \Sigma^i(L_{\mathrm{reg}})$$

on the regular part of the state manifold.

3.2. Applicable domains. Consider symmetric tensors $\tau \in S^{2k}(Q)$, $k \in \mathbb{N}$, and say that they are *positive,* $\tau > 0$, if $\langle \tau, p^{\otimes 2k}\rangle > 0$, for any tensor $p \in Q^* = P$.

Let now A be a random operator and $L \subset \mathbf{J}^1$ be the corresponding state Legendrian manifold. Consider the regular parts $L_{\mathrm{reg}}^{(2k)} \subset \mathbf{J}^{2k}$ of the prolongations of L.

We say that a point $x_{2k} \in L_{\mathrm{reg}}^{(2k)}$ is *applicable* if for any positive polynomial

$$f(X) = X^{\otimes 2k} + \binom{2}{2k} a_{2k-2} X^{\otimes(2k-2)} + \cdots + \binom{l}{2k} a_l X^{\otimes l} + \cdots + a_0,$$

where $a_i \in S^{2k-i}(Q)$, $X \in P$, the mean of the random operator $f(A)$, $\langle f(A)\rangle_\lambda \in S^{2k}(Q)$ is positive at the point x_k.

We will call the set of all applicable points $x_{2k} \in L_{\mathrm{reg}}^{(2k)}$ applicable domain of order $2k$ and denote it by $L_{\mathrm{app}}^{(2k)}$.

REMARK 1. *We have used only polynomials $f(X)$ with the condition $a_{2k-1} = 0$ in the definition of positivity. The shift $X \mapsto X + a_{k-1}$ shows that the above definition of positivity is equivalent to the definition, where we use arbitrary positive polynomials with the condition $a_{2k-1} \neq 0$.*

Let us analyze the conditions of the positivity for the cases $k = 1$ and $k = 2$.

Thus, let $k = 1$, and

$$f(X) = X^{\otimes 2} + a_2, \quad a_2 \in S^{2k}(Q),$$

be a positive tensor, i. e. $a_2 > 0$.

Then

$$\langle f(A)\rangle_\lambda - f(\mu) = c_2,$$

and it is positive for all positive polynomial f if and only if $c_2 \ge 0$.

For the case $k = 2$, we have polynomial of the form

$$f(X) = X^{\otimes 4} + 6a_2 X^{\otimes 2} + 4a_1 X + a_0,$$
$$a_2 \in S^2(Q), \quad a_1 \in S^3(Q), \quad a_0 \in S^4(Q).$$

The conditions of their positivity are the following (see [**10**]):

$$a_2 > 0, \quad a_0 a_2 - a_1^2 - a_2^3 > 0.$$

We have

$$\langle f(A)\rangle_\lambda - f(\mu) = c_4 + 4\mu c_3 + 6(\mu^2 + a_2)c_2.$$

Moreover, we have $f(\mu) \geq 0$ because of positivity f.

Let us consider the difference $\langle f(A)\rangle_\lambda - f(\mu) = 6c_2\mu^2 + 4c_3\mu + c_4 + 6a_2c_2$ as a quadratic polynomial in μ.

Therefore, the difference $\langle f(A)\rangle_\lambda - f(\mu)$ is a positive polynomial if

$$c_2 > 0,$$
$$16c_3^2 - 24c_2\left(c_4 + 6a_2c_2\right) < 0,$$

Rewrite the last inequality in the following form

$$\left(3c_2c_4 - 2c_3^2\right) + 18a_2c_2 > 0.$$

As we have seen, tensors c_2, a_2 are positive but a_2 can be arbitrary small.

Therefore, we guarantee positivity of the mean $\langle f(A)\rangle_\lambda$ if and only if $c_2 > 0$ and the following inequality for symmetric tensors

$$3c_2c_4 - 2c_3^2 > 0 \tag{3}$$

holds.

Thus, we get the following description of applicable domains.

THEOREM 2. *The domain $L_{\text{app}}^{(2)}$ consist of points where the symmetric differential 2-form c_2 is positive and the domain $L_{\text{app}}^{(4)}$ consists of the points where symmetric differential 2-form c_2 is positive and symmetric* 6*-form* $3c_2c_4 - 2c_3^2 > 0$ *is positive too.*

REMARK 3. *Thus, the second central moment c_2 defines the structures of Riemannian manifolds on the applicable domains. The third and fourth central moments are not arbitrary symmetric differential forms of degree 3 and 4, but they have to satisfy the inequality* (3).

REMARK 4. *The constructive conditions for the positivity of symmetric* 6*-forms can also be found in the paper* [**10**].

3.3. Phase transitions. Let $L^{(1)} \subset P \times Q \times \mathbb{R} = \mathbf{J}^1$ be a Legendrian manifold that represents a state manifold for a measurement procedure.

The first type of singularity that we study is the singularity of the projection $\pi_1 \colon L \to Q$, that is the set

$$\operatorname{Sing} L = \{x \in L, \dim\ker d_x\pi_1 \neq 0\}.$$

Removing this singularity, we get a Legendrian manifold $L_{\text{reg}} = L \setminus \operatorname{Sing} L$, that can be seen as the graph of 1-jet of a multivalued function H (Gibbs energy) over the domain $Q \setminus \pi_1(\operatorname{Sing} L)$, where the set $\pi_1(\operatorname{Sing} L) \subset Q$ is called in thermodynamics *coexistence surface.*

Taking now prolongations $L_{\text{reg}}^{(k)} \subset \mathbf{J}^k$, we get state manifolds of orders $k \geq 2$. In this procedure, we meet another type of singularity.

To clarify these phenomena, we consider the following "basic" case.

Namely, let us consider a covering $\alpha \colon M \to B$ of finite degree n, where M and B are smooth manifolds.

Let $\pi_1(M)$ and $\pi_1(B)$ be fundamental groups of the manifolds M and B respectively. Then it is known (see, for example, [**4**]), that the manifold M is a $\pi_1(B)$-manifold, and $\pi_1(B)$-orbits are exactly the fibres of the projection α and they are isomorphic to the coset $\pi_1(B)/\pi_1(M)$ of the group $\pi_1(B)$ by the subgroup $\alpha_*(\pi_1(M)) \subset \pi_1(B)$.

Moreover, if we identify functions on B with functions on M, then the subgroup $\alpha_*(\pi_1(M))$ acts in the trivial way on $C^\infty(B)$ and $f \in C^\infty(M)$ is a fixed element of $\alpha_*(\pi_1(M))$-action, then $f \in C^\infty(B)$.

Using this observation, we take a point $b \in B$ and let the fibre $\alpha^{-1}(b)$ consist of points $\{b_1, \dots, b_n\}$.

Take a function $h \in C^\infty(M)$ and consider the following polynomial in a parameter z:

$$p_h(z,b) = \prod_{i=1}^{n} (z - h(b_i)). \tag{4}$$

THEOREM 5. *Let $h \in C^\infty(M)$ and*

$$p_h(z,b) = z^n + a_{n-1}(b)z^{n-1} + \cdots + a_1(b)z + a_0(b).$$

Then, $a_i \in C^\infty(B)$.

PROOF. Assume, that the fundamental group $\pi_1(B)$ acts in the trivial way on z. Then, by definition 4, this polynomial is $\pi_1(B)$-invariant. Therefore, $a_i(b)$ are also $\pi_1(B)$-invariants and $a_i \in C^\infty(B)$. □

Let us consider now the case when $M \subset L^{(k)}_{\text{app}} \subset \mathbf{J}^k$, $\alpha = \pi_k$ and $B = \pi_k(M) \subset Q$ is a domain.

Let $(q_1, \dots, q_N)$ be coordinates in Q, and $(q_1, \dots, q_N, u, u_1, \dots, u_N, \dots, u_\sigma, \dots)$, $\sigma = (\sigma_1, \dots, \sigma_N)$ be the standard coordinates in the jet bundles $\mathbf{J}^l$, $l = 0, 1, \dots, k, \dots$

Then the prolongation of M on the level of $(k+1)$-jets is described by the values of functions $u_{\sigma+1_i} = \frac{du_\sigma}{dq_i}$ on B. Here we denoted by $\frac{d}{dq_i}$ the total derivative along q_i.

On the other hand, due to the above theorem, we have relations of the form

$$p_\sigma(u_\sigma) = 0, \tag{5}$$

for all σ, $|\sigma| = k$, where p_σ are the following polynomials

$$p_\sigma(z) = z^n + a_{\sigma,n-1}z^{n-1} + \cdots + a_{\sigma,0},$$

with coefficients in $C^\infty(B)$.

Taking total derivatives of (5) we get the following equations

$$u_{\sigma+1_i}\frac{\partial p_\sigma}{\partial z}(u_\sigma) + \frac{\partial p_\sigma}{\partial q_i}(u_\sigma) = 0, \tag{6}$$

for all multi-indices σ of length $|\sigma| = k$.

Here,

$$\frac{\partial p_\sigma}{\partial q_i}(u_\sigma) = \frac{\partial a_{\sigma,n-1}}{\partial q_i}z^{n-1} + \cdots + \frac{\partial a_{\sigma,0}}{\partial q_i}.$$

We call points on M, where the discriminants of all polynomial p_σ, $|\sigma| = k$ do not vanish, *prolongable points.*

Remark that at such points all solutions $u_{\sigma+1_i}$ of the equations (6) are smooth, and therefore these points belong to the image of the mapping $\pi_{k+1,k}\colon L^{(k+1)} \to L^{(k)}$.

In the case when the discriminant of some polynomial p_σ, $|\sigma| = k$, equals zero at a point $b \in B$, equations (6) have a smooth solution $u_{\sigma+1_i}$ if and only if the polynomials $\frac{\partial p_\sigma}{\partial q_i}$ are proportional to $\frac{\partial p_\sigma}{\partial z}$.

Thus, we extend the notion of prolongable point by adding the latter points.

Namely, we call *prolongable* such points on $M \subset L^{(k)}$, where the discriminants of all polynomial p_σ, $|\sigma| = k$, are different from zero or if a discriminant of some polynomial p_σ equals zero then common roots of polynomials p_σ and $\frac{\partial p_\sigma}{\partial z}$ are also roots of polynomials $\frac{dp_\sigma}{dq_i}$.

Remark that under these prolongability conditions prolongable points are precisely the points of M that belong to the image of the mapping $\pi_{k+1,k}\colon L^{(k+1)} \to L^{(k)}$.

It is worth to note also that the above condition on polynomials p_σ, $\frac{\partial p_\sigma}{\partial z}$ and $\frac{dp_\sigma}{dq_i}$ to have common roots is too strong and rarely realizes in practice.

Going back to the probabilistic interpretation, we will say that non-prolongable points correspond to states where (central) moments of order k are blowing up i.e. tends to infinity in some directions.

Thus, we have two types of singular phenomena (or so called phase transition), the first one (first order phase transitions) occurs on the level $k = 1$ and corresponds to singularities of the projection $\pi_1\colon L^{(1)} \to Q$, and the second ones ($k$-th order phase transitions) do occur in orders $k \geq 2$, where the non-prolongable points appear. They correspond to blow up of k-moments of processes.

3.4. The Van der Waals gas. The conservation energy law for real gases has the form $\theta = 0$, where the differential 1-form θ is the following

$$\theta = d\sigma - T^{-1}\, d\varepsilon - pT^{-1}\, dv,$$

where σ, ε, v are the specific entropy, the specific internal energy and the specific volume, correspondingly

They are extensive variables, but the corresponding intensive variables are temperature T and pressure P.

The Van der Waals equation of state is the Lagrangian manifold $L \subset \mathbb{R}^4(\varepsilon, v, T, p)$ given by the following equations [**11**]:

$$p = \frac{8T}{3v-1} - \frac{3}{v^2}, \quad \varepsilon = \frac{4n}{3}T - \frac{3}{v}.$$

If we introduce the function

$$G = \sigma - \frac{pv - \varepsilon}{T},$$

then the state equation $L^{(1)}$ takes the form

$$G_T = -\frac{v}{T}, \quad G_p = \frac{\varepsilon + pv}{T}.$$

In the jet coordinates (T, p, u, u_1, u_2) on $\mathbf{J}^1$ we have

$$u = G, \quad u_1 = \frac{\varepsilon + pv}{T}, \quad u_2 = -\frac{v}{T},$$

and the state equation $L^{(1)}$ finally takes the form:

$$\mathcal{E}_1 = 3pT^3u_2^3 + T^2(p+8T)u_2^2 + 9Tu_2 + 3 = 0,$$
$$\mathcal{E}_2 = pT^2u_2^2 - 4T^2\left(n - \frac{3u_1}{4}\right)u_2 - 9 = 0.$$

We get its first prolongation by taking the first total derivatives of the above equations.

Thus we get:

$$\begin{aligned}
\mathcal{E}_{11} &= -9pT^2u_2^3 - (9pT^3u_{12} + 2pT + 24T^2)u_2^2 - (2pT^2u_{12} + 16T^3u_{12} + 9)u_2 - 9Tu_{12},\\
\mathcal{E}_{12} &= -9pT^3u_2^2u_{22} - 3T^3u_2^3 - 2pT^2u_2u_{2,2} - 16T^3u_2u_{2,2} - T^2u_2^2 - 9Tu_{22},\\
\mathcal{E}_{21} &= -3T^3(pu_{12} + u_{11})u_2^2 - 18u_2 - 9Tu_{12},\\
\mathcal{E}_{22} &= -3T^2u_2^3 - 3T^2(pu_{22} + u_{12})u_2^2 - 9u_{22}.
\end{aligned}$$

Eliminating u_2 from $\mathcal{E}_1$ and $\mathcal{E}_{11}$, we get a polynomial of degree 3 with coefficients being polynomials in T and p. The discriminant D_{11} of this polynomial equals

$$D_{11} = T^{12}f_{11}^2\Delta,$$

where

$$f_{11} = p^4 + 36p^3t + 480p^2t^2 + 2816pt^3 + 6144t^4 + 108p^3 + 324p^2t - 15552pt^2 - \\ 89856t^3 + 4374p^2 - 61236pt + 723168t^2 + 78732p - 1180980t + 531441,$$

and

$$\Delta = -p^3 - (24T + 54)p^2 - (192T^2 - 1080T + 729)p - 512T^3 + 432T^2$$

is the discriminant of the polynomial $\mathcal{E}_1$ in u_2.

In the same way, eliminating u_2 from $\mathcal{E}_{12}$ and $\mathcal{E}_1$ we get a polynomial of degree 3 in u_{22} with the discriminant

$$D_{12} = -559872T^8f_{12}^2\Delta,$$

where

$$f_{12} = p^4 + 36p^3T + 480p^2T^2 + 2816pT^3 + 6144T^4 + 108T^3 + 324p^2T - 15552pT^2 - \\ 89856T^3 + 4374p^2 - 61236pT + 723168T^2 + 78732p - 1180980T + 531441.$$

Equations $\mathcal{E}_{21}$ and $\mathcal{E}_{22}$ give us nothing new and we see that the first order phase transitions occur on the discriminant set $\Delta^{-1}(0) \subset \mathbb{R}^2(T,p)$ (see more details in [**10**]) and the second order phase transitions occurs on the set $f_{11}^{-1}(0) \cup f_{12}^{-1}(0) \subset \mathbb{R}^2(T,p)$.

Below, we picture the first order phase transitions

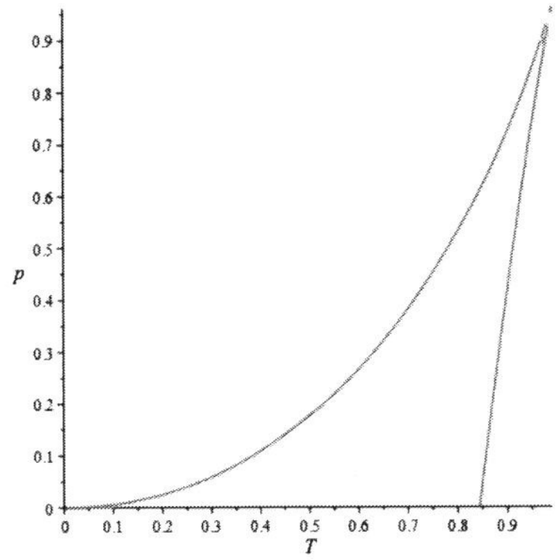

FIGURE 1. First-order phase transitions.

and the curves, where the second order phase transitions occur.

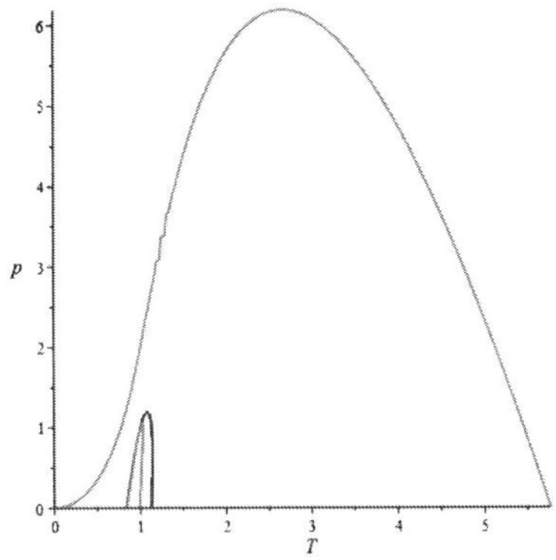

FIGURE 2. Second-order phase transitions.

The last picture shows all phase transitions together

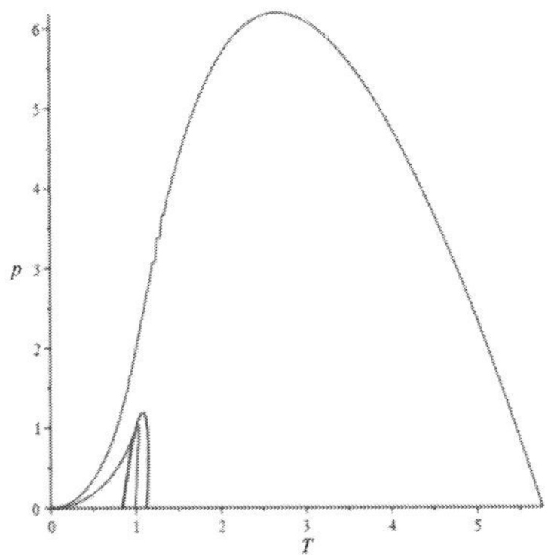

FIGURE 3. Phase transitions of the first and second orders.

4. Random endomorphisms

At first, we remind some constructions from multilinear algebra in a suitable for us form.

Let V be a finite-dimensional vector space over $\mathbb{R}$ and $\dim V = n$.

The following two graded algebras associated with V will be important for us.

Namely, the tensor algebra $T(V) = \oplus_{k=0} V^{\otimes k}$, and the exterior algebra $\Lambda(V) = \oplus_{k=0}^{n} \Lambda^k(V)$.

Moreover, any linear operator $A \in \operatorname{End} V$ generates two grade homomorphisms $T(A)\colon T(V) \to T(V)$ and $\Lambda(A)\colon \Lambda(V) \to \Lambda(V)$, k-th components of which we will denote by $A^{\otimes k}$ and $\Lambda^k(A)$ respectively.

Then the operator $A^{\otimes k}\colon V^{\otimes k} \to V^{\otimes k}$ is a representation of the element $A^{\otimes} \in \operatorname{End}(V)^{\otimes}$ and we will identify them.

Denote by $\operatorname{Alt}_k\colon V^{\otimes k} \to \Lambda^k V$ the skew-symmetrization operator and by $\iota_k\colon \Lambda^k V \to V^{\otimes k}$ the natural embedding of vector spaces.

Then, the k-th exterior power $\Lambda^k(A)$ of the operator A is the restriction of the tensor degree $A^{\otimes k}$ in the following sense: $\Lambda^k(A) = \operatorname{Alt}_k \circ A^{\otimes} \circ \iota_k$.

In these notation, the characteristic polynomial $\operatorname{ch}_A(\lambda) = \det(A - \lambda \operatorname{id}) = \sum_{k=0}^{n} p_k \lambda^k$ has the following coefficients:

$$p_k = (-1)^{n-k} \operatorname{Tr}\left(\Lambda^{n-k} A\right).$$

Consider now the random operator

$$A\colon \Omega \to \operatorname{End} V,$$

and let μ_α, $\alpha \in \operatorname{End} V$, be the extreme probability measure (2).

Then coefficients $p_k(A)$ of the characteristic polynomial $\operatorname{ch}_A(\lambda)$ are random variables, and their means $\langle p_k(A)\rangle_\alpha$ are the following

$$\begin{aligned}\langle p_k(A)\rangle_\alpha &= (-1)^{n-k} \operatorname{Tr}\left(\left\langle \Lambda^{n-k} A\right\rangle_\alpha\right) = (-1)^{n-k} \operatorname{Tr}\left(\left\langle \operatorname{Alt}_k \circ A^{\otimes k} \circ \iota_k\right\rangle_\alpha\right) = \\ &(-1)^{n-k} \operatorname{Tr}\left(\operatorname{Alt}_k \circ \left\langle A^{\otimes k}\right\rangle_\alpha \circ \iota_k\right) = (-1)^{n-k} \operatorname{Tr}\left(\operatorname{Alt}_k \circ \mu_k(\alpha) \circ \iota_k\right).\end{aligned}$$

On the other hand, the characteristic polynomial of the mean $\langle A\rangle_\alpha$ has the coefficients $p_k\left(\langle A\rangle_\alpha\right)$:

$$p_k\left(\langle A\rangle_\alpha\right) = (-1)^{n-k} \operatorname{Tr}\left(\operatorname{Alt}_k \circ \langle A\rangle_\alpha^{\otimes k} \circ \iota_k\right) = (-1)^{n-k} \operatorname{Tr}\left(\operatorname{Alt}_k \circ \mu_1(\alpha)^{\otimes k} \circ \iota_k\right).$$

Summarizing, we get the following result.

THEOREM 6. *The difference $\Delta_\alpha(A)$ between the mean of characteristic polynomial $\langle \operatorname{ch}_A(\lambda)\rangle_\alpha$ and the characteristic polynomial of the mean $\operatorname{ch}_{\langle A\rangle_\alpha}(\lambda)$ equals*

$$\Delta_\alpha(A) = \sum_{k=1}^{n-1} (-1)^{n-k} \operatorname{Tr}\left(\operatorname{Alt}_k \circ c_k(\alpha) \circ \iota_k\right) \lambda^k \tag{7}$$

Assume now, that the state Legendrian manifold is given

$$L^{(1)} \subset \mathbf{J}^1 = \operatorname{End} V \times \operatorname{End} V \times \mathbb{R},$$

where the first multiplier $\operatorname{End} V = Q$ is formed by intensive quantities α, that label extreme measures, and the second corresponds to the extensive quantities, that are means of random operators $A(\omega)$ regarding the extreme measure.

Central moments c_k are symmetric forms on the prolongations $L^{(k)} \subset \mathbf{J}^k$. Therefore, formula (7) gives us the polynomial $\Delta(A)$ of degree $(n-1)$ in λ on the

$(n-1)$ prolongation $L^{(n-1)} \subset \mathbf{J}^{n-1}$ of the state manifold. In particular, the phase transitions up to the order $(n-1)$ exert an essential influence on the behavior of the characteristic polynomial.

References

[1] Shun-ichi Amari and Hiroshi Nagaoka, *Methods of information geometry*, Translations of Mathematical Monographs, vol. 191, American Mathematical Society, Providence, RI; Oxford University Press, Oxford, 2000. Translated from the 1993 Japanese original by Daishi Harada, DOI 10.1090/mmono/191. MR1800071

[2] V. I. Arnol′d, *Singularities of Legendre varieties, of evolvents and of fronts at an obstacle*, Ergodic Theory Dynam. Systems **2** (1982), no. 3-4, 301–309 (1983), DOI 10.1017/S0143385700001620. MR721725

[3] Paul Ehrenfest, *Phasenumwandlungen im ueblichen und erweiterten Sinn, classifiziert nach den entsprechenden Singularitaeten des thermodynamischen Potentiales*, NV Noord-Hollandsche Uitgevers Maatschappij, 1933.

[4] Allen Hatcher, *Algebraic topology*, Cambridge University Press, Cambridge, 2002. MR1867354

[5] I. S. Krasil′shchik, V. V. Lychagin, and A. M. Vinogradov, *Geometry of jet spaces and nonlinear partial differential equations*, Advanced Studies in Contemporary Mathematics, vol. 1, Gordon and Breach Science Publishers, New York, 1986. Translated from the Russian by A. B. Sosinskiĭ. MR861121

[6] Boris Kruglikov and Valentin Lychagin, *Geometry of differential equations*, Handbook of global analysis, Elsevier Sci. B. V., Amsterdam, 2008, pp. 725–771, 1214, DOI 10.1016/B978-044452833-9.50015-2. MR2389645

[7] L. D. Landau and E. M. Lifshitz, *Statistical physics*, Course of Theoretical Physics, Vol. 5, Pergamon Press, Ltd., London-Paris; Addison-Wesley Publishing Company, Inc., Reading, Mass., 1958. Translated from the Russian by E. Peierls and R. F. Peierls. MR0136378

[8] V. V. Lychagin, *Geometric theory of singularities of solutions of nonlinear differential equations* (Russian), Problems in geometry, Vol. 20 (Russian), Itogi Nauki i Tekhniki, Akad. Nauk SSSR, Vsesoyuz. Inst. Nauchn. i Tekhn. Inform., Moscow, 1988, pp. 207–247. Translated in J. Soviet Math. **51** (1990), no. 6, 2735–2757. MR966202

[9] Eivind Schneider, *Differential invariants in thermodynamics*, Nonlinear PDEs, their geometry, and applications, Tutor. Sch. Workshops Math. Sci., Birkhäuser/Springer, Cham, 2019, pp. 223–232. MR3932303

[10] Eivind Schneider, *Thermodynamics as a theory of measurement*, Journal of Geometry and Physics **172** (2022), 104430 (en).

[11] V. V. Lychagin and M. D. Roop, *Critical phenomena in filtration processes of real gases*, Lobachevskii J. Math. **41** (2020), no. 3, 382–399, DOI 10.1134/s1995080220030129. MR4131613

[12] Frank Nielsen, *The many faces of information geometry*, Notices Amer. Math. Soc. **69** (2022), no. 1, 36–45, DOI 10.1090/noti2403. MR4353351

[13] H. Eugene Stanley, *Introduction to phase transitions and critical phenomena*, The International series of monographs on physics, Clarendon Press, Oxford, 1971.

V. A. Trapeznikov Institute of Control Sciences of Russian Academy of Sciences 65 Profsoyuznaya street, Moscow 117997, Russia

Email address: `valentin.lychagin@uit.no`

Contemporary Mathematics
Volume **788**, 2023
https://doi.org/10.1090/conm/788/15827

Metrics admitting projective and c-projective vector fields

Gianni Manno, Jan Schumm, and Andreas Vollmer

In memory of A. M. Vinogradov, whose Mathematics was and will be of great impact.

ABSTRACT. Vector fields on a smooth manifold with an affine connection are called *projective* (resp. *c-projective*) if their local flows preserve geodesics (resp. J-planar curves). In this expository paper, metrics with such vector fields are discussed. Emphasis is put on Lie's classical 1882 problem of finding a local description of surfaces with projective vector fields, known results on this problem, as well as on various extensions in the literature.

1. Introduction

Let M be a smooth manifold of dimension n equipped with a connection ∇. In the present paper, all connections are torsion-free and affine. A curve $\gamma : I \subseteq \mathbb{R} \mapsto \gamma(t) \in M$ such that

$$\nabla_{\dot\gamma}\dot\gamma = \alpha(t)\dot\gamma \tag{1.1}$$

for some function $\alpha \in C^\infty(I)$, is called an *unparametrized geodesic* of ∇.

DEFINITION 1.1. Two connections on the same manifold M are *projectively equivalent* if they share the same unparametrized geodesics. The collection of all connections projectively equivalent to a given connection ∇ is called the *projective class* of ∇, denoted by $[\nabla]_{\mathfrak{p}}$, and it is a *projective structure* on M.

DEFINITION 1.2. A *projective transformation* is a (local) diffeomorphism of M that sends unparametrized geodesics into unparametrized geodesics. A vector field on M is *projective* if its (local) flow acts by projective transformations.

The projective vector fields of a given connection ∇ form a Lie algebra [**26**], called the *projective Lie algebra*, which we denote by $\mathfrak{p}(\nabla)$.

The above definitions can be given also in the case that M is equipped with a *metric* g, i.e., a symmetric, non-degenerate $(0, 2)$-tensor field of arbitrary signature,

---

2020 *Mathematics Subject Classification.* Primary 53B10; Secondary 53A20, 53B20.

The authors acknowledge support through the project PRIN 2017 "Real and Complex Manifolds: Topology, Geometry and holomorphic dynamics", the project "Connessioni proiettive, equazioni di Monge-Ampère e sistemi integrabili" Istituto Nazionale di Alta Matematica (INdAM) and the MIUR grant "Dipartimenti di Eccellenza 2018-2022 (E11G18000350001)". The first author acknowledges the "Finanziamento alla Ricerca (53\RBA17MANGIO)". The first author is a member of GNSAGA of INdAM. The second author was a research fellow at INdAM.

Corresponding author: Gianni Manno.

©2023 by the authors

by considering its Levi-Civita connection. In this case, the projective class of g will be denoted by $[g]_{\mathfrak{p}}$ and its Lie algebra of projective vector fields by $\mathfrak{p}(g)$.

If g is positive definite, it is called a *Riemannian* metric. A metric is of *constant curvature* if its sectional curvatures coincide and are constant. In particular, if M is 2-dimensional, this means that the Gaussian curvature is constant.

DEFINITION 1.3. A projective vector field v is *homothetic* (for the metric g) if the Lie derivative of g along v satisfies $\mathcal{L}_v g = \lambda g$ for some constant $\lambda \in \mathbb{R}$. If v is not homothetic, it is called an *essential* projective vector field.

In the 1880s, Sophus Lie posed the following problem for 2-dimensional surfaces [**25, 26**]:[1]

Lie's 1882 problem: It is being required to determine the form of the arc element of any surface whose geodesic curves admit an infinitesimal transformation.

In fact, Lie distinguishes two problems, dubbed Lie's First Problem and Second Problem, respectively. The first is the one stated above, and asks to describe metrics whose geodesic curves admit (at least) one infinitesimal transformation. The other asks to describe metrics whose geodesic curves admit *at least* two independent infinitesimal transformations.[2] For details we refer to Section 3 below. The distinction made by Lie turns out to be much deeper than it might initially seem. Indeed we shall see that the solution techniques found in the literature are fundamentally different in either case.

Lie's formulation of the problem appears ambiguous on various levels, and we would like to mention at least the following aspects here. Indeed, Lie merely asks us to *determine* ("bestimmen") the metrics. This might be understood as the task to find a list of candidates (which is not required to be sharp, i.e. might contain metrics not satisfying the criteria). However, a more thorough answer would provide a sharp list of metrics (such that any entry has the desired properties). Lastly, one might even ask for a proper classification, e.g. up to projective or isometric transformations.

Another ambiguity arises as the projective Lie algebra action could show *singularities*, i.e. it can have orbits of nonconstant dimension. While Lie's classical problem is well understood today around generic points, such singularities are much less well understood (some partial results are obtained in [**30**]).

A last, and less deep, ambiguity is that Lie explicitly asks us only for the metrics as such. However, one would naturally ask, as well, for the full and explicit projective vector fields that these metrics admit. For simplicity of exposition, we focus on the following interpretation of Lie's problem.

Lie's classical problem: Determine a sharp list of local forms of 2-dimensional metrics admitting a projective Lie algebra of dimension greater than zero, along with the admitted projective vector fields.

[1]German original: "Es wird verlangt, die Form des Bogenelementes einer jeden Fläche zu bestimmen, deren geodätische Curven eine infinitesimale Transformation gestatten".

[2]Lie's Second Problem (original German version in [**25**]): "Man soll die Form des Bogenelementes einer jeden Fläche bestimmen, deren geodätische Curven mehrere infinitesimale Transformationen gestatten." A rough translation is: "One should determine the form of the arc element of any surface whose geodesic curves admit several infinitesimal transformations."

The idea of Lie's problem is, of course, not restricted to the 2-dimensional case. The analogous problem for any dimension $n \geq 2$ can be stated and studied as well:

Extended Lie problem: Determine a sharp list of local forms of n-dimensional metrics ($n \geq 3$) admitting a projective Lie algebra of dimension greater than zero, along with the admitted projective vector fields.

We are going to come back to this problem in Section 4. Before proceeding to a detailed exposition of the established solution of Lie's classical problem, we mention that Lie's problem also has a natural analog in the realm of Kähler geometry. In this context one needs to study infinitesimal transformations which preserve the so-called J-planar curves, a generalization of geodesics for complex manifolds. Detailed definitions and a synopsis of some established results are given in Section 5.

2. Projective structures, their metrizability and degree of mobility

In the present section the Einstein summation convention is applied unless otherwise specified. Condition (1.1) reads, in a system of coordinates $(y^1, \dots, y^n)$ of M,

$$\ddot{y}^k + \Gamma^k_{ij}\dot{y}^i\dot{y}^j = \alpha(t)\dot{y}^k\,.$$

By eliminating the function $\alpha(t)$, a system of $n-1$ ordinary differential equations (ODEs) describing *unparametrized geodesics* is obtained:

$$\begin{aligned} (2.1)\quad y^k_{xx} + \Gamma^k_{11} + \sum_{i=2}^{n}(2\Gamma^k_{1i} - \delta^k_i\Gamma^1_{11})y^i_x \\ + \sum_{i,j=2}^{n}(\Gamma^k_{ij} - \delta^k_i\Gamma^1_{1j} - \delta^k_j\Gamma^1_{1i})y^i_x y^j_x - \sum_{i,j=2}^{n}\Gamma^1_{ij}\, y^i_x y^j_x y^k_x\,, \end{aligned}$$

where $k \in \{2, \dots, n\}$ and

$$(y^1, y^2, \dots, y^n) = (x, y^2, \dots, y^n)\,.$$

Since the equivalence class $[\nabla]_{\mathfrak{p}}$ of connections that are projectively equivalent to ∇ is represented by (2.1), we call (2.1) the *projective connection associated to* ∇.

Note that the System (2.1) is of special type, i.e. of the form

$$(2.2)\qquad y^k_{xx} = f^k_{11} + \sum_{i=2}^{n} f^k_{1i}y^i_x + \sum_{i,j=2}^{n} f^k_{ij}y^i_x y^j_x + \sum_{i,j=2}^{n} f_{ij}\, y^i_x y^j_x y^k_x\,,$$

where $k \in \{2, \dots, n\}$ and where f^k_{ab} and f_{ij} are functions of $(x, y^2, \dots, y^n)$.

DEFINITION 2.1. A *projective connection* on a smooth manifold M is a system of ODEs of type (2.2).

REMARK 2.2. Essentially, a projective connection on M realizes a projective structure on M in terms of a system of ODEs. A straightforward computation shows that the class of ODEs (2.2) is closed under *point transformations*, i.e., transformations

$$(2.3)\qquad (x, y^2, \dots, y^n) \to (u, v^2, \dots, v^n)$$

with u, v^2, $\dots$, v^n functions of $(x, y^2, \dots, y^n)$.

DEFINITION 2.3. Two systems of type (2.2) are called *(point-)equivalent* if there exists a (point-)transformation (2.3) mapping one into the other.

One might now ask oneself under which conditions a projective structure is *metrizable*, by which we mean the following:

Metrizability problem: Given a connection ∇, decide whether there exists a Levi-Civita connection ∇^g (for a metric g) such that $\nabla^g \in [\nabla]_{\mathfrak{p}}$.

From an analytic viewpoint this is equivalent, due to (2.1) and (2.2), to proving the existence of functions $g_{ij} = g_{ij}(y^1, \ldots, y^n)$, i.e. of a metric tensor g, satisfying the following system:

$$\begin{cases} \Gamma^k_{11} = -f^k_{11}\,, & k \in \{2, \ldots, n\} \\ 2\Gamma^k_{1i} - \delta^k_i \Gamma^1_{11} = -f^k_{1i}\,, & k \in \{2, \ldots, n\} \\ \Gamma^k_{ij} - \delta^k_i \Gamma^1_{1j} - \delta^k_j \Gamma^1_{1i} = -f^k_{ij}\,, & k \in \{2, \ldots, n\} \\ \Gamma^1_{ij} = f_{ij} \end{cases} \tag{2.4}$$

where $i, j \in \{2, \ldots, n\}$ and

$$\Gamma^k_{ij} = \frac{1}{2} g^{kh} \left(\frac{\partial g_{jh}}{\partial y^i} + \frac{\partial g_{ih}}{\partial y^j} - \frac{\partial g_{ij}}{\partial y^h} \right), \quad i, j, k \in \{1, \ldots, n\},$$

are the *Christoffel symbols*. System (2.4) is a non-linear system of partial differential equations (PDEs), but it can be linearized thanks to the following proposition.

PROPOSITION 2.4 ([**19**]). A metric g on an n-dimensional manifold lies in the projective class of a given connection ∇ if and only if σ^{ij} defined by

$$\sigma^{ij} = |\det(g)|^{\frac{1}{n+1}} g^{ij} \tag{2.5}$$

is a solution of the linear system

$$\nabla_a \sigma^{bc} - \frac{1}{n+1} (\delta^c_a \nabla_i \sigma^{ib} + \delta^b_a \nabla_i \sigma^{ic}) = 0 \tag{2.6}$$

with

$$\nabla_a \sigma^{bc} = \sigma^{bc}_{,a} + \Gamma^b_{ad} \sigma^{dc} + \Gamma^c_{ad} \sigma^{db} - \frac{2}{n+1} \Gamma^d_{da} \sigma^{bc},$$

where Γ^k_{ij} are the Christoffel symbols.

REMARK 2.5. The tensor σ whose components are (2.5) is a *weighted tensor field* on M and as such an element of $S^2(M) \otimes (\Lambda^n(M))^{\frac{2}{n+1}}$, where $S^2(M)$ denotes the space of symmetric $(2,0)$-tensor fields on M. In dimension 2, specifically, linearization can also be achieved using the weighted tensor field $a = |\det(g)|^{-\frac{2}{3}} g$ instead of $\sigma = |\det(g)|^{\frac{1}{3}} g^{-1}$. Note that the components of a and σ are related by the classical adjugate. Indeed, in the papers treating the 2-dimensional case, i.e. [**13, 33, 39**], the weighted tensor field a is frequently used rather than σ.

DEFINITION 2.6. The vector space of the solutions to System (2.6) is going to be denoted by Σ. The *degree of mobility* of a metric g is the dimension of Σ.

Important consequences of Proposition 2.4 are the following: firstly, System (2.6) is projectively invariant as it depends on the projective class $[\nabla]_{\mathfrak{p}}$ only. Thus, if v is a projective vector field, the Lie derivative $\mathcal{L}_v$ sends solutions of (2.6) to solutions of (2.6), i.e., it is a linear map on Σ:

$$\mathcal{L}_v : \Sigma \to \Sigma$$

with

$$\mathcal{L}_v \sigma^{ij} = |\det(g)|^{\frac{1}{n+1}} \left(\frac{g^{mn}\mathcal{L}_v g_{mn}}{n+1} g^{ij} + \mathcal{L}_v g^{ij} \right). \tag{2.7}$$

Secondly, if σ_1 and σ_2 are solutions to (2.6), also $c_1\sigma_1 + c_2\sigma_2$ is a solution for any $c_1, c_2 \in \mathbb{R}$. Therefore, if g_1 and g_2 are projectively equivalent metrics on an n-dimensional manifold M, then

$$\frac{\left(c_1|\det(g_1)|^{\frac{1}{n+1}} g_1^{-1} + c_2|\det(g_2)|^{\frac{1}{n+1}} g_2^{-1}\right)^{-1}}{\det\left(c_1|\det(g_1)|^{\frac{1}{n+1}} g_1^{-1} + c_2|\det(g_2)|^{\frac{1}{n+1}} g_2^{-1}\right)} \tag{2.8}$$

is a projectively equivalent metric to g_1 (and g_2) for any $c_1, c_2 \in \mathbb{R}$, whenever it is defined.

Important papers concerning the metrizability of projective structures are [**12**, **18**]. The paper [**12**] finds differential invariants responsible for the metrizability of a given 2-dimensional projective structure and then explicit obstructions to the existence of a Levi-Civita connection within a given projective class. Indeed the authors studied the overdetermined system (2.6) by employing the prolongation method. Analogous results are contained in [**18**], where the authors face the problem of metrizability of 3-dimensional projective structures.

We conclude this section by introducing, for any pair of projectively equivalent metrics $g, \hat{g}$ on M, the $(1,1)$-tensor field

$$L = L(g, \hat{g}) = \left| \frac{\det(\hat{g})}{\det(g)} \right|^{\frac{1}{n+1}} \hat{g}^{-1} g = \hat{\sigma}\sigma^{-1}, \tag{2.9}$$

which has been an important tool to solve both Lie's classical as well as Lie's extended problem

3. Lie's classical problems

Now we review Lie's second problem (Section 3.1) and Lie's first problem (Section 3.2), which have been solved using quite different techniques. To start with, some clarifications about Lie's problem are in order. As we already pointed out, the set of projective vector fields form a finite dimensional Lie algebra: in what follows we assume that the action of the projective Lie algebra has orbits of constant dimension. Then one can use the results by S. Lie. He found all possible Lie algebras that can arise from projective vector fields of a 2-dimensional projective connection [**25**]. Moreover, Lie obtained the possible realizations in terms of vector fields on (open neighborhoods in) a 2-dimensional manifold [**26**], see also [**27, 46**]. In Section 3.1 and 3.2 below we shall explain how to obtain a *local* description of metrics admitting at least one projective vector field. Global aspects of Lie's theorem, in the case of Riemannian metrics, can be found in [**34–37**].

We start treating Lie's second problem that, from a chronological viewpoint, was the first to be solved [**13**]. Historically, it was believed that A.V. Aminova had solved Lie's second problem in [**1, 2**]. Some years later a gap in her proof was pointed out and filled by [**13**]. This gap will be addressed below. Indeed, Aminova obtained an exhaustive (but not sharp) list of 2-dimensional metrics admitting nonzero projective vector fields, some of which are however only described via ODEs.

3.1. Lie's second problem: 2-dimensional metrics g with $\dim \mathfrak{p}(g) \geq 2$. We now discuss the aforementioned gap pointed out in [**13**]. Briefly speaking, it occurred due to a misinterpretation of a statement by G. Kœnigs in the 1890s, see [**16**, p. 374]. This statement is summarized in the following theorem.

THEOREM 3.1. *A 2-dimensional metric whose geodesic flow admits three independent quadratic integrals (in momenta) is a metric of a surface of revolution.*[3]

From Theorem 3.1 it follows that the metric admits a Killing vector field or, equivalently, a linear integral of the geodesic flow.

The issue here is how the number of independent integrals is counted. Indeed, Kœnigs does not include the Hamiltonian obtained from the metric into the count of the integrals, essentially counting integrals of motion up to the addition of multiples of the Hamiltonian. Instead, Aminova applies the following interpretation:

NON-THEOREM 3.1. *If the geodesic flow of a 2-dimensional metric admits three independent quadratic integrals in momenta (where in the tally the Hamiltonian is included), then the metric is a metric of a surface of revolution.*

The following counter-example shows that this interpretation is not viable.

EXAMPLE 3.2. All quadratic first integrals in the momenta p, q of the metric $g = (4x^2 + y^2 + 1)(dx^2 + dy^2)$ are linear combinations of

$$\begin{cases} H = \frac{p^2+q^2}{4x^2+y^2+1} \\ F_1 = \frac{y^2p^2-(4x^2+1)q^2}{4x^2+y^2+1} \\ F_2 = \frac{-xy^2p^2+y(4x^2+y^2+1)pq-x(4x^2+2y^2+1)q^2}{4x^2+y^2+1} \end{cases}$$

It is not difficult to confirm that this metric admits no non-zero Killing vector field.

Kœnigs' theorem, in an enriched form, therefore would be as follows:

THEOREM 3.3. *The geodesic flow of a 2-dimensional metric admits* 1, 2, 3, 4 *or* 6 *linearly independent quadratic first integrals in momenta (including the Hamiltonian). If the metric admits* 4 *linearly independent quadratic first integrals, then it admits a non-zero linear first integral (or, equivalently, a non-zero Killing vector field). If the metric admits* 6 *linearly independent quadratic first integrals, then it is of constant curvature.*

The following theorem provides a link between quadratic integrals of the geodesic flow of a metric g and its projective class $[g]_{\mathfrak{p}}$. In particular, in dimension 2, the degree of mobility (see Definition 2.6) coincides with the dimension of the vector space of quadratic integrals. The insight goes back, at least, to the end of the 19th century and is mentioned by Levi-Civita in [**24**, page 276], who attributes it to Painlevé [**45**, Ch. 2] (see also [**48**]).

THEOREM 3.4. *Let g and $\hat{g}$ be metrics on an n-dimensional manifold M. If they are projectively equivalent, then the function*

$$K : v \in TM \to K(v,v) := \left(\frac{\det(g)}{\det(\hat{g})}\right)^{\frac{2}{n+1}} g(v,v)$$

is a quadratic first integral in velocities of g. If $n = 2$ the above implication is an equivalence.

[3] French original: "Tout ds^2 dont les géodésiques admettent trois intégrales quadratiques indépendantes convient à une surface de révolution".

The statement of Non-Theorem 3.1 appears crucial in the proof in [**1**, **2**]. Indeed, one can construct a quadratic first integral starting from a projective vector field (see Proposition 5 of [**31**] and also [**21**, **43**]). Then, if a metric admits two independent projective vector fields, it would admit, in view of Non-Theorem 3.1 of the above reasoning, also a Killing vector field. This, of course, would be a huge simplification of Lie's second problem as the components of the metric tensor to be found would depend on one variable only. The aforementioned gap was observed and remedied by R. L. Bryant and coworkers in [**13**]. Indeed, the implication

$$\begin{aligned} &g \text{ is a 2-dimensional metric with } \dim \mathfrak{p}(g) \geq 2 \\ &\Longrightarrow \text{existence of a Killing vector field of } g \end{aligned}$$

is actually true, but so far only an easy a posteriori confirmation exists using the complete list of Theorem 3.5 below. Therefore, the list of metrics obtained by Aminova in [**1**, **2**] is correct and obtained from that in Theorem 3.5 below after suitable changes of the coordinates. However, the list of metrics of [**1**, **2**] contains some functions given in terms of certain differential equations, contrary to that of Theorem 3.5, proved in Sections 3.1.1–3.1.4.

THEOREM 3.5 ([**13**]). *Let g be a 2-dimensional metric on a manifold M and assume $\dim \mathfrak{p}(g) \geq 2$. Then, almost every point of M has a neighborhood U such that the restriction of g to U has constant curvature, or such that there exists a coordinate system (x, y) on U in which the metric g takes one of the following forms:*

(1) *Metrics with* $\dim \mathfrak{p}(g) = 2$.
  (a) $\varepsilon_1 e^{(b+2)\,x}\, dx^2 + \varepsilon_2 e^{b\,x}\, dy^2$, *where* $b \in \mathbb{R} \setminus \{-2, 0, 1\}$ *and* $\varepsilon_i \in \{-1, 1\}$ *are constants,*
  (b) $a\left(\frac{e^{(b+2)\,x} dx^2}{(e^{b\,x}+\varepsilon_2)^2} + \varepsilon_1 \frac{e^{b\,x} dy^2}{e^{b\,x}+\varepsilon_2}\right)$, *where* $a \in \mathbb{R} \setminus \{0\}$, $b \in \mathbb{R} \setminus \{-2, 0, 1\}$, *and* $\varepsilon_i \in \{-1, 1\}$ *are constants, and*
  (c) $a\left(\frac{e^{2\,x} dx^2}{x^2} + \varepsilon \frac{dy^2}{x}\right)$, *where* $a \in \mathbb{R} \setminus \{0\}$, *and* $\varepsilon \in \{1, -1\}$ *are constants.*

(2) *Metrics with* $\dim \mathfrak{p}(g) = 3$.
  (a) $\varepsilon_1 e^{3x} dx^2 + \varepsilon_2 e^{x} dy^2$, *where* $\varepsilon_i \in \{-1, 1\}$ *are constants,*
  (b) $a\left(\frac{e^{3x} dx^2}{(e^{x}+\varepsilon_2)^2} + \varepsilon_1 \frac{e^{x} dy^2}{(e^{x}+\varepsilon_2)}\right)$, *where* $a \in \mathbb{R} \setminus \{0\}$, $\varepsilon_i \in \{-1, 1\}$ *are constants, and*
  (c) $a\left(\frac{dx^2}{(2x^2+cx+\varepsilon_2)^2 x} + \varepsilon_1 \frac{x dy^2}{(2x^2+cx+\varepsilon_2)}\right)$, *where* $a > 0$, $\varepsilon_i \in \{-1, 1\}$, $c \in \mathbb{R}$ *are constants.*

No two distinct metrics from this list are isometric.

3.1.1. *Preliminary results and ideas for the solution of Lie's second problem.* In the 2-dimensional case, Systems (2.1) and (2.2) read, respectively,

$$y_{xx} = -\Gamma^2_{11} + (\Gamma^1_{11} - 2\Gamma^2_{12}) y_x - (\Gamma^2_{22} - 2\Gamma^1_{12}) y_x^2 - \Gamma^1_{22} y_x^3$$

and

$$y_{xx} = A_0(x, y) + A_1(x, y) y_x + A_2(x, y) y_x^2 + A_3(x, y) y_x^3 \,. \tag{3.1}$$

S. Lie proved [**25**] that the algebras that are the projective Lie algebras of 2-dimensional metrics, or, equivalently, the symmetry Lie algebras of the ODEs (3.1), can only be the following ones:

$$0\,, \quad \mathbb{R}\,, \quad \mathfrak{s} : [X, Y] = X\,, \quad \mathfrak{sl}(2, \mathbb{R})\,, \quad \mathfrak{sl}(3, \mathbb{R})\,, \tag{3.2}$$

where $\mathfrak{s}$ is the 2-dimensional non-commutative Lie algebra. Note that $\mathfrak{s}$ is a subalgebra of both $\mathfrak{sl}(2,\mathbb{R})$ and $\mathfrak{sl}(3,\mathbb{R})$. Since in this section we are interested in metrics admitting at least a 2-dimensional Lie algebra of projective vector fields, we shall concentrate on the last three ones. Lie also showed that an ODE of type (3.1) admitting $\mathfrak{sl}(3,\mathbb{R})$ as symmetry Lie algebra is equivalent (see Definition 2.3) to

$$y_{xx} = 0\,, \tag{3.3}$$

i.e., to the ODE whose solutions are the lines of $\mathbb{R}^2$. In other words, all 2-dimensional metrics admitting an 8-dimensional Lie algebra of projective vector fields are of constant curvature. Liouville proved [**29**] that

$$(L_1 dx + L_2 dy) \otimes (dx \wedge dy) \tag{3.4}$$

where

$$\begin{aligned} L_1 &= 2A_{1xy} - A_{2xx} - 3A_{0yy} - 6A_0A_{3x} \\ &\qquad - 3A_3A_{0x} + 3A_0A_{2y} + 3A_2A_{0y} + A_1A_{2x} - 2A_1A_{1y} \\ L_2 &= 2A_{2xy} - A_{1yy} - 3A_{3xx} + 6A_3A_{0y} \\ &\qquad + 3A_0A_{3y} - 3A_3A_{1x} - 3A_1A_{3x} - A_2A_{1y} + 2A_2A_{2x} \end{aligned}$$

is a differential invariant of the projective connection (3.1): the quantity (3.4) vanishes if and only if (3.1) is equivalent to (3.3) (see also [**15**, **49**]). Note also that in the list (3.2) there is a dimensional "gap", since $\mathfrak{sl}(2,\mathbb{R})$ is 3-dimensional and $\mathfrak{sl}(3,\mathbb{R})$ 8-dimensional. This means that an ODE (3.1) admitting more than three independent projective vector fields automatically admits $\mathfrak{sl}(3,\mathbb{R})$ as symmetry Lie algebra, implying that such an ODE is equivalent to (3.3).

It remains to investigate 2-dimensional metrics admitting $\mathfrak{s}$ and $\mathfrak{sl}(2,\mathbb{R})$ (cf. list (3.2)) as projective Lie algebras. To this end, a result by Lie [**26**–**28**] concerning all possible realizations of the algebra $\mathfrak{s}$ as an algebra of vector fields on $\mathbb{R}^2$ is very helpful. He showed that, in a neighborhood of almost every point,[4] there exist coordinates (x, y) such that vector fields X, Y satisfying $[X, Y] = X$ are, up to the trivial case $X = 0$, described by

- *Non-Transitive case:* $X = e^y\partial_y$, $Y = -\partial_y$;
- *Transitive case:* $X = \partial_y$, $Y = \partial_x + y\partial_y$.

3.1.2. *The non-transitive case.* By a direct computation one confirms that the most general projective connection (3.1) admitting $e^y\partial_y$ and ∂_y as infinitesimal symmetries is

$$y_{xx} = A_1(x)y_x + y_x^2\,. \tag{3.5}$$

Since the Liouville invariant (3.4) of (3.5) vanishes, (3.5) admits a symmetry Lie algebra isomorphic to $\mathfrak{sl}(3,\mathbb{R})$, i.e., (3.5) is equivalent to (3.3). We conclude that 2-dimensional metrics g such that $\dim \mathfrak{p}(g) = 2$ with $\mathfrak{p}(g)$ acting non-transitively are of constant curvature.

[4] We recall the assumption that the orbits of the projective Lie algebra action are of constant dimension.

3.1.3. *The transitive case with* $\dim \mathfrak{p}(g) = 2$. By a direct computation one confirms that the most general projective connection (3.1) admitting ∂_y and $\partial_x + y\partial_y$ as infinitesimal symmetries is

$$y_{xx} = K_0 e^x + K_1 y_x + K_2 e^{-x} y_x^2 + K_3 e^{-2x} y_x^3 , \quad K_i \in \mathbb{R} . \tag{3.6}$$

In this case, the linear system (2.6) reads as follows:

$$\begin{cases} \sigma^{22}_{,x} - \frac{2}{3} K_1 \sigma^{22} - 2K_0 e^x \sigma^{12} = 0 \\ \\ \sigma^{22}_{,y} - 2\sigma^{12}_{,x} - \frac{4}{3} K_2 e^{-x} \sigma^{22} - \frac{2}{3} K_1 \sigma^{12} + 2 K_0 e^x \sigma^{11} = 0 \\ \\ -2\sigma^{12}_{,y} + \sigma^{11}_{,x} - 2 K_3 e^{-2x} \sigma^{22} + \frac{2}{3} K_2 e^{-x} \sigma^{12} + \frac{4}{3} K_1 \sigma^{11} = 0 \\ \\ \sigma^{11}_{,y} + 2 K_3 e^{-2x} \sigma^{12} + \frac{2}{3} K_2 e^{-x} \sigma^{11} = 0 \end{cases} \tag{3.7}$$

Since we are supposing $\dim \mathfrak{p}(g) = 2$, we can further assume (see page 454 of [**13**])

$$K_2 \neq 0 \text{ and } K_1 = K_3 = 0 , \quad \text{or}$$
$$K_2 = 0, K_3 \neq 0 \text{ and if } K_0 = 0 \text{ then } K_1 \notin \{\tfrac{1}{2}, 2\}$$

As the coefficients of the above linear system do not depend on y, if $\sigma = \sigma^{ij}$ is a solution to (3.7), also $\frac{\partial^k \sigma}{\partial y^k}$ is a solution ($\forall k \in \mathbb{N}$), implying that any linear combination $\sum_{k=0} c_k \frac{\partial^k \sigma}{\partial y^k}$ is a solution ($c_k \in \mathbb{R}$). We stress that, according to Theorems 3.3 and 3.4, the dimension of the solution space to (3.7) can be equal to $0, 1, 2, 3, 4, 6$ (but 0 would imply non-metrizability). There is another restriction:

PROPOSITION 3.6 ([**13**, p. 453]). *A 2-dimensional metric g with $\dim \mathfrak{p}(g) = 2$ has degree of mobility less than four.*

Due to Proposition 3.6, four independent solutions to System (3.7) cannot exist, so there must exist constants c_i, with $(c_0, c_1, c_2, c_3) \neq (0, 0, 0, 0)$ such that

$$c_0 \sigma + c_1 \frac{\partial \sigma}{\partial y} + c_2 \frac{\partial^2 \sigma}{\partial y^2} + c_3 \frac{\partial^3 \sigma}{\partial y^3} = 0 . \tag{3.8}$$

One finds solutions to (3.8), depending on the multiplicity of the roots of its characteristic polynomial, and then one substitutes them into System (3.5). A long and tedious computation (see [**13**, pages 453-455]) shows that $\frac{\partial \sigma}{\partial y} = 0$, i.e., in view of formula (2.5), the metrics we are looking for admit ∂_x as a Killing vector field. This leads to obtain the list of metrics of Theorem 3.5 with $\dim \mathfrak{p}(g) = 2$.

3.1.4. *The transitive case with* $\dim \mathfrak{p}(g) = 3$. In this case, one can find a basis X, Y, Z of projective vector fields satisfying the following commutation relations:

$$[X, Y] = X , \quad [X, Z] = Y , \quad [Y, Z] = Z . \tag{3.9}$$

We can then exploit the realization of X and Y, in terms of vector fields, given in Section 3.1.1 for the transitive case, $X = \partial_y$ and $Y = \partial_x + y\partial_y$, in order to find also a coordinate description of the vector field Z. Indeed, conditions (3.9) translate into a system of PDEs, with the components of Z as unknown functions, that can be easily integrated:

$$Z = (y + C_1 e^x)\partial_x + \left(\frac{y^2}{2} + C_2 e^{2x} \right) \partial_y . \tag{3.10}$$

The only possibility for the projective connection (3.6) to admit an infinitesimal symmetry of type (3.10) and, at the same time, to admit *exactly* a 3-dimensional symmetry Lie algebra, is that $K_0 = K_2 = 0$, $K_1 = \frac{1}{2}$, $K_3 \neq 0$, $C_1 = C_2 = 0$. To sum up, we have to consider the projective connection

$$y_{xx} = \frac{1}{2}y_x + K_3 e^{-2x} y_x^3 \,, \quad K_3 \neq 0 \tag{3.11}$$

with the following vector fields as a basis of the projective Lie algebra:

$$X = \partial_y \,, \quad Y = \partial_x + y\partial_y \,, \quad Z = y\partial_x + \frac{y^2}{2}\partial_y \,. \tag{3.12}$$

The following proposition ([**13**, page 456]) is crucial for obtaining the list of metrics of Theorem 3.5 in the case $\dim \mathfrak{p}(g) = 3$.

PROPOSITION 3.7. A 2-dimensional metric g such that $\dim \mathfrak{p}(g) \geq 3$ admits a non-zero Killing vector field.

Without loss of generality, we may suppose that X or Y or Z of (3.12) is a Killing vector field for the metric g, thus obtaining three distinct cases. We discuss only the case in which $X = \partial_y$ is a Killing vector field for the metric g, as the other two cases can be treated similarly. In the case under consideration, the metric coefficients g_{ij} depend only on x, so that also σ^{ij} (see (2.5)) depend only on x. Substituting $\sigma^{ij} = \sigma^{ij}(x)$ and $A_0 = A_2 = 0$, $A_1 = \frac{1}{2}$, $A_3 = K_3 e^{-2x}$ (cf. (3.11)) into System (3.7), one obtains a system that yields

$$g = e^{3x} C_2 (-C_1 e^x + 2C_2 K_3)^2 \, dx^2 - \frac{e^x}{C_2^2(-C_1 e^x + 2C_2 K_3)} \, dy^2 \,, \quad C_i \in \mathbb{R} \,.$$

By virtue of an appropriate change of coordinate, the metric g is transformed into either a metric 2a or 2b in Theorem 3.5. The metrics 2c in Theorem 3.5 are obtained by a similar consideration of the remaining cases.

3.2. Lie's first problem. We now consider 2-dimensional metrics with a projective Lie algebra of dimension exactly 1, i.e. we require that $\dim \mathfrak{p}(g) = 1$. Thereby we follow the chronological order: historically Lie's second problem has been solved before Lie's first problem. The techniques employed in these solutions are also quite different. Indeed one could follow the same strategy as in [**13**], outlined above, but in the case under consideration the projective connection (3.1) would depend on univariate functions rather than constants, contrary to (3.6), making the general solution to System (3.7) harder to obtain. As mentioned in the beginning of Section 3, Aminova [**1, 2**] had already found a non-sharp list of normal forms for 2-dimensional metrics g with $\dim \mathfrak{p}(g) = 1$ (involving unsolved ODEs).

The hypothesis $\dim \mathfrak{p}(g) = 1$ implies that, if v is a non-zero projective vector field, then any other projective vector field will be a scalar multiple of v. We recall that, due to Theorems 3.3 and 3.4, the degree of mobility of an arbitrary 2-dimensional metric satisfies

$$\dim(\Sigma) \in \{1, 2, 3, 4, 6\} \,.$$

Each case is now discussed individually. We begin with $\dim(\Sigma) = 6$. Due to Theorems 3.3 and 3.4, the metric is then of constant curvature. It therefore admits an 8-dimensional Lie algebra of projective vector fields and the projective Lie algebra is isomorphic to $\mathfrak{sl}(3, \mathbb{R})$.

The next possible case is $\dim(\Sigma) = 4$. By virtue of Theorems 3.3 and 3.4, a Killing vector field exists. In a neighborhood of a point p such that $v_p \neq 0$, there exist coordinates (x, y) such that $v = \partial_x$: in this system of coordinates

$$g = E(y)dx^2 + 2F(y)dxdy + G(y)dy^2 \,.$$

Of course, this is a non-sharp description as, for instance, in the case of constant E, F and G this metric is of constant curvature and hence does not meet the hypothesis.

We now turn to the case when $\dim(\Sigma) = 1$, in which all metrics projectively equivalent to the metric g are proportional to g by a non-zero factor, and therefore v necessarily is homothetic for any such multiple. In suitable local coordinates we achieve, around points where v does not vanish, $v = \partial_x$ and

$$g = e^{\lambda x}\left(E(y)\,dx^2 + 2\,F(y)\,dxdy + G(y)\,dy^2\right)\,,$$

which is a non-sharp description (for instance, again, if E, F and G are constants then g admits another projective vector field, namely ∂_y).

We are thus left with two cases, $\dim(\Sigma) \in \{2, 3\}$. As pointed out in Section 2, the Lie derivative $\mathcal{L}_v$ along v is a linear operator on Σ, see (2.7). Furthermore, the operator $\mathcal{L}_v$ is non-degenerate [**39**, Lemma 3].

– If $\dim(\Sigma) = 2$, then since $\dim \mathfrak{p}(g) = 1$, we can rescale v by a non-zero constant factor and choose an appropriate basis of Σ such that $\mathcal{L}_v$ assumes one of the following Jordan forms (for a constant λ):

$$\begin{pmatrix} 1 & 1 \\ 0 & 1 \end{pmatrix}, \quad \begin{pmatrix} \lambda & -1 \\ 1 & \lambda \end{pmatrix}, \quad \begin{pmatrix} \lambda & 0 \\ 0 & 1 \end{pmatrix}. \tag{3.13}$$

– If $\dim(\Sigma) = 3$, there similarly exists an appropriate basis of Σ and a rescaling of v such that $\mathcal{L}_v$ assumes ones of the following Jordan forms:

$$\begin{pmatrix} 1 & 1 & 0 \\ 0 & 1 & 1 \\ 0 & 0 & 1 \end{pmatrix}, \quad \begin{pmatrix} 1 & 1 & 0 \\ 0 & 1 & 0 \\ 0 & 0 & \mu \end{pmatrix}, \quad \begin{pmatrix} \lambda & -1 & 0 \\ 1 & \lambda & 0 \\ 0 & 0 & \mu \end{pmatrix}, \quad \begin{pmatrix} \lambda & 0 & 0 \\ 0 & 1 & 0 \\ 0 & 0 & \mu \end{pmatrix} \tag{3.14}$$

for constants λ, μ. We observe that the non-degenerate operator $\mathcal{L}_v$ always possesses a 2-dimensional invariant subspace $\mathcal{V}$. Therefore, up to a non-zero constant factor, the restriction of $\mathcal{L}_v$ to $\mathcal{V}$ is described by one of the matrices (3.13).

The previous discussion reduces Lie's first problem, at least partially, to the following: find $\mathcal{L}_v$-invariant subspaces $\Sigma' \subseteq \Sigma$ of dimension 2 such that $\dim \mathfrak{p}(g) = 1$, then try to extend them to the full space Σ. Two pre-existing results make this approach feasible: first, we can assume $\mathcal{L}_v$ to have Jordan form as in (3.13). Second, the basis of Σ', in suitable local coordinates on the manifold, can be taken in canonical form according to the following proposition.

PROPOSITION 3.8 ([**5, 10, 17**]). *Let g_1 and g_2 be projectively equivalent, non-proportional 2-dimensional metrics. Then, in a neighborhood of almost every point, there are coordinates (x, y) such that the metrics assume one of the following three normal forms:*

	Liouville	Complex-Liouville	Jordan-Block
g_1	$(f_1 - f_2)(dx^2 \pm dy^2)$	$(\overline{h(z)} - h(z))\,(d\overline{z}^2 - dz^2)$	$(1 + xf_2')dxdy$
g_2	$\left(\frac{1}{f_1} - \frac{1}{f_2}\right)\left(\frac{dx^2}{f_1} \pm \frac{dy^2}{f_2}\right)$	$\left(\frac{1}{\overline{h(z)}} - \frac{1}{h(z)}\right)\left(\frac{d\overline{z}^2}{\overline{h(z)}} - \frac{dz^2}{h(z)}\right)$	$\frac{1+xf_2'}{f_2^4}\,(-2f_2\,dxdy$ $+(1 + xf_2')\,dy^2)$

Here, $f_1 = f_1(x)$ and $f_2 = f_2(y)$ are functions of one variable only, and in the complex Liouville case we use coordinates $z = x + iy$, $\overline{z} = x - iy$.

The strategy for the solution of Lie's first problem pursued in [**39**] is as follows. Let us assume $\dim(\Sigma) = 2$. Let $\{\sigma_1, \sigma_2\}$ be a basis of Σ. The equations

$$\mathcal{L}_v \begin{pmatrix} \sigma_1 \\ \sigma_2 \end{pmatrix} = A \begin{pmatrix} \sigma_1 \\ \sigma_2 \end{pmatrix}$$

where A is one of the matrices in (3.13), translate into a system of PDEs with the metric coefficients of g_1, g_2 and the components of v as unknown functions, because of (2.5) and (2.7). By substituting g_1 and g_2 with those contained in the table of Proposition 3.8, we obtain (for each considered matrix (3.13)) three distinct cases, each of which is a system of 6 PDEs that can be integrated. Once g_1 and g_2 are obtained from the above computation, Formula (2.8) gives all metrics in the projective class $[g_1]_\mathfrak{p}$ of g_1 (and, of course, g_2).

If $\dim \Sigma = 3$, a similar reasoning allows one to find a suitable basis of Σ, such that $\mathcal{L}_v$ is described by one of the matrices (3.14), starting from a basis of the invariant subspace $\mathcal{V}$ that is inherited from the previous case.

The next question is how to obtain a list of pairwise non-isometric metrics lying in the same projective class. This question was answered in [**33**]. The idea is the following. Since v is a projective vector field, its local flow preserves the equivalence class $[g_1]_\mathfrak{p}$: two metrics in $[g_1]_\mathfrak{p}$ linked by the local flow of v are thus isometric. An additional tool for distinguishing non-isometric metrics in the same projective class is to study invariants of tensor (2.9).

The full proof is rather tedious and technical. We therefore refer the interested reader to the literature [**2, 32, 39**]. The result can be summarized as follows: if a metric g satisfies $\dim \mathfrak{p}(g) = 1$, then locally it is either of the form

$$g = e^{\mu x}(E(y)dx^2 + 2F(y)\,dxdy + G(y)\,dy^2)\,, \qquad \mu \in \mathbb{R}\,,$$

and the projective vector fields are multiples of ∂_x, or it can be transformed into exactly one metric from an (extensive) list of normal forms, see [**32**]. We also refer the reader to the concise table in Appendix A of [**31**].

4. Higher dimensional extensions

Having discussed Lie's classical problem of metrics with infinitesimal symmetries of their geodesics, we shall now turn our attention to extensions of this problem. The most natural extension is to ask for metrics in higher dimension $n \geq 3$ that have projective vector fields.

While Lie's classical problem (see page 1) is solved, the same cannot be said about higher dimensions. Various partial solutions have been obtained, however. Typically, these works impose additional restrictions on the type of metrics. In particular, Levi-Civita's metrics have received some attention, which were obtained by

Levi-Civita when studying pairs of metrics that share the same unparametrized geodesics; a detailed introduction to these metrics can be found in the next subsection. We present two major results from the literature.

Firstly, *in arbitrary dimension $n \geq 3$, Levi-Civita metrics of the most basic type* have been investigated by Aleksandr Solodovnikov [**47**]. He obtained a description in terms of ordinary differential equations which we review in Section 4.2. For a given metric, these equations determine its projective vector fields. On the other hand, they also allow one to find metrics with projective vector fields.

Secondly, the latter point has been executed for *Levi-Civita metrics of arbitrary type in dimension $n = 3$* [**33**]. Two of the authors thus obtained a sharp description (see Section 4.1), not only for Levi-Civita metrics of basic type, but for all Levi-Civita metrics in dimension 3. In particular, this suffices to solve Lie's problem for Riemannian metrics in dimension $n = 3$ (see Section 4.4). For Levi-Civita metrics of non-basic type this also reveals a splitting-gluing phenomenon effectively reducing the problem to Lie's classical, 2-dimensional problem.

These situations will be elaborated upon in the dedicated sections 4.2 as well as 4.3 and 4.4, respectively. We stress that the Einstein convention is not used. The next section is dedicated to Levi-Civita metrics in general.

4.1. Levi-Civita metrics. In his 1896 work [**24**], Levi-Civita considers pairs of non-proportional, Riemannian metrics $g, \hat{g}$ that share the same unparametrized geodesics. His aim is to find local canonical forms for such pairs. He relies on previous work by Appel [**4**], Painlevé [**45**] and Liouville [**29**]. For the purposes here, the most important result of Levi-Civita is that in a sufficiently small neighborhood, there exist local coordinates to put the metrics g and $\hat{g}$ into a canonical form:

PROPOSITION 4.1 ([**7, 8, 38**]). *Let g be a Riemannian, and $\hat{g}$ a metric projectively equivalent and non-proportional to g. Then (almost everywhere, in a neighborhood) local coordinates exist such that $g, \hat{g}$ assume the form*

$$g = \sum_{i=1}^{r} P_i (dx_i^1)^2 + \sum_{i=r+1}^{m} \left[P_i \sum_{\alpha_i,\beta_i}^{k_i} (h_i(x_i))_{\alpha_i,\beta_i} dx_i^{\alpha_i} dx_i^{\beta_i} \right] \tag{4.1a}$$

$$\hat{g} = \sum_{i=1}^{r} P_i \rho_i (dx_i^1)^2 + \sum_{i=r+1}^{m} \left[P_i \rho_i \sum_{\alpha_i,\beta_i}^{k_i} (h_i(x_i))_{\alpha_i,\beta_i} dx_i^{\alpha_i} dx_i^{\beta_i} \right] \tag{4.1b}$$

where the functions $(h_i)_{\alpha\beta}$ depend on a subset (x_i) of the coordinates only, denoted by $(x_i^1, \dots, x_i^{k_i})$, and where

$$P_i = \pm \prod (f_i - f_j) \qquad \text{and} \qquad \rho_i = \frac{1}{f_i \prod_\alpha f_\alpha^{k_\alpha}},$$

with f_i denoting the eigenvalue of $L = L(g, \hat{g})$ (see (2.9)) for the eigendistribution $\mathcal{D}_i$. The numbers k_i are larger or equal to 2, and the dimension of the manifold is

$$n = r + \sum_{i=r+1}^{m} k_i.$$

The description is sharp in the sense that the metrics g, $\hat{g}$ of this form are indeed projectively equivalent. Note that these normal forms include three major "building components": Univariate functions $f_i(x_i)$, constants ρ_μ and lower-dimensional (but otherwise arbitrary) metrics h_μ. The building components f_i and ρ_μ of g, $\hat{g}$ take on a specific meaning: they are eigenvalues of L. We remark that the corresponding eigenspaces yield integrable distributions [**7**]. The metrics h_μ correspond to constant eigenvalues ρ_μ of the endomorphism L.

DEFINITION 4.2. We say that a Levi-Civita metric is of *basic type* if L has only 1-dimensional eigenspaces.

While projectively equivalent, the metrics g, $\hat{g}$ of Levi-Civita type do not necessarily admit projective vector fields. However, they are natural candidates for such metrics: let us now consider a metric g that admits a projective vector field v. If v is not homothetic, i.e. $\mathcal{L}_v g$ is not proportional to g, the construction introduced in Section 2 allows us to write down, using (2.5),

$$\hat{\sigma} = \mathcal{L}_v\sigma = \mathcal{L}_v(|\det(g)|^{\frac{1}{n+1}} g^{-1})$$

which solves (2.6). If $\hat{\sigma}$ is non-degenerate, then

$$\hat{g} = |\det(\hat{\sigma})|^{-1}\hat{\sigma}^{-1}$$

is a metric, such that $(g, \hat{g})$ is a pair of metrics satisfying Levi-Civita's hypothesis. If $\hat{\sigma}$ is degenerate, we replace it by a suitable linear combination of $\hat{\sigma}$ and σ to obtain such a metric. This confirms: Riemannian metrics with a projective vector field that is not homothetic are of Levi-Civita's type.

We conclude the section by mentioning that Levi-Civita's metrics are not confined to Riemannian geometry. Indeed, using the canonical form (4.1) we obtain many examples $g, \hat{g}$ of projectively equivalent metrics that are of mixed signature.

4.2. Solodovnikov's solution for basic Levi-Civita metrics. We now present a result obtained by A. Solodovnikov in the 1950s, which characterizes the Riemannian Levi-Civita metrics of basic type with homothetic or non-homothetic projective vector fields. As the signature turns out not to be crucial for the reasoning, we shall consider metrics of the more general form

$$g = \sum_{i=1}^{n}\prod_{j=1}^{n} (f_j(x_j) - f_i(x_i))\, dx_i^2 \qquad (n \geq 3)\,, \tag{4.2}$$

with $\hat{g}$ having the corresponding form via (4.1b). As explained in detail in [**7**, **8**], the projective vector fields for such metrics are of the form

$$v = \sum_i v_i(x_i)\partial_{x_i}\,.$$

Briefly speaking the reason for this form of v is that the distributions associated to eigenspaces of L are integrable, cf. (2.9), see also [**7**]. The ordinary differential equations contained in the following theorem were obtained by A. Solodovnikov.

THEOREM 4.3. *Let g be a metric (4.2) with non-zero projective vector field v.* *(i) If v is homothetic, then there are constants $c, a \in \mathbb{R}$ such that*

$$\begin{aligned} v_i(x_i)f_i'(x_i) &= f_i(x_i)^2 + cf_i(x_i) + a \\ 2v_i'(x_i) &= (3-n)f_i(x_i) - c(n-1) \end{aligned}$$

(ii) If v is not homothetic, then there are constants $c, \ell, \lambda \in \mathbb{R}$ such that

$$\begin{aligned} v_i(x_i) f_i'(x_i) &= c f_i(x_i) + \ell \\ 2v_i'(x_i) &= -c(n-1) + \lambda \end{aligned}$$

PROOF. One can show that, in the adapted coordinates we are using, each component v_i of the projective vector field depends on the variable x_i only (in other words, the eigendistributions of L are integrable). This implies $\nabla_i v_j = 0$ if $i \neq j$. Solodovnikov then shows that if v is not homothetic, then without loss of generality we have

$$\nabla_i v_i = (f_i + \sum_j f_j) \prod_{j \neq i} |f_i - f_j|$$

(achieving this form might require a common translation $f_i \to f_i + \mu$, $\mu \in \mathbb{R}$, which does not modify the metric g). He obtains the following solution:

$$2\frac{\partial v_i}{\partial x_i} = -\sum_{j \neq i} \frac{\frac{\partial f_j}{\partial x_j} v_j - \frac{\partial f_i}{\partial x_i} v_i}{f_j - f_i} + 2f_i + \sum_{j \neq i} f_j \,.$$

Since the left hand side depends on x_i only, this must hold also for the expression on the right hand side. In particular, for $j \neq i$,

$$-\frac{\frac{\partial f_j}{\partial x_j} v_j - \frac{\partial f_i}{\partial x_i} v_i}{f_j - f_i} + 2f_i = F_i^j(x_i)\,,$$

should be univariate. It necessarily follows that $c_{ij} = f_i - F_i^j$ is a constant and that $c_{ij} = c_{ji}$. Writing this out, we have that

$$\tfrac{\partial f_i}{\partial x_i} - f_i^2 - c_{ij} f_i = \tfrac{\partial f_j}{\partial x_j} - f_j^2 - c_{ji} f_j =: a$$

is a constant, and, indeed, a does not depend on the choice of i, j. Resubstituting into the formula for $\frac{\partial v_i}{\partial x_i}$, the claim is obtained for the case of a non-homothetic vector field. The equations for homothetic vector fields can be developed similarly. $\square$

The theorem establishes a characterization of metrics (4.2) with projective vector fields. For given functions $f_i(x_i)$, employing it directly allows one to check if a projective vector field exists. More importantly, however, we view these equations as ODEs for both v and f (ignoring labels as the equations are identical for all blocks). This allows us to find in explicit terms all the metrics with homothetic and non-homothetic vector fields, along with the vector fields themselves. As a metric can admit both kinds of projective vector fields simultaneously, the solutions should be expected to overlap. We follow this train of thought in the next section where we discuss 3-dimensional Levi-Civita metrics.

4.3. Levi-Civita metrics in dimension $n = 3$. The current section reviews the work by two of the authors [**33**], addressing the case of Levi-Civita metrics in dimension $n = 3$. Writing out formula (4.1a), we obtain the following two cases of Levi-Civita metrics in dimension 3:

Type 111 $\quad g = (X - Y)(X - Z)dx^2 + (Y - X)(Y - Z)dy^2 + (Z - X)(Z - Y)dz^2$

Type 21 $\quad g = (X - \rho)dx^2 + (\rho - X)h$

where $X = X(x)$, $Y = Y(y)$, $Z = Z(z)$ and where

$$-h = h_{11}(y,z)dy^2 + 2h_{12}(y,z)dydz + h_{22}(y,z)dz^2\,.$$

Redefining $\zeta(x) = X(x) - \rho$, we write the type-21 metric in more compact form as

$$g = \zeta(x)(dx^2 + h)\,.$$

The task is now to find the metrics among these that admit projective vector fields. This is particularly easy for the first case, as we simply have to solve Solodovnikov's equations. Note that these simplify in the case $n = 3$. As the equations are of Riccati type, the solutions are readily obtained. For example, one solution is

$$\begin{aligned} g = {} & k_1(\tanh(x) - \tanh(y))(\tanh(x) - \tanh(z))dx^2 \\ & + k_2(\tanh(y) - \tanh(x))(\tanh(y) - \tanh(z))dy^2 \\ & + k_3(\tanh(z) - \tanh(x))(\tanh(z) - \tanh(y))dz^2\,. \end{aligned}$$

The full list of solutions can be found in [**33**, Lemmas 10 and 11].

Now let us turn to the 21-type. One can show, again using that eigenspaces correspond to integrable distributions, that if v is projective for a 21-type metric, $v = (\alpha(x), u^2(y,z), u^3(y,z))$.

PROPOSITION 4.4. *If $v = \alpha(x)\partial_x + u$ is a projective vector field of a 21-type Levi-Civita metric, then u is homothetic for h. Conversely, if u is homothetic for h with $\mathcal{L}_u h = -Ch$, then v is homothetic if and only if*

$$\begin{aligned} \zeta\left(\alpha\zeta'' - \alpha'\zeta' - C\zeta'\right) &= \alpha\zeta'^2 \\ \zeta\left(-2\alpha''\zeta - \alpha\zeta'' + \alpha'\zeta'\right) &= \alpha\zeta'^2 \end{aligned}$$

This reduces the problem effectively to one dimension lower, as it permits us to use the solution of Lie's classical 2-dimensional problem, and then construct all 3-dimensional 21-type metrics from these by solving the above system of differential equations. The next section gives a qualitative impression of the results obtained.

4.4. A sharp description of Riemannian metrics in $n = 3$. We note the following easy lemma.

LEMMA 4.5. *Let g be a Riemannian metric of dimension $n = 3$ that admits a non-vanishing projective vector field v. Then:*

- *If v is homothetic, then locally, around almost every point of the manifold, there are local coordinates such that $v = \partial_x$ and $g = e^{\mu x}h(y,z)$;*
- *If v is not homothetic, then g is of constant curvature, or a Levi-Civita metric of either type 111 or type 21.*

PROOF. The first part of the claim is obvious by the rectification theorem. For the second part, assume v is not homothetic. Let $\sigma = |\det(g)|^{\frac{1}{n+1}} g^{-1}$ and then let $\hat\sigma = \mathcal{L}_v\sigma$. If $\hat\sigma$ is non-degenerate, then $\hat g = |\det\hat\sigma|^{-1}\hat\sigma^{-1}$ is a metric such that $(g, \hat g)$ is a Levi-Civita pair. If $\hat\sigma$ is degenerate, we replace $\hat\sigma$ by a linear combination of $\hat\sigma$ and σ and then obtain $\hat g$ in the obvious, analogous way. □

The lemma immediately gives us a sharp description of 3-dimensional Riemannian metrics with projective vector fields.

THEOREM 4.6. *Let g be a Riemannian metric of $n = 3$ with a non-vanishing projective vector field. Then, locally, in a small neighborhood of almost any point, it falls into exactly one of the following cases:*

(1) *The metric g is of constant curvature and after a local change of coordinates its Lie algebra of projective vector fields is isomorphic to $\mathfrak{sl}(n+1, \mathbb{R})$.*

(2) *The metric g has non-constant curvature and admits a homothetic vector field and, after a local change of coordinates,*
$$g = e^{\mu x} h(y, z)$$
(recall: we exclude cases of constant curvature as these have already been covered).

(3) *The metric g has non-constant curvature and admits no homothetic vector field. Then, after a local change of coordinates,*
$$\begin{aligned} g = {} & k_1(f(x) - f(y))(f(x) - f(z))e^{bx}dx^2 \\ & + k_2(f(y) - f(x))(f(y) - f(z))e^{by}dy^2 \\ & + k_3(f(z) - f(x))(f(z) - f(y))e^{bz}dz^2 \,, \end{aligned}$$
where $(b \in \mathbb{R})$
$$(i)\ f : t \to \tanh(t)\,, \quad (ii)\ f : t \to \tan(t)\,, \quad (iii)\ f : t \to \frac{1}{t}\,, b = 2\,.$$
The projective vector field is then a multiple of $\partial_x + \partial_y + \partial_z$.

(4) *The metric g has non-constant curvature and admits a non-homothetic vector field and, after a local change of coordinates,*
$$g = k\, f'(x)(dx^2 + h)\,,$$
where $k \in \mathbb{R} \setminus \{0\}$ and where h does not admit any non-zero homothetic vector field. Moreover,
$$(i)\ f(x) = \frac{1}{x} \qquad (ii)\ f(x) = \tan(x) \qquad (iii)\ f(x) = \tanh(x)\,,$$
and the respective projective vector field is a multiple of $f(x)\partial_x$.

PROOF. The first two cases are, of course, classical. As outlined in the previous section, if g has no homothetic vector fields, it is necessarily of Levi-Civita type 111 or 21, see (4.1). Therefore there is an endomorphism L with at least one non-constant eigenvalue of multiplicity one. In the 111 case, we integrate Solodovnikov's ODE system. In the case of 21-type metrics one proceeds analogously. □

We conclude the section by commenting on the case when both non-homothetic and homothetic vector fields exist. We proceed along the degree of mobility, and in the case of 21-type Levi-Civita metrics also the dimension of the homothetic algebra of h.

Due to a theorem by Matveev-Kiosak [**22**], the degree of mobility is either $D = 1$ or $D = 2$, if the metric is of non-constant curvature. If $D = 1$, then we are in the first case. So assume $D = 2$, which guarantees that g is Levi-Civita. Let us begin with the 111-type. One finds that L has at least one constant eigenvalue, and thus that the projective Lie algebra of g is generically 1-dimensional. However, there are also cases with an algebra of dimension 2 or 3, and the details can be found in [**33**].

Finally, let us assume $D = 2$ and that the metric is of 21-type. Then h has to have non-constant Gaussian curvature (as one can show that otherwise g would have constant curvature already). The projective vector field of g is necessarily of the form $v = \alpha(x)\partial_x + u$ where u is a homothetic vector field of h. Therefore we arrive at the following cases:

- $\zeta = const.$ Then the dimension $\dim \mathfrak{p}(g) = \dim \mathfrak{p}(h) + 2$ is at least 2 and at most 5.
- h has constant curvature. Then $\dim \mathfrak{p}(g) \in \{3, 4\}$.
- $\dim \mathfrak{p}(h) = 1$. Then $\dim \mathfrak{p}(g) = 2$.
- $\dim \mathfrak{p}(h) = 1$. Then $\dim \mathfrak{p}(g) \in \{2, 3\}$.

Here, we do not have space to list the metrics realizing the cases in (2) of Theorem 4.6. But these can be found in terms of local canonical forms in [**33**], see Propositions 9, 10 and 11 of the reference (111-case), as well as in Corollary 3 and Theorem 3 (21-case).

THEOREM 4.7. *Let g be a local Riemannian metric of $n = 3$, not of constant curvature, with a non-vanishing, homothetic vector field v. Then, locally, in a small neighborhood of almost any point, the projective Lie algebra of g is of dimension*

$$1 \leq \dim \mathfrak{p}(g) \leq 5\,,$$

where each number is realized. In particular, if $\dim \mathfrak{p}(g) = 1$, all projective vector fields are multiples of v. In the other extreme, $\dim \mathfrak{p}(g) = 5$, the metric h is of constant curvature and $g = k(h + dz^2)$ $(k \neq 0)$. Then $v = u + (k_1 z + k_0)\partial_z$ with constants $k_0, k_1 \in \mathbb{R}$.

5. Complex Lie problem

The complex analog of projective geometry is called *c-projective geometry.* This field studies what is now known as *J-planar curves* which were treated in 1954 by Otsuki and Tashiro [**44**] after S. Bochner [**6**, Theorem 2] had shown in 1947 that two metrics which are Kähler w.r.t. the same complex structure J and are geodesically equivalent must be affinely equivalent (i.e. their Levi-Civita connections coincide). A comprehensive overview of the topic can be found in [**14**].

Let M be an n-dimensional complex manifold with complex structure J (here n is the complex dimension, thus the real dimension of M is $2n$) equipped with a *complex connection* ∇, i.e., a connection such that $\nabla J = 0$. A *J-planar curve* is a curve $\gamma : I \subseteq \mathbb{R} \to M$ such that

$$\nabla_{\dot\gamma}\dot\gamma = \alpha(t)\dot\gamma + \beta(t)J(\dot\gamma) \tag{5.1}$$

for some smooth functions $\alpha, \beta \in C^\infty(I)$. It is a natural generalization of unparametrized geodesics. Note that Equation (5.1) is equivalent to $\nabla_{\dot\gamma}\dot\gamma \wedge \dot\gamma \wedge J(\dot\gamma) = 0$ whereas unparametrized geodesics are characterized by $\nabla_{\dot\gamma}\dot\gamma \wedge \dot\gamma = 0$.

DEFINITION 5.1. Two complex connections on the same complex manifold (M, J) are *c-projectively equivalent* if they share the same J-planar curves. The set of all connections c-projectively equivalent to a given connection ∇ is called the *c-projective class* of ∇.

DEFINITION 5.2. A *c-projective transformation* is a local diffeomorphism of M that sends J-planar curves into J-planar curves. A *c-projective vector field* is a vector field on M whose local flow acts by c-projective transformations.

The c-projective vector fields of a complex connection form a Lie algebra.

The above definitions can also be given in the case that M is equipped with a *Kähler* metric g by considering its Levi-Civita connection. A metric is said to be of *constant holomorphic curvature* if the sectional curvature, restricted to J-invariant planes, is constant on M. We can then state the Lie problem in the complex case:

Complex Lie problem: Determine a sharp list of local forms of $2n$-dimensional Kähler metrics admitting a non-zero c-projective vector field.

Already in the case $2n = 4$ many different cases can occur. Assume that g and $\hat{g}$ are c-projectively equivalent and consider the $(1,1)$-tensor (this is the c-projective analog of (2.9))

$$A = \left|\frac{\det(\hat{g})}{\det(g)}\right|^{\frac{1}{2(n+1)}} \hat{g}^{-1} g\,.$$

Then for any $t \in \mathbb{R}$ the vector field

$$J \operatorname{grad} \sqrt{\det(t\mathbb{1} - A)}$$

is Killing for g and $\hat{g}$; these *canonical Killing vector fields* [**11**, **14**, **23**] are the examples of c-projective vector fields that are easiest to find. A list of local normal forms for c-projectively equivalent metrics was published in [**11**].

EXAMPLE 5.3 (c-projectively equivalent metrics). Let g and $\hat{g}$ be c-projectively equivalent metrics of dimension $2n = 4$ and let A have two real non-constant eigenvalues ρ, σ. Then, in a neighbourhood of almost any point, there exist coordinates (x, y, s, t) such that $\rho = \rho(x)$, $\sigma = \sigma(y)$ and the metric g, the Kähler form $\omega(\cdot,\cdot) = g(J\cdot,\cdot)$ and the endomorphism A take the forms

$$\begin{aligned}
g &= (\rho - \sigma)(dx^2 + \epsilon dy^2) + \frac{1}{\rho - \sigma}\left[(\rho')^2(ds + \sigma dt)^2 + \epsilon(\sigma')^2(ds + \rho dt)^2\right] \\
\omega &= d((\rho + \sigma)ds + \rho\sigma dt) \\
A &= \rho\partial_x \otimes dx + \sigma\partial_y \otimes dy + (\rho + \sigma)\partial_s \otimes ds + \rho\sigma\partial_s \otimes dt - \partial_t \otimes ds
\end{aligned}$$

where $\epsilon = 1$ in the case of positive signature and $\epsilon = -1$ in the case of split signature. J and $\hat{g}$ can be computed from the given data via the formulae for ω and A mentioned above. This is the c-projective analog of the Liouville case of Proposition 3.8.

For the full c-projective analog of Proposition 3.8 see [**9**, Theorem 3.1] and for the generalization to all dimensions [**11**, Theorem 1.6]. The classification of the metrics in [**11**, Theorem 1.6] w.r.t. the existence of canonical Killing vector fields is evident but some metrics may admit non-canonical Killing vector fields as well as other c-projective vector fields. Due to its significance regarding other problems, in particular the Yano-Obata conjecture [**20**, **40**], existing research has been focused on the following problem.

Complex Lie problem for non-affine c-projective vector fields: Determine a sharp list of local forms of $2n$-dimensional Kähler metrics admitting a non-affine c-projective vector field.

In the case of dimension $2n = 4$ there is a characterization in terms of the degree of mobility:

PROPOSITION 5.4. Let (M, g, J) be a connected Kähler manifold of real dimension 4 (of arbitrary signature) and of non-constant holomorphic sectional curvature. Then the degree of mobility is at most 2 [**3**, **20**].

If the degree of mobility is one, then only homothetic and Killing vector fields are possible. The case of degree of mobility three cannot occur and if the degree of mobility attains the maximal value four, then the metric is of constant holomorphic sectional curvature [**41**]. In the case that the degree of mobility is two, a list of metrics admitting a non-affine c-projective vector field is given in [**9**, Theorems 1.2 and 1.5].

EXAMPLE 5.5 ([**9**, Theorem 1.2, case L1]). The Kähler structure defined by

$$g = (x-y)(c_1^2 dx^2 + \epsilon c_2^2 dy^2) + \frac{1}{x-y}\left[\frac{1}{c_1^2}(ds + y dt)^2 + \epsilon \frac{1}{c_2^2}(ds + x dt)^2\right]$$
$$\omega = d\left[(x+y)\, ds + xy\, dt\right]$$

admits the non-affine c-projective vector field $v = \partial_x + \partial_y - t\partial_s$.

The list in [**9**, Theorems 1.2 and 1.5] is complete in the sense that for any metric admitting a non-affine c-projective vector field in a neighborhood of almost any point there exist coordinates such that the metric takes one of the given forms. But the list is not sharp in the sense that some of the metrics contained are diffeomorphic to others in the list, and some of them have constant holomorphic sectional curvature.

In real dimension $2n \geq 6$ Proposition 5.4 is true only under the additional assumption of closedness [**20**], but fails without the additional assumption [**41**]. The results in [**42**, §2.2] allow the construction of (local) metrics with degree of mobility ≥ 3 and non-constant holomorphic sectional curvature. For Riemannian Kähler metrics that admit a c-projectively equivalent metric that is not affinely equivalent the possible degrees of mobility and the possible dimensions of the space of non-affine c-projective vector fields have been determined in [**41**].

References

[1] A. V. Aminova, *A Lie problem, projective groups of two-dimensional Riemann surfaces, and solitons* (Russian), Izv. Vyssh. Uchebn. Zaved. Mat. **6** (1990), 3–10; English transl., Soviet Math. (Iz. VUZ) **34** (1990), no. 6, 1–9. MR1076673

[2] A. V. Aminova, *Projective Transformations of Pseudo-Riemannian Manifolds*, Journal of Mathematical Sciences **113** (2003), no. 3, 367–470.

[3] Vestislav Apostolov, David M. J. Calderbank, Paul Gauduchon, and Christina W. Tønnesen-Friedman, *Hamiltonian 2-forms in Kähler geometry. II. Global classification*, J. Differential Geom. **68** (2004), no. 2, 277–345. MR2144249

[4] Paul Appell, *Sur des transformations de mouvements* (French), J. Reine Angew. Math. **110** (1892), 37–41, DOI 10.1515/crll.1892.110.37. MR1580273

[5] E. Beltrami, *Risoluzione del problema: riportare i punti di una superficie sopra un piano in modo che le linee geodetiche vengano rappresentate da linee rette*, Ann. Mat. **1** (1865), no. 7, 185–204.

[6] S. Bochner, *Curvature in Hermitian metric*, Bull. Amer. Math. Soc. **53** (1947), 179–195, DOI 10.1090/S0002-9904-1947-08778-4. MR19983

[7] Alexey V. Bolsinov and Vladimir S. Matveev, *Splitting and gluing lemmas for geodesically equivalent pseudo-Riemannian metrics*, Trans. Amer. Math. Soc. **363** (2011), no. 8, 4081–4107, DOI 10.1090/S0002-9947-2011-05187-1. MR2792981

[8] Alexey V. Bolsinov and Vladimir S. Matveev, *Local normal forms for geodesically equivalent pseudo-Riemannian metrics*, Trans. Amer. Math. Soc. **367** (2015), no. 9, 6719–6749, DOI 10.1090/S0002-9947-2014-06416-7. MR3356952

[9] Alexey V. Bolsinov, Vladimir S. Matveev, Thomas Mettler, and Stefan Rosemann, *Four-dimensional Kähler metrics admitting c-projective vector fields*, J. Math. Pures Appl. (9) **103** (2015), no. 3, 619–657, DOI 10.1016/j.matpur.2014.07.005. MR3310270

[10] Alexey V. Bolsinov, Vladimir S. Matveev, and Giuseppe Pucacco, *Normal forms for pseudo-Riemannian 2-dimensional metrics whose geodesic flows admit integrals quadratic in momenta*, J. Geom. Phys. **59** (2009), no. 7, 1048–1062, DOI 10.1016/j.geomphys.2009.04.010. MR2536861

[11] Alexey V. Bolsinov, Vladimir S. Matveev, and Stefan Rosemann, *Local normal forms for c-projectively equivalent metrics and proof of the Yano-Obata conjecture in arbitrary signature. Proof of the projective Lichnerowicz conjecture for Lorentzian metrics* (English, with English and French summaries), Ann. Sci. Éc. Norm. Supér. (4) **54** (2021), no. 6, 1465–1540, DOI 10.24033/asens.2487. MR4371392

[12] Robert Bryant, Maciej Dunajski, and Michael Eastwood, *Metrisability of two-dimensional projective structures*, J. Differential Geom. **83** (2009), no. 3, 465–499. MR2581355

[13] Robert L. Bryant, Gianni Manno, and Vladimir S. Matveev, *A solution of a problem of Sophus Lie: normal forms of two-dimensional metrics admitting two projective vector fields*, Math. Ann. **340** (2008), no. 2, 437–463, DOI 10.1007/s00208-007-0158-3. MR2368987

[14] David M. J. Calderbank, Michael G. Eastwood, Vladimir S. Matveev, and Katharina Neusser, *C-projective geometry*, Mem. Amer. Math. Soc. **267** (2020), no. 1299, v+137, DOI 10.1090/memo/1299. MR4194892

[15] E. Cartan, *Sur les variétés à connexion projective*, Bull. Soc. Math. France **52** (1924), 205–241. MR1504846

[16] G. Darboux, E. Picard, G.X.P. Koenigs, and E.M.P. Cosserat, *Leçons sur la théorie générale des surfaces et les applications géométriques du calcul infinitésimal.*, Cours de géométrie de la Faculté des sciences, Gauthier-Villars, (1896).

[17] Ulisse Dini, *Sopra un problema che si presenta nella teoria generale delle rappresentazioni geografiche di una superficie su di un'altra*, Annali di Matematica Pura ed Applicata (1867-1897) **3** (1869), no. 1, 269–293.

[18] Maciej Dunajski and Michael Eastwood, *Metrisability of three-dimensional path geometries*, Eur. J. Math. **2** (2016), no. 3, 809–834, DOI 10.1007/s40879-016-0095-3. MR3536153

[19] Michael Eastwood and Vladimir Matveev, *Metric connections in projective differential geometry*, Symmetries and overdetermined systems of partial differential equations, IMA Vol. Math. Appl., vol. 144, Springer, New York, 2008, pp. 339–350, DOI 10.1007/978-0-387-73831-4_16. MR2384718

[20] A. Fedorova, V. Kiosak, V. S. Matveev, and S. Rosemann, *The only Kähler manifold with degree of mobility at least 3 is* $(\mathbb{C}P(n), g_{Fubini\text{-}Study})$, Proc. Lond. Math. Soc. (3) **105** (2012), no. 1, 153–188. MR2948791

[21] Hitosi Hiramatu, *Riemannian manifolds admitting a projective vector field*, Kodai Math. J. **3** (1980), no. 3, 397–406. MR604484

[22] Volodymyr Kiosak and Vladimir S. Matveev, *Proof of the projective Lichnerowicz conjecture for pseudo-Riemannian metrics with degree of mobility greater than two*, Comm. Math. Phys. **297** (2010), no. 2, 401–426, DOI 10.1007/s00220-010-1037-4. MR2651904

[23] Kazuyoshi Kiyohara and Peter Topalov, *On Liouville integrability of h-projectively equivalent Kähler metrics*, Proc. Amer. Math. Soc. **139** (2011), no. 1, 231–242, DOI 10.1090/S0002-9939-2010-10576-2. MR2729086

[24] T. Levi-Civita, *Sulle trasformazioni delle equazioni dinamiche*, Ann. Mat. Pura Appl. Ser 2a **24** (1896), 255–300.

[25] Sophus Lie, *Untersuchungen über geodätische Curven* (German), Math. Ann. **20** (1882), no. 3, 357–454, DOI 10.1007/BF01443601. MR1510173

[26] Sophus Lie, *Classification und Integration von gewöhnlichen Differentialgleichungen zwischen xy, die eine Gruppe von Transformationen gestatten*, Archiv for Mathematik og Naturvidenskab. Christiana. **8** (1883), 187–288.

[27] Sophus Lie, *Classification und Integration von gewöhnlichen Differentialgleichungen zwischen xy, die eine Gruppe von Transformationen gestatten* (German), Math. Ann. **32** (1888), no. 2, 213–281, DOI 10.1007/BF01444068. MR1510512

[28] Sophus Lie, *Vorlesungen über Differentialgleichungen mit bekannten infinitesimalen Transformationen* (ger), Teubner, (1891).

[29] R. Liouville, *Sur les invariants de certaines équations différentielles et sur leurs applications*, Journal de l'École Polytechnique **59** (1889), 7–76.

[30] Tianyu Ma, *Geodesic rigidity of Levi-Civita connections admitting essential projective vector fields*, Geom. Dedicata **205** (2020), 147–166, DOI 10.1007/s10711-019-00469-7. MR4076823

[31] Gianni Manno and Andreas Vollmer, *(Super-)integrable systems associated to 2-dimensional projective connections with one projective symmetry*, J. Geom. Phys. **145** (2019), 103476, 22, DOI 10.1016/j.geomphys.2019.07.007. MR3989667

[32] Gianni Manno and Andreas Vollmer, *Normal forms of two-dimensional metrics admitting exactly one essential projective vector field* (English, with English and French summaries), J. Math. Pures Appl. (9) **135** (2020), 26–82, DOI 10.1016/j.matpur.2020.01.003. MR4066105

[33] Gianni Manno and Andreas Vollmer, *3-dimensional Levi-Civita metrics with projective vector fields* (English, with English and French summaries), J. Math. Pures Appl. (9) **163** (2022), 473–517, DOI 10.1016/j.matpur.2022.05.012. MR4438907

[34] V. S. Matveev, *Solodovnikov's theorem in dimension two* (Russian), Dokl. Akad. Nauk **396** (2004), no. 1, 25–27. MR2115905

[35] Vladimir S. Matveev, *Die Vermutung von Obata für Dimension 2*, Arch. Math. (Basel) **82** (2004), no. 3, 273–281. MR2053631

[36] Vladimir S. Matveev, *Lichnerowicz-Obata conjecture in dimension two*, Comment. Math. Helv. **80** (2005), no. 3, 541–570, DOI 10.4171/CMH/25. MR2165202

[37] Vladimir S. Matveev, *Proof of the projective Lichnerowicz-Obata conjecture*, J. Differential Geom. **75** (2007), no. 3, 459–502. MR2301453

[38] Vladimir S. Matveev, *Geodesically equivalent metrics in general relativity*, J. Geom. Phys. **62** (2012), no. 3, 675–691, DOI 10.1016/j.geomphys.2011.04.019. MR2876790

[39] Vladimir S. Matveev, *Two-dimensional metrics admitting precisely one projective vector field*, Math. Ann. **352** (2012), no. 4, 865–909, DOI 10.1007/s00208-011-0659-y. MR2892455

[40] Vladimir S. Matveev and Stefan Rosemann, *Proof of the Yano-Obata conjecture for h-projective transformations*, J. Differential Geom. **92** (2012), no. 2, 221–261. MR2998672

[41] Vladimir S. Matveev and Stefan Rosemann, *Conification construction for Kähler manifolds and its application in c-projective geometry*, Adv. Math. **274** (2015), 1–38. MR3318143

[42] J. Mikeš, *Holomorphically projective mappings and their generalizations*, J. Math. Sci. (New York) **89** (1998), no. 3, 1334–1353, DOI 10.1007/BF02414875. Geometry, 3. MR1619720

[43] J. E. R. O'Connor and G. E. Prince, *Finding collineations of Kimura metrics*, Gen. Relativity Gravitation **30** (1998), no. 1, 69–82. MR1601561

[44] Tominosuke Ōtsuki and Yoshihiro Tashiro, *On curves in Kaehlerian spaces*, Math. J. Okayama Univ. **4** (1954), 57–78. MR66024

[45] P. Painlevé, *Mémoire sur la transformation des équations de la dynamique*, Journal de Mathématiques Pures et Appliquées **10** (1894), 5–92.

[46] Yu. R. Romanovskiĭ, *Calculation of local symmetries of second-order ordinary differential equations by Cartan's equivalence method*, Mat. Zametki **60** (1996), no. 1, 75–91, 159. MR1431461

[47] A. S. Solodovnikov, *Projective transformations of Riemannian spaces* (Russian), Uspehi Mat. Nauk (N.S.) **11** (1956), no. 4(70), 45–116. MR0084826

[48] Peter Topalov and Vladimir S. Matveev, *Geodesic equivalence via integrability*, Geom. Dedicata **96** (2003), 91–115, DOI 10.1023/A:1022166218282. MR1956835

[49] Ar. Tresse, *Détermination des invariants ponctuels de l'équation différentielle ordinaire du second ordre* (1896).

Politecnico di Torino, via Duca degli Abruzzi 24, 10129 Torino (Italy)
Email address: `giovanni.manno@polito.it`

Politecnico di Torino, via Duca degli Abruzzi 24, 10129 Torino (Italy)
Email address: `jsmaths@gmx.net`

Politecnico di Torino, via Duca degli Abruzzi 24, 10129 Torino (Italy) and University of Hamburg, Bundesstr. 55, 20146 Hamburg (Germany)
Email address: `andreas.vollmer@uni-hamburg.de`

Contemporary Mathematics
Volume **788**, 2023
https://doi.org/10.1090/conm/788/15828

Lax representations via twisted extensions of infinite-dimensional Lie algebras: some new results

Oleg I. Morozov

ABSTRACT. We find new integrable partial differential equations with Lax representations generated by extensions of Lie algebras of the Kac–Moody type as well as the Lie algebra of Hamiltonian vector fields on $\mathbb{R}^2$.

1. Introduction

Lax representations provide the basic construction that allows implementation of a number of techniques for studying nonlinear partial differential equations (PDEs), whence they are considered as the key feature indicating integrability thereof, see [**1, 5, 12, 20, 29, 32, 34, 36, 37**] and references therein. Therefore the problem of finding intrinsic conditions that ensure existence of a Lax representation for a given PDE is of great importance in the theory of integrable systems. In the recent papers [**22**] – [**26**] we propose an approach to tackle this problem. We have shown there that for a number of PDEs including the potential Khokhlov–Zabolotskaya equation, the Boyer–Finley equation, the hyper-CR equation of Einstein–Weyl structures, the reduced quasiclassical self-dual Yang–Mills equation, the 4D Martínez Alonso–Shabat equation, the 4D universal hierarchy equation, and other equations, their known Lax representations can be inferred from non-triviality of the second twisted cohomology groups of the Lie algebras of contact symmetries of the PDEs. Moreover, we have shown that the technique allows one to find new Lax representations of some PDEs.

The aim of the present paper is to make a further progress in understanding of the relationship between the structure theory of infinite-dimensional Lie algebras and the theory of integrable systems. In particular, we construct new examples of deriving Lax representations of PDEs from twisted extensions of some Lie algebras.

In a number of above-mentioned examples the symmetry algebras of the PDEs have the form of the semi-direct sum $\mathfrak{s}_\diamond \ltimes \mathfrak{q}_{N,\varepsilon}$ of a finite-dimensional Lie algebra $\mathfrak{s}_\diamond$ and the infinite-dimensional Lie algebra $\mathfrak{q}_{N,\varepsilon}$ of the Kac–Moody type, [**6**], that is, the deformation of the tensor product $\mathfrak{q}_{N,0} = \mathbb{R}_N[s] \otimes \mathfrak{w}$, where $\mathbb{R}_N[s]$ is the commutative unital algebra of the truncated polynomials in the (formal) variable s of degree N and $\mathfrak{w}$ is the Lie algebra of the polynomial vector fields on $\mathbb{R}$, see Section 3 for definition of $\mathfrak{q}_{N,\varepsilon}$. The second twisted cohomology groups of the Lie

2020 *Mathematics Subject Classification.* Primary 58H05, 58J70, 35A30, 37K05, 37K10.
Key words and phrases. Twisted cohomology, Maurer–Cartan forms, symmetries of differential equations, Lax representations.

©2023 American Mathematical Society

algebras $\mathfrak{s}_\diamond \ltimes \mathfrak{q}_{N,\varepsilon}$ from our examples turn out to be nontrivial, and the nontrivial twisted 2-cocycles generate twisted extensions of these Lie algebras. Linear combinations of the Maurer–Cartan forms of the twisted extensions provide the Wahlquist–Estabrook forms that generate the Lax representations of the PDEs.

In examples from [**22**]–[**26**] we have $N \geq 3$, and the natural question is whether there exist Lie algebras of the form $\mathfrak{s}_\diamond \ltimes \mathfrak{q}_{N,\varepsilon}$ with $N < 3$ and nontrivial second twisted cohomology groups whose twisted extensions generate Lax representations of some PDEs. In Sections 4, 5, and 6 we present three systems (4.3) - (4.4), (5.5) - (5.6), and (6.2) - (6.3) whose Lax representations are generated by extensions of the Lie algebras $\mathfrak{q}_{1,-1}$, $\mathfrak{q}_{1,-1/2}$, and $\mathfrak{q}_{2,-1}$, respectively. Equation (5.5) can be considered as a 3D generalization of the generalized 2D Hunter–Saxton equation [**3, 9, 10, 27**] with the special value of the parameter. We compare the structure of the symmetry algebras of the obtained systems with the structure of the Lie algebras that generate the Lax representations.

Paper [**24**] shows that for a number of known 4D integrable equations their Lax representations can be derived from the twisted extensions of the symmetry algebras, which turn out to be of the form $\mathfrak{s}_\diamond \ltimes (\mathbb{R}_N[s] \otimes \mathbb{R}[t] \otimes \mathfrak{w})$. In [**26**] we apply this technique to a Lie algebra of such a form and construct a new 4D integrable equation. In the present paper we address an interesting question of finding new examples of integrable PDEs whose symmetry algebras include the Lie algebra $\mathfrak{h}$ of Hamiltonian vector fields on $\mathbb{R}^2$ as a subalgebra. In Section 7 we present such an example. We derive equation (7.4) from the Lax representation defined by the extensions of the Lie algebra $\mathfrak{h} \oplus \mathfrak{w}$. Other examples of integrable systems whose symmetry algebras include $\mathfrak{h}$ as 'building blocks' are given by the family of heavenly equations [**4, 7, 18, 33**] and their 'symmetric deformations' [**19**].

2. Preliminaries and notation

2.1. Symmetries and Lax representations. The presentation in this subsection closely follows [**13**]—[**17**] and [**35**]. All our considerations are local. Let $\pi\colon \mathbb{R}^n \times \mathbb{R}^m \to \mathbb{R}^n$, $\pi\colon (x^1, \dots, x^n, u^1, \dots, u^m) \mapsto (x^1, \dots, x^n)$, be a trivial bundle, and $J^\infty(\pi)$ be the bundle of its jets of the infinite order. The local coordinates on $J^\infty(\pi)$ are $(x^i, u^\alpha, u^\alpha_I)$, where $I = (i_1, \dots, i_n)$ are multi-indices with $i_k \geq 0$, and for every local section $f\colon \mathbb{R}^n \to \mathbb{R}^n \times \mathbb{R}^m$ of π the corresponding infinite jet $j_\infty(f)$ is a section $j_\infty(f)\colon \mathbb{R}^n \to J^\infty(\pi)$ such that $u^\alpha_I(j_\infty(f)) = \dfrac{\partial^{\#I} f^\alpha}{\partial x^I} = \dfrac{\partial^{i_1+\dots+i_n} f^\alpha}{(\partial x^1)^{i_1} \dots (\partial x^n)^{i_n}}$. We put $u^\alpha = u^\alpha_{(0,\dots,0)}$. Also, we will simplify notation in the following way: e.g., in the case of $n = 3$, $m = 1$ we denote $x^1 = t$, $x^2 = x$ $x^3 = y$, and $u^1_{(i,j,k)} = u_{t\dots tx\dots xy\dots y}$ with i times t, j times x, and k times y.

The vector fields

$$D_{x^k} = \frac{\partial}{\partial x^k} + \sum_{\#I \geq 0} \sum_{\alpha=1}^{m} u^\alpha_{I+1_k} \frac{\partial}{\partial u^\alpha_I}, \qquad k \in \{1, \dots, n\},$$

$(i_1, \dots, i_k, \dots, i_n) + 1_k = (i_1, \dots, i_k + 1, \dots, i_n)$, are called *total derivatives.* They commute everywhere on $J^\infty(\pi)$: $[D_{x^i}, D_{x^j}] = 0$.

The *evolutionary vector field* associated to an arbitrary vector-valued smooth function $\varphi\colon J^\infty(\pi) \to \mathbb{R}^m$ is the vector field

$$\mathbf{E}_\varphi = \sum_{\#I\geq 0}\sum_{\alpha=1}^{m} D_I(\varphi^\alpha)\,\frac{\partial}{\partial u_I^\alpha}$$

with $D_I = D_{(i_1,\dots i_n)} = D_{x^1}^{i_1}\circ\dots\circ D_{x^n}^{i_n}$.

A system of PDEs $F_r(x^i, u_I^\alpha) = 0$ of the order $s \geq 1$ with $\#I \leq s$, $r \in \{1,\dots,R\}$ for some $R \geq 1$, defines the submanifold $\mathcal{E} = \{(x^i, u_I^\alpha) \in J^\infty(\pi) \mid D_K(F_r(x^i, u_I^\alpha)) = 0,\ \#K \geq 0\}$ in $J^\infty(\pi)$.

A function $\varphi\colon J^\infty(\pi) \to \mathbb{R}^m$ is called a *(generator of an infinitesimal) symmetry* of equation $\mathcal{E}$ when $\mathbf{E}_\varphi(F) = 0$ on $\mathcal{E}$. The symmetry φ is a solution to the *defining system*

$$\ell_{\mathcal{E}}(\varphi) = 0, \tag{2.1}$$

where $\ell_{\mathcal{E}} = \ell_F|_{\mathcal{E}}$ with the matrix differential operator

$$\ell_F = \left(\sum_{\#I\geq 0}\frac{\partial F_r}{\partial u_I^\alpha}\,D_I\right).$$

The *symmetry algebra* $\mathrm{Sym}(\mathcal{E})$ of equation $\mathcal{E}$ is the linear space of solutions to (2.1) endowed with the structure of a Lie algebra over $\mathbb{R}$ by the *Jacobi bracket* $\{\varphi,\psi\} = \mathbf{E}_\varphi(\psi) - \mathbf{E}_\psi(\varphi)$. The *algebra of contact symmetries* $\mathrm{Sym}_0(\mathcal{E})$ is the Lie subalgebra of $\mathrm{Sym}(\mathcal{E})$ defined as $\mathrm{Sym}(\mathcal{E})\cap C^\infty(J^1(\pi))$. Symmetries with generators of the form $\varphi^\alpha = \eta^\alpha - \sum_i \xi^i\,u_i^\alpha$, $\eta^\alpha, \xi^i \in C^\infty(J^0(\pi))$, are referred to as *point symmetries*. They correspond to vector fields $\sum_i \xi^i\,\partial_{x^i} + \sum_\alpha \eta^\alpha\,\partial_{u^\alpha}$ on $J^0(\pi)$.

Let the linear space $\mathcal{W}$ be either $\mathbb{R}^N$ for some $N \geq 1$ or $\mathbb{R}^\infty$ endowed with local coordinates w^a, $a \in \{1,\dots,N\}$ or $a \in \mathbb{N}$, respectively. Variables w^a are called *pseudopotentials* [**36**]. Locally, a *differential covering* of $\mathcal{E}$ is a trivial bundle $\varpi\colon J^\infty(\pi)\times\mathcal{W} \to J^\infty(\pi)$ equipped with *extended total derivatives*

$$\widetilde{D}_{x^k} = D_{x^k} + \sum_a T_k^a(x^i, u_I^\alpha, w^b)\,\frac{\partial}{\partial w^a}$$

such that $[\widetilde{D}_{x^i}, \widetilde{D}_{x^j}] = 0$ for all $i \neq j$ whenever $(x^i, u_I^\alpha) \in \mathcal{E}$. Define the partial derivatives of w^a by $w^s_{x^k} = \widetilde{D}_{x^k}(w^s)$. This yields the over-determined system of PDEs

$$w^a_{x^k} = T_k^a(x^i, u_I^\alpha, w^b) \tag{2.2}$$

which is compatible whenever $(x^i, u_I^\alpha) \in \mathcal{E}$. System (2.2) is referred to as the *covering equations* or the *Lax representation* of equation $\mathcal{E}$.

Dually, the differential covering is defined by the *Wahlquist–Estabrook forms*

$$\tau^a = dw^a - \sum_{k=1}^{m} T_k^a(x^i, u_I^\alpha, w^b)\,dx^k \tag{2.3}$$

as follows: when w^a and u^α are considered to be functions of x^1, …, x^n, forms (2.3) are equal to zero if and only if system (2.2) holds.

2.2. Twisted cohomology of Lie algebras. For a Lie algebra $\mathfrak{g}$ over $\mathbb{R}$, its representation $\rho\colon \mathfrak{g} \to \mathrm{End}(V)$, and $k \ge 1$ let $C^k(\mathfrak{g}, V) = \mathrm{Hom}(\Lambda^k(\mathfrak{g}), V)$ be the space of all k–linear skew-symmetric mappings from $\mathfrak{g}$ to V. Then the Chevalley–Eilenberg differential complex

$$V = C^0(\mathfrak{g}, V) \xrightarrow{d} C^1(\mathfrak{g}, V) \xrightarrow{d} \cdots \xrightarrow{d} C^k(\mathfrak{g}, V) \xrightarrow{d} C^{k+1}(\mathfrak{g}, V) \xrightarrow{d} \ldots$$

is generated by the differential $d\colon \theta \mapsto d\theta$ such that

$$\begin{aligned} &d\theta(X_1, \ldots, X_{k+1}) \\ &= \sum_{q=1}^{k+1}(-1)^{q+1}\rho(X_q)\left(\theta(X_1, \ldots, \hat{X}_q, \ldots, X_{k+1})\right) \\ &+ \sum_{1\le p<q\le k+1} (-1)^{p+q+1}\theta([X_p, X_q], X_1, \ldots, \hat{X}_p, \ldots, \hat{X}_q, \ldots, X_{k+1}). \end{aligned} \tag{2.4}$$

The cohomology groups of the complex $(C^*(\mathfrak{g}, V), d)$ are referred to as the *cohomology groups of the Lie algebra* $\mathfrak{g}$ *with coefficients in the representation* ρ. For the trivial representation $\rho_0\colon \mathfrak{g} \to \mathbb{R}$, $\rho_0\colon X \mapsto 0$, the cohomology groups are denoted by $H^*(\mathfrak{g})$.

Consider a Lie algebra $\mathfrak{g}$ over $\mathbb{R}$ with non-trivial first cohomology group $H^1(\mathfrak{g})$ and take 1-form $\alpha \neq 0$ on $\mathfrak{g}$ such that $d\alpha = 0$. Then for each $c \in \mathbb{R}$ define the *twisted differential* $d_{c\alpha}\colon C^k(\mathfrak{g}, \mathbb{R}) \to C^{k+1}(\mathfrak{g}, \mathbb{R})$ by the formula

$$d_{c\alpha}\theta = d\theta - c\,\alpha \wedge \theta.$$

From $d\alpha = 0$ it follows that $d_{c\alpha}^2 = 0$. The cohomology groups of the complex

$$C^1(\mathfrak{g}, \mathbb{R}) \xrightarrow{d_{c\alpha}} \cdots \xrightarrow{d_{c\alpha}} C^k(\mathfrak{g}, \mathbb{R}) \xrightarrow{d_{c\alpha}} C^{k+1}(\mathfrak{g}, \mathbb{R}) \xrightarrow{d_{c\alpha}} \ldots$$

are referred to as the *twisted cohomology groups* [**30**,**31**] of $\mathfrak{g}$ and denoted by $H^*_{c\alpha}(\mathfrak{g})$.

3. Lie algebras of the Kac–Moody type and their extensions

Consider the Lie algebra $\mathfrak{q}_{N,0} = \mathbb{R}_N[s] \otimes \mathfrak{w}$, where $\mathbb{R}_N[s] = \mathbb{R}[s]/\langle s^{N+1} = 0\rangle$ is the commutative unital algebra of the truncated polynomials of the formal variable s of degree N, and $\mathfrak{w} = \langle V_k \mid k \ge 0\rangle$, $V_k = \dfrac{1}{k!}\,t^k\,\partial_t$, is the Lie algebra of polynomial vector fields on $\mathbb{R}$ referred to as the (one-sided) Witt algebra. Algebra $\mathfrak{q}_{N,0}$ admits the deformation[1] generated by cocycle $\Psi \in H^2(\mathfrak{q}_{N,0}, \mathfrak{q}_{N,0})$,

$$\Psi(s^p \otimes V_m, s^q \otimes V_n) = \begin{cases} \dfrac{p\,n - q\,m}{m+n}\dbinom{m+n}{m}\, s^{p+q} \otimes V_{m+n-1}, & m+n \ge 1, p+q \le N, \\ 0, & \text{otherwise.} \end{cases}$$

For each $\varepsilon \neq 0$ this cocycle defines new bracket $[\cdot,\cdot]_\varepsilon = [\cdot,\cdot] + \varepsilon\,\Psi(\cdot,\cdot)$ on the linear space $\langle s^p \otimes V_m \,|\, p \le N, m \ge 0\rangle$. We denote the resulting Lie algebra as $\mathfrak{q}_{N,\varepsilon}$. In other words, the Lie algebra $\mathfrak{q}_{N,\varepsilon}$ is isomorphic to the linear space of functions $f(t,s) = f_0(t) + s\,f_1(t) + \cdots + s^N\,f_N(t)$, $f_k \in \mathbb{R}[t]$, equipped with the bracket

$$[f, g]_\varepsilon = f\,g_t - g\,f_t + \varepsilon\,s\,(f_s\,g_t - g_s\,f_t) \tag{3.1}$$

such that there holds $s^k = 0$ for $k > N$. Arguments similar to [**6**] show that $\mathfrak{q}_{N,\varepsilon}$ is a proper subalgebra of the affine Kac–Moody Lie algebra $\mathfrak{g}(A^{(1)}_M)$ with the

[1] For the full description of deformations of the Lie algebra $\mathfrak{q}_{N,0}$ see [**38**].

generalized Cartan matrix $A_M^{(1)}$ [**11**] for some $M \geq N$. Therefore the Lie algebras $\mathfrak{q}_{N,\varepsilon}$ are referred to as Lie algebras of the Kac–Moody type.

Consider the dual 1-forms $\theta_{k,m}$ to the basis elements $s^p \otimes V_n$ of $\mathfrak{q}_{N,\varepsilon}$, that is, the linear maps $\theta_{k,m} \colon \mathfrak{q}_{N,\varepsilon} \to \mathbb{R}$ such that $\theta_{k,m}(s^p \otimes V_n) = \delta_{k,p}\, \delta_{m,n}$. Define the formal series

$$\Theta = \sum_{k=0}^{N} \sum_{m=0}^{\infty} \frac{h_0^k h_1^m}{m!} \theta_{k,m} \tag{3.2}$$

with the formal parameters h_0 and h_1 such that $h_0^k = 0$ when $k > N$ and $dh_0 = dh_1 = 0$. Then (2.4) and (3.1) entail the *structure equations*

$$d\Theta = \Theta_{h_1} \wedge (\Theta + \varepsilon\, h_0\, \Theta_{h_0}) \tag{3.3}$$

of the Lie algebra $\mathfrak{q}_{N,\varepsilon}$. Here and below we use notation $\Theta_{h_i} = \partial_{h_i}\Theta$ for the partial derivatives of the formal series of 1-forms with respect to the formal parameters h_i.

For each $N \in \mathbb{N}$ and $\varepsilon \in \mathbb{R}$ the map $D_0 \colon s^p \otimes V_n \mapsto p\, s^p \otimes V_n$ is an outer derivation of $\mathfrak{q}_{N,\varepsilon}$. Denote by $\mathfrak{a}_0 \ltimes \mathfrak{q}_{N,\varepsilon}$ the associated one-dimensional 'right' extension, [**8**, §1.4.4], of $\mathfrak{q}_{N,\varepsilon}$. As a vector space $\mathfrak{a}_0 \ltimes \mathfrak{q}_{N,\varepsilon} = \langle Z_0 \rangle \oplus \mathfrak{q}_{N,\varepsilon}$, and the bracket on $\mathfrak{q}_{N,\varepsilon}$ is extended to Z_0 by the formula $[Z_0, s^p \otimes V_n]_\varepsilon = D_0(s^p \otimes V_n) = p\, s^p \otimes V_n$. Let $\alpha_0 \colon \mathfrak{a}_0 \ltimes \mathfrak{q}_{N,\varepsilon} \to \mathbb{R}$ be the dual form to $-Z_0$, that is, $\alpha_0(Z_0) = -1$ and $\alpha_0(s^p \otimes V_n) = 0$. Then the structure equations for the Lie algebra $\mathfrak{a}_0 \ltimes \mathfrak{q}_{N,\varepsilon}$ read

$$\left\{ \begin{array}{lcl} d\Theta & = & \Theta_{h_1} \wedge (\Theta + \varepsilon\, h_0\, \Theta_{h_0}) + h_0\, \alpha_0 \wedge \Theta_{h_0}, \\ d\alpha_0 & = & 0. \end{array} \right.$$

For some values of N and ε the Lie algebras $\mathfrak{a}_0 \ltimes \mathfrak{q}_{N,\varepsilon}$ admit further right extensions. In [**22**–**25**] we have shown examples of integrable PDEs whose Lax representations can be inferred from such extensions when $N \geq 3$. In the next three sections we consider integrable PDEs that are related to extensions of the Lie algebras $\mathfrak{a}_0 \ltimes \mathfrak{q}_{1,-1}$, $\mathfrak{a}_0 \ltimes \mathfrak{q}_{1,-1/2}$, and $\mathfrak{a}_0 \ltimes \mathfrak{q}_{2,-1}$.

As it was shown in [**23**], for $N \in \mathbb{N}$ and $\varepsilon = -1/r$ with $r \in \{1, \ldots, N\}$ the Lie algebra $\mathfrak{a}_0 \ltimes \mathfrak{q}_{N,\varepsilon}$ admits the right extension $\mathfrak{a}_1 \ltimes \mathfrak{q}_{N,\varepsilon}$ generated by the outer derivation $D_1 \colon \mathfrak{a}_0 \ltimes \mathfrak{q}_{N,\varepsilon} \to \mathfrak{a}_0 \ltimes \mathfrak{q}_{N,\varepsilon}$ with

$$D_1(s^p \otimes V_k) = \left\{ \begin{array}{ll} k\, s^{p+r} \otimes V_{k-1}, & p + r \leq N, k \geq 1, \\ 0, & \text{otherwise.} \end{array} \right.$$

We have $[D_0, D_1] = D_0 \circ D_1 - D_1 \circ D_0 = r\, D_1$. Then $\mathfrak{a}_1 \ltimes \mathfrak{q}_{N,-1/r}$ as a vector space is $\langle Z_1 \rangle \oplus (\mathfrak{a}_0 \ltimes \mathfrak{q}_{N,-1/r})$, with the extension of the bracket of $\mathfrak{q}_{N,-1/r}$ given by $[Z_0, Z_1]_{-1/r} = -r\, Z_1$ and $[Z_1, s^p \otimes V_k]_{-1/r} = D_1(s^p \otimes V_k)$. Consider the dual form α_1 to the vector $-Z_1$, that, put $\alpha_1(Z_1) = -1$, $\alpha_1(Z_0) = \alpha_1(s^p \otimes V_k) = 0$. Then the structure equations for the Lie algebra $\mathfrak{a}_1 \ltimes \mathfrak{q}_{N,-1/r}$ get the form

$$\left\{ \begin{array}{lcl} d\Theta & = & \Theta_{h_1} \wedge \left(\Theta - \dfrac{h_0}{r}\, \Theta_{h_0} - h_0^r\, \alpha_1 \right) + h_0\, \alpha_0 \wedge \Theta_{h_0}, \\ d\alpha_0 & = & 0, \\ d\alpha_1 & = & r\, \alpha_0 \wedge \alpha_1. \end{array} \right. \tag{3.4}$$

These equations yield $H^1(\mathfrak{a}_1 \ltimes \mathfrak{q}_{N,-1/r}) = \langle \alpha_0 \rangle$ and $[\alpha_0 \wedge \alpha_1] \in H^2_{r\alpha_0}(\mathfrak{a}_1 \ltimes \mathfrak{q}_{N,-1/r})$, therefore the Lie algebra $\mathfrak{a}_1 \ltimes \mathfrak{q}_{N,-1/r}$ admits the twisted extension $\mathfrak{a}_2 \ltimes \mathfrak{q}_{N,-1/r}$ with the structure equations obtained by appending equation

$$d\alpha_2 = r\, \alpha_0 \wedge \alpha_2 + \alpha_0 \wedge \alpha_1 \tag{3.5}$$

to system (3.4).

Furthermore, we have $[\alpha_1 \wedge \alpha_2] \in H^2_{2r\alpha_0}(\mathfrak{a}_2 \ltimes \mathfrak{q}_{N,-1/r})$, therefore the Lie algebra $\mathfrak{a}_2 \ltimes \mathfrak{q}_{N,-1/r}$ admits the twisted extension $\mathfrak{a}_3 \ltimes \mathfrak{q}_{N,-1/r}$ whose structure equations are obtained by appending equation

$$d\alpha_3 = 2\,r\,\alpha_0 \wedge \alpha_3 + \alpha_1 \wedge \alpha_2 \tag{3.6}$$

to system (3.4), (3.5). This process can be repeated: $\mathfrak{a}_k \ltimes \mathfrak{q}_{N,-1/r}$ admits the twisted extension $\mathfrak{a}_{k+1} \ltimes \mathfrak{q}_{N,-1/r}$ defined by the twisted 2-cocycle $[\alpha_1 \wedge \alpha_k] \in H^2_{kr\alpha_0}(\mathfrak{a}_k \ltimes \mathfrak{q}_{N,-1/r})$. In other words, the extended Lie algebra $\mathfrak{a}_{k+1} \ltimes \mathfrak{q}_{N,-1/r}$ has the structure equations

$$\left\{\begin{array}{lcl} d\Theta & = & \Theta_{h_1} \wedge \left(\Theta - \dfrac{h_0}{r}\,\Theta_{h_0} - h_0^r\,\alpha_1\right) + h_0\,\alpha_0 \wedge \Theta_{h_0}, \\ d\alpha_0 & = & 0, \\ d\alpha_1 & = & r\,\alpha_0 \wedge \alpha_1, \\ d\alpha_{m+1} & = & m\,r\,\alpha_0 \wedge \alpha_{m+1} + \alpha_1 \wedge \alpha_m, \qquad m \in \{1, \ldots, k\}. \end{array}\right. \tag{3.7}$$

4. Equation associated to $\mathfrak{a}_2 \ltimes \mathfrak{q}_{1,-1}$

Consider the Lie algebra $\mathfrak{a}_3 \ltimes \mathfrak{q}_{1,-1}$ defined by the structure equations (3.7) with $r = 1$ and $k = 2$, that is, by system

$$\left\{\begin{array}{lcl} d\Theta & = & \Theta_{h_1} \wedge (\Theta - h_0\,\Theta_{h_0} - h_0\,\alpha_1) + h_0\,\alpha_0 \wedge \Theta_{h_0}, \\ d\alpha_0 & = & 0, \\ d\alpha_1 & = & \alpha_0 \wedge \alpha_1, \\ d\alpha_2 & = & \alpha_0 \wedge \alpha_2 + \alpha_0 \wedge \alpha_1, \\ d\alpha_3 & = & 2\,\alpha_0 \wedge \alpha_3 + \alpha_0 \wedge \alpha_2. \end{array}\right. \tag{4.1}$$

Frobenius' theorem allows one to integrate equations (4.1) step by step. In particular, we have

$$\alpha_0 = \frac{dq}{q}, \quad \alpha_1 = q\,dy, \quad \alpha_2 = q\,(dw + \ln q\;dy), \quad \alpha_3 = q^2\,(dv - w\,dy),$$

$$\theta_{0,0} = a\,dt, \quad \theta_{1,0} = q\,(dx - \ln a\;dy + u\,dt), \quad \theta_{1,1} = q\,a^{-1}\,(du - b\,dy - p\,dt),$$

where $q \neq 0$, $a \neq 0$, t, x, y, u, v, w are free parameters ('constants of integration'). We do not need explicit expressions for the other forms $\theta_{i,j}$ in what follows.

We proceed by imposing the requirement for the linear combination $\theta_{1,1} - \theta_{1,0} = q\,a^{-1}\,(du - a\,dx - (b - a\,\ln a)\,dy - (p + a\,u)\,dt)$ to be a multiple of the contact form $du - u_t\,dt - u_x\,dx - u_y\,dy$ on the bundle of jets of sections of the bundle $\pi\colon \mathbb{R}^4 \to \mathbb{R}^3$, $\pi\colon (t,x,y,u) \mapsto (t,x,y)$, that is, we take $a = u_x$, $b = u_y + u_x \ln u_x$, $p = u_t - u\,u_x$. Then we consider the linear combination $\tau = \alpha_3 - \theta_{1,0} = q^2\,(dv - q^{-1}(u\,dt + dx + (q\,w - \ln u_x)\,dy)$ and put $q = v_x^{-1}$. This yields

$$\tau = v_x^{-2}\,(dv - u\,v_x\,dt - v_x\,dx - (w - v_x\,\ln u_x)\,dy).$$

The restriction of form τ on the bundle of sections of the bundle $\mathbb{R}^6 \to \mathbb{R}^3$, $(t,x,y,u,v,w) \mapsto (t,x,y)$, gives the over-determined system of PDEs

$$\left\{\begin{array}{lcl} v_t & = & u\,v_x, \\ v_y & = & w - v_x\,\ln u_x. \end{array}\right. \tag{4.2}$$

The integrability condition $(v_t)_y = (v_y)_t$ thereof gives two equations

$$u_{tx} = u\, u_{xx} - u_x^2 \ln u_x - u_x\, u_y \tag{4.3}$$

and

$$w_t = u\, w_x. \tag{4.4}$$

Thus system (4.2) provides a Lax representation of system (4.3), (4.4). We can find the Lax representation of equation (4.3) by proceeding as follows. Equation (4.4) is a copy of the first equation in (4.2). We consider w as a new pseudopotential and include equation (4.4) into the Lax representation for equation (4.3) by adding the copy $w_y = w_1 - w_x \ln u_x$ of the second equation from (4.2) with additional function w_1, and then repeat this process. In other words, we rename $v = v_0$, $w = v_1$, then add the sequence of functions v_k, $k \geq 2$, and consider the infinite system

$$\left\{ \begin{array}{lcl} v_{k,t} & = & u\, v_{k,x}, \\ v_{k,y} & = & v_{k+1} - v_{k,x} \ln u_x. \end{array} \right. \qquad k \geq 0. \tag{4.5}$$

The compatibility condition of this system coincides with equation (4.3). System (4.5) can be written in the finite form by introducing the function

$$r = \mathrm{e}^{-\lambda\, y} \sum_{k=0}^{\infty} \lambda^k\, v_k \tag{4.6}$$

where $\sum_{k=0}^{\infty} \lambda^k\, v_k$ is a formal series with respect to a formal parameter λ. Then we have

$$\left\{ \begin{array}{lcl} r_t & = & u\, r_x, \\ r_y & = & -r_x \ln u_x. \end{array} \right. \tag{4.7}$$

This system provides a Lax representation for equation (4.3).

Straightforward computations[2] give the following statement:

PROPOSITION 1. *The contact symmetry algebra of equation* (4.3) *is generated by the vector fields*

$$A_0\, \partial_t - A_0'\, y\, \partial_x - (A_0'\, u - A_0''\, y)\, \partial_u, \quad A_1\, \partial_t - A_1'\, \partial_u, \quad x\, \partial_x + y\, \partial_y + u\, \partial_u, \quad \partial_y, \tag{4.8}$$

where $A_i = A_i(t)$ *are arbitrary smooth functions of* t. *Restricting these functions to polynomials gives the Lie algebra isomorphic to* $\mathfrak{a}_2 \ltimes \mathfrak{q}_{1,-1}$. *The contact symmetry algebra of system* (4.3), (4.4) *is obtained by appending the vector field* $B\, \partial_w$ *with arbitrary smooth function* $B = B(y, w)$ *to the vector fields* (4.8). □

5. 3D generalized Hunter–Saxton equation

The Lie algebra $\mathfrak{a}_0 \ltimes \mathfrak{q}_{1,-1/2}$ admits the outer derivation $D_2(s^p\, f(t)) = s^{p+1}\, f''(t)$, that is,

$$D_2(s^p \otimes V_k) = \left\{ \begin{array}{ll} k\,(k-1)\, s \otimes V_{k-2}, & p = 0 \text{ and } k \geq 2, \\ 0, & p = 1 \text{ or } k \in \{0, 1\}. \end{array} \right.$$

This derivation produces the right extension $\mathfrak{b}_1 \ltimes \mathfrak{q}_{1,-1/2} = \langle Z_1 \rangle \ltimes (\mathfrak{a}_0 \ltimes \mathfrak{q}_{1,-1/2})$ of $\mathfrak{a}_0 \ltimes \mathfrak{q}_{1,-1/2}$, where $[Z_1, V_k]_{-1/2} = D_2(V_k)$ and $[Z_0, Z_1]_{-1/2} = Z_1$. Denote by α_1

[2]We carried out computations of generators of contact symmetries in the *Jets* software [**2**].

the dual form to $-Z_1$, that is, put $\alpha_1(Z_1) = -1$, $\alpha_1(Z_0) = 0$, $\alpha_1(V_k) = 0$. Then the structure equations for $\mathfrak{b}_1 \ltimes \mathfrak{q}_{1,-1/2}$ acquire the form

$$
\left\{
\begin{array}{lcl}
d\Theta & = & \Theta_{h_1} \wedge (\Theta - 2\,h_0\,\Theta_{h_0}) + h_0\,\alpha_0 \wedge \Theta_{h_0} - h_0\,\alpha_1 \wedge \Theta_{h_1 h_1}, \\
d\alpha_0 & = & 0, \\
d\alpha_1 & = & \alpha_0 \wedge \alpha_1.
\end{array}
\right. \tag{5.1}
$$

The Lie algebra $\mathfrak{b}_1 \ltimes \mathfrak{q}_{1,-1/2}$ admits the sequence of twisted extensions $\mathfrak{b}_k \ltimes \mathfrak{q}_{1,-1/2}$, $k \ge 2$, as it was described when constructing system (3.7). In this section we need the second extension from this series. Specifically, we have $[\alpha_0 \wedge \alpha_1] \in H^2_{\alpha_0}(\mathfrak{b}_1 \ltimes \mathfrak{q}_{1,-1/2})$, hence we append equation

$$
d\alpha_2 = \alpha_0 \wedge \alpha_2 + \alpha_0 \wedge \alpha_1, \tag{5.2}
$$

to system (5.1) and get the structure equations of $\mathfrak{b}_2 \ltimes \mathfrak{q}_{1,-1/2}$. Then $[\alpha_1 \wedge \alpha_2] \in H^2_{2\alpha_0}(\mathfrak{b}_2 \ltimes \mathfrak{q}_{1,-1/2})$, and the structure equations of $\mathfrak{b}_3 \ltimes \mathfrak{q}_{1,-1/2}$ are obtained by adding equation

$$
d\alpha_3 = 2\,\alpha_0 \wedge \alpha_3 + \alpha_1 \wedge \alpha_2. \tag{5.3}
$$

to system (5.1), (5.2).

Successively integrating equations (5.1), (5.2), and (5.3) we get

$$
\alpha_0 = \frac{dq}{q}, \quad \alpha_1 = q\,dy, \quad \alpha_2 = q\,(dw + \ln q\;dy), \quad \alpha_3 = q^2\,(dv - w\,dy),
$$

$$
\theta_{0,0} = a\,dt, \quad \theta_{0,1} = \frac{da}{a} + p_1\,dt, \quad \theta_{0,2} = \frac{dp_1}{a} + p_2\,dt,
$$

$$
\theta_{1,0} = q\,a^{-1}\,(dx + p_1\,dt + u\,dt), \quad \theta_{1,1} = q\,a^{-2}\,(du - p_1\,dx + (a\,p_2 - p_1^2)\,dy + p_3\,dt),
$$

where $q \neq 0$, $a \neq 0$, t, x, y, u, v, w, p_1, p_2, p_3 are free parameters. By altering notation as $p_1 = u_x$, $p_2 = a^{-1}\,(u_x^2 - u_y)$, and $p_3 = -u_t$ we obtain $\theta_{1,1} = q\,a^{-2}\,(du - u_t\,dt - u_x\,dx - u_y\,dy)$. Then we consider the linear combination $\tau = \alpha_3 - \theta_{1,0} = q^2\,\left(dv - a^{-1}\,q^{-1}\,(u\,dt + dx + (a\,q\,w + u_x)\,dy)\right)$, put $a = q^{-1}\,v_x^{-1}$ and obtain

$$
\tau = q^2\,(dv - u\,v_x\,dt - v_x\,dx - (w + u_x\,v_x)\,dy).
$$

Upon restriction to the sections of the bundle $\mathbb{R}^6 \to \mathbb{R}^3$, $(t,x,y,u,v,w) \mapsto (t,x,y)$ this form produces the over-determined system

$$
\left\{
\begin{array}{lcl}
v_t & = & u\,v_x, \\
v_y & = & w + u_x\,v_x,
\end{array}
\right. \tag{5.4}
$$

which is compatible by virtue of two equations

$$
u_{tx} = u\,u_{xx} - u_x^2 - u_y \tag{5.5}
$$

and

$$
w_t = u\,w_x. \tag{5.6}
$$

The symmetry reduction of equation (5.5) with respect to $u_y = 0$ coincides with the generalized Hunter– Saxton equation [**3, 9, 10, 27**]

$$
u_{tx} = u\,u_{xx} + \beta\,u_x^2 \tag{5.7}
$$

with the special value $\beta = -1$ of parameter β. Hence (5.5) can be considered as a three-dimensional generalization of the particular case $u_{tx} = u\,u_{xx} - u_x^2$ of equation (5.7).

Similar to the constructions in 4, we can find the Lax representation of equation (5.5) by renaming $v = v_0$, $w = v_1$ and including equation (5.6) into the infinite system

$$(5.8)\qquad \begin{cases} v_{k,t} = u\, v_{k,x}, \\ v_{k,y} = v_{k+1} + u_x\, v_{k,x}, \end{cases} \qquad k \ge 0.$$

System (5.8) is compatible by virtue of equation (5.5). Series (4.6) allows one to write (5.8) in the form

$$(5.9)\qquad \begin{cases} r_t = u\, r_x, \\ r_y = u_x\, r_x. \end{cases}$$

We have

PROPOSITION 2. *The contact symmetry algebra of equation* (5.5) *admits generating vector fields*

$$A_0\, \partial_t + (A_0''\, y - A_0'\, x)\, \partial_x - (2\, A_0'\, u - A_0''\, x + A_0'''\, y)\, \partial_u, \quad A_1\, \partial_x - A_1'\, \partial_u,$$
$$x\, \partial_x + y\, \partial_y + u\, \partial_u, \quad \partial_y,$$

where $A_i = A_i(t)$ *are arbitrary smooth functions of* t*. Restriction of these functions to* $\mathbb{R}[t]$ *gives the Lie algebra isomorphic to* $\mathfrak{b}_2 \ltimes \mathfrak{q}_{1,-1/2}$*. The contact symmetry algebra of system* (5.5), (5.6) *has additional generating vector field* $B\, \partial_w$ *with arbitrary smooth function* $B = B(y, w)$. □

6. Equation associated to $\mathfrak{a}_2 \ltimes \mathfrak{q}_{2,-1}$

The structure equations of the Lie algebra $\mathfrak{a}_3 \ltimes \mathfrak{q}_{2,-1}$ have the form (4.1), where now the formal series Θ is given by formula (3.2) with $N = 2$. We take the same forms $\alpha_0, \ldots, \alpha_3$, and $\theta_{0,0}$ as in Section 4, and put

$$\theta_{1,0} = q\, (db - \ln a\, dy - p_1\, dt), \quad \theta_{2,0} = -q^2\, a^{-1}\, (dx + p_1\, dy - p_2\, dt).$$

Then the linear combination

$$\theta_{1,0} + \theta_{2,0} + \alpha_2 = q\, \left(db + dw + \frac{a\, p_1 + q\, p_2}{a}\, dt - \frac{q}{a}\, dx - \frac{a\, (\ln a - \ln q) + q\, p_1}{a}\, dy \right)$$

after altering notation $b = u - w$, $q = a\, u_x$, $p_1 = (u_y + \ln u_x)\, u_x^{-1}$, and $p_2 = -(u_t\, u_x + u_y + \ln u_x)\, u_x^{-2}$ acquires the form $\theta_{1,0} + \theta_{2,0} + \alpha_2 = q\, (du - u_t\, dt - u_x\, dx - u_y\, dy)$, while for the form

$$\begin{aligned} \tau &= \alpha_3 + \theta_{2,0} \\ &= a^2\, u_x^2\, \left(dv - \frac{1}{a} \left(dx + \frac{u_t\, u_x + u_y + \ln u_x}{u_x^2}\, dt + \frac{a\, u_x\, w + u_y + \ln u_x}{u_x}\, dy \right) \right) \end{aligned}$$

after renaming $a = v_x^{-1}$ we have

$$\tau = \frac{u_x^2}{v_x^2} \left(dv - v_x \left(dx + \frac{u_t\, u_x + u_y + \ln u_x}{u_x^2}\, dt \right) + \frac{u_x\, w + (u_y + \ln u_x)\, v_x}{u_x}\, dy \right).$$

Restricting this onto sections of the bundle $(t, x, y, u, v, w) \mapsto (t, x, y)$ produces system

$$(6.1)\qquad \begin{cases} v_t = \dfrac{u_t\, u_x + u_y + \ln u_x}{u_x^2}\, v_x, \\[2ex] v_y = \dfrac{u_y + \ln u_x}{u_x}\, v_x + w, \end{cases}$$

which is compatible by virtue of equations

$$u_{yy} = u_{tx} - \frac{(u_y + \ln u_x)^2 + u_t\,u_x}{u_x^2}\,u_{xx} + \frac{2\,(u_y + \ln u_x) - 1}{u_x}\,u_{xy} \tag{6.2}$$

and

$$w_t = \frac{u_t\,u_x + u_y + \ln u_x}{u_x^2}\,w_x. \tag{6.3}$$

The last equation is a copy of the first equation in (6.1). Therefore we can take w as a new pseudopotential, rename $v = v_0$, $w = v_1$, add the infinite sequence of pseudopotentials v_j, $j \ge 2$, and consider infinite system

$$\left\{\begin{array}{lcl} v_{k,t} & = & \dfrac{u_t\,u_x + u_y + \ln u_x}{u_x^2}\,v_{k,x}, \\ v_{k,y} & = & \dfrac{u_y + \ln u_x}{u_x}\,v_{k,x} + v_{k+1}, \end{array}\right. \qquad k \ge 0. \tag{6.4}$$

This system is compatible by virtue of equation (6.2) and thus defines a Lax representation thereof. Introducing series (4.6), we rewrite (6.4) in the form of another Lax representation

$$\left\{\begin{array}{lcl} r_t & = & \dfrac{u_t\,u_x + u_y + \ln u_x}{u_x^2}\,r_x, \\ r_y & = & \dfrac{u_y + \ln u_x}{u_x}\,r_x \end{array}\right. \tag{6.5}$$

for equation (6.2).

PROPOSITION 3. *The following vector fields*

$$A_0\,\partial_t - \left(A_0'\,x + \frac{1}{2}\,A_0''\,y^2\right)\partial_x - A_0'\,y\,\partial_u, \quad A_1'\,y\,\partial_x + A_1\,\partial_u, \quad A_2\,\partial_x,$$
$$2\,x\,\partial_x + y\,\partial_y + (u+y)\,\partial_u, \quad \partial_y.$$

with arbitrary smooth functions $A_i = A_i(t)$ are generators for the contact symmetry algebra of equation (6.2). Restricting these functions to polynomials gives the Lie algebra isomorphic to $\mathfrak{a}_2 \ltimes \mathfrak{q}_{2,-1}$. The contact symmetry algebra of system (6.2), (6.3) is obtained by appending the vector field $B\,\partial_w$ with arbitrary smooth function $B = B(y, w)$. □

7. Equation associated to $\mathfrak{h} \oplus \mathfrak{w}$

In this section we construct a Lie algebra that includes the algebra of polynomial Hamiltonian vector fields

$$\mathfrak{h} = \left\langle W_{m,n} = n\,t^m\,x^{n-1}\,\partial_t - m\,t^{m-1}\,x^n\,\partial_x \mid m, n \in \mathbb{N}_0,\ m^2 + n^2 \ne 0\right\rangle$$

on $\mathbb{R}^2$ as a subalgebra and has a twisted extension that generates an integrable PDE. To find such a Lie algebra we proceed as follows. The Lie algebra $\mathfrak{h}$ admits the grading $\mathrm{gr}(W_{m,n}) = m + n - 2$ that defines the outer derivation $D\colon W_{m,n} \mapsto (m+n-2)\,W_{m,n}$ and the right extension $\langle Z\rangle \ltimes \mathfrak{h}$ with $[Z, W_{m,n}] = D(W_{m,n})$. Denote the dual forms to $W_{p,q}$ and $-Z$ as $\theta_{i,j}$ and α, so $\theta_{i,j}(W_{m,n}) = \delta_{i,m}\,\delta_{j,n}$, $\theta_{i,j}(Z) = 0$, $\alpha(W_{m,n}) = 0$, and $\alpha(Z) = -1$. Further, take $\mathfrak{w} = \langle V_k = \frac{1}{k!}\,y^k\,\partial_y \mid k \ge 0\rangle$ and consider the direct sum of the Lie algebras

$$\widetilde{\mathfrak{h}} = (\langle Z\rangle \ltimes \mathfrak{h}) \oplus \mathfrak{w}.$$

Denote by ω_m the dual forms for V_k, so $\omega_m(V_k)=\delta_{m,k}$, $\omega_m(W_{k,n})=\omega_m(Z)=0$, and $\theta_{m,n}(V_k)=\alpha(V_k)=0$. Put

$$\Theta=\sum_{m\ge 0,\, n\ge 0,\, m^2+n^2\neq 0}\frac{h_1^m\,h_2^n}{m!\,n!}\,\theta_{m,n},\qquad \Omega=\sum_{k\ge 0}\frac{h_3^k}{k!}\,\omega_k,$$

where h_i are the formal parameters such that $dh_i=0$. Then the structure equations for the Lie algebra $\widetilde{\mathfrak{h}}$ acquire the form

$$\text{(7.1)}\qquad \left\{\begin{array}{lcl} d\Theta &=& -\Theta_{h_1}\wedge\Theta_{h_2}-\alpha\wedge(h_1\,\Theta_{h_1}+h_2\,\Theta_{h_2}-2\,\Theta)\,,\\ d\Omega &=& \Omega_{h_3}\wedge\Omega,\\ d\alpha &=& 0.\end{array}\right.$$

This system entails $H^1(\widetilde{\mathfrak{h}})=\langle\alpha\rangle$ and $[\theta_{1,0}\wedge\theta_{0,1}]\in H^2_{2\alpha}(\widetilde{\mathfrak{h}})$. Hence the Lie algebra $\widetilde{\mathfrak{h}}$ admits the twisted extension defined by appending equation

$$\text{(7.2)}\qquad d\sigma=2\,\alpha\wedge\sigma+\theta_{1,0}\wedge\theta_{0,1}$$

to system (7.1).

Integrating equations (7.1), (7.2) step by step we obtain

$$\theta_{1,0}=a_{11}\,dt+a_{12}\,dx,\qquad \theta_{0,1}=a_{21}\,dt+a_{22}\,dx,$$

$$\alpha=\frac{1}{2}\,\frac{dq}{q},\qquad q=\det\left(\begin{array}{cc} a_{11} & a_{12}\\ a_{21} & a_{22}\end{array}\right),$$

$$\theta_{2,0}=\frac{1}{q}\,(a_{12}\,da_{11}-a_{11}\,da_{12})+b_1\,dt+b_2\,dx,$$

$$\theta_{0,2}=\frac{1}{q}\,(a_{22}\,da_{21}-a_{21}\,da_{22})+b_3\,dt+b_4\,dx,$$

$$\theta_{1,1}=\frac{1}{2\,q}\,(a_{22}\,da_{11}-a_{11}\,da_{22}+a_{12}\,da_{21}-a_{21}\,da_{12})+b_5\,dt+b_6\,dx,$$

$$\omega_0=p\,dy,\qquad \sigma=q\,(dv-x\,dt),$$

where $a_{i,j}$, b_1, $\dots$, b_4, p, v are free parameters such that $q\neq 0$, $p\neq 0$, and

$$b_5=\frac{1}{q}\,(-a_{21}\,a_{22}\,b_1+a_{21}^2\,b_2+a_{11}\,a_{12}\,b_3-a_{11}^2\,b_4),$$

$$b_6=\frac{1}{q}\,(-a_{22}^2\,b_1+a_{21}\,a_{22}\,b_2+a_{12}^2\,b_3-a_{11}\,a_{12}\,b_4),$$

Put $a_{21}=u\,a_{22}$, $b_3=-a_{22}^2\,q^{-1}\,u_t$, $b_4=-a_{22}^2\,q^{-1}\,u_x$, $p=-a_{22}^2\,q^{-1}\,u_y$. This yields $\theta_{0,2}-\omega_0=a_{22}^2\,q^{-1}\,(du-u_t\,dt-u_x\,dx-u_y\,dy)$. We consider the linear combination

$$\tau=\sigma+\theta_{1,0}-\tfrac{1}{2}\,\omega_0=a_{22}\,A\left(dv+\frac{dx}{A}-\frac{(x\,A-u)\,dt}{A}-\frac{u_y\,dy}{2\,A^2}\right).$$

where $A=a_{11}-u\,a_{12}$. By altering notation $A=-v_x^{-1}$ we get

$$\tau=-\frac{a_{22}}{v_x}\,\left(dv-v_x\,dx-(x+u\,v_x)\,dt-\tfrac{1}{2}\,u_y\,v_x^2\,dy\right).$$

Then upon restricting form τ to the sections of the bundle $\mathbb{R}^5\to\mathbb{R}^3$, $(t,x,y,u,v)\mapsto(t,x,y)$, we obtain the over-determined system

$$\text{(7.3)}\qquad \left\{\begin{array}{lcl} v_t &=& x+u\,v_x,\\ v_y &=& \tfrac{1}{2}\,u_y\,v_x^2\end{array}\right.$$

The compatibility condition of this system $(v_t)_y = (v_y)_t$ holds if and only

$$u_{ty} = u\,u_{xy} - 2\,u_x\,u_y. \tag{7.4}$$

Therefore system (7.3) defines a Lax representation for equation (7.4).

We notice that equation (7.4) is invariant with respect to translations along x, while the Lax representation (7.3) does not admit such translations, cf [**21**, Th. 4] and [**23**, Example 7].

The next proposition can be readily verified by straightforward computations.

PROPOSITION 4. *The contact symmetry algebra of equation* (7.4) *has generators*

$$A_x\,\partial_t - A_t\,\partial_x + (A_{tt} - 2\,u\,A_{tx} + u^2\,A_{xx})\,\partial_u, \quad B\,\partial_y, \quad t\,\partial_t + x\,\partial_x,$$

where $A = A(t,x)$ and $B = B(y)$ are arbitrary smooth functions of their arguments. Restriction of these functions to polynomials $A = t^m\,x^n$, $B = y^k$, $m, n, k \in \mathbb{N}_0$, $m^2 + n^2 \neq 0$, produces the Lie algebra isomorphic to $\widetilde{\mathfrak{h}}$. □

8. Concluding remarks

In the present paper we have used the method of twisted extensions to derive new integrable PDEs related to some infinite-dimensional Lie algebras. In the obtained examples as well as in some examples in [**22**]–[**26**] the integrable equations were constructed starting from certain extensions of the Lie algebras of the Kac–Moody type or the Lie algebra of Hamiltonian vector fields on $\mathbb{R}^2$. In examples of Sections 4, 5, and 6 the symmetry algebras of the obtained integrable systems turn out to be wider than the initial infinite-dimensional Lie algebras used in the construction, while the specific form of equations (4.4), (5.6), and (6.3) allowed us to find the infinite-component Lax representations (4.5), (5.8), and (6.4) as well as the one-component Lax representations (4.7), (5.9), and (6.5) for equations (4.3), (5.5), and (6.2), respectively. The polynomial parts of the symmetry algebras of equations (4.3), (5.5), and (6.2) coincide with the infinite-dimensional Lie algebras whose twisted extensions were used to construct systems (4.2), (5.4), and (6.1).

We hope further examples will enlighten relations between the structure theory of Lie algebras and integrable PDEs. In particular, it is important to address the following issues in the future research:

- to find other examples of integrable systems related to the Lie algebras of the Kac–Moody type with small values of N,
- to consider extensions of the general deformations of the Lie algebras $\mathbb{R}_N[s] \otimes \mathfrak{w}$, cf. [**38**],
- to construct other integrable systems whose symmetry algebras are extensions of the Lie algebras of Hamiltonian vector fields,
- to generalize the technique used in the present paper to the Lie–Rinehart algebras, cf. [**28**].

Acknowledgments

The author is very grateful to I.S. Krasil′shchik for useful discussions. The author thanks E.V. Ferapontov, M.V. Pavlov, and the anonymous referee for important remarks.

References

[1] M. J. Ablowitz and P. A. Clarkson, *Solitons, nonlinear evolution equations and inverse scattering*, London Mathematical Society Lecture Note Series, vol. 149, Cambridge University Press, Cambridge, 1991, DOI 10.1017/CBO9780511623998. MR1149378

[2] H. Baran, M. Marvan. Jets: A software for differential calculus on jet spaces and diffieties. Available on-line at `http://jets.math.slu.cz`

[3] Richard Beals, David H. Sattinger, and Jacek Szmigielski, *Inverse scattering solutions of the Hunter-Saxton equation*, Appl. Anal. **78** (2001), no. 3-4, 255–269, DOI 10.1080/00036810108840938. MR1883537

[4] C. P. Boyer and P. Winternitz, *Symmetries of the self-dual Einstein equations. I. The infinite-dimensional symmetry group and its low-dimensional subgroups*, J. Math. Phys. **30** (1989), no. 5, 1081–1094, DOI 10.1063/1.528379. MR992582

[5] Alan Coley, Decio Levi, Robert Milson, Colin Rogers, and Pavel Winternitz (eds.), *Bäcklund and Darboux transformations. The geometry of solitons*, CRM Proceedings & Lecture Notes, vol. 29, American Mathematical Society, Providence, RI, 2001, DOI 10.1090/crmp/029. MR1870397

[6] D. David, N. Kamran, D. Levi, and P. Winternitz, *Subalgebras of loop algebras and symmetries of the Kadomtsev-Petviashvili equation*, Phys. Rev. Lett. **55** (1985), no. 20, 2111–2113, DOI 10.1103/PhysRevLett.55.2111. MR811845

[7] B. Doubrov and E. V. Ferapontov, *On the integrability of symplectic Monge-Ampère equations*, J. Geom. Phys. **60** (2010), no. 10, 1604–1616, DOI 10.1016/j.geomphys.2010.05.009. MR2661158

[8] D. B. Fuks, *Cohomology of infinite-dimensional Lie algebras*, Contemporary Soviet Mathematics, Consultants Bureau, New York, 1986. Translated from the Russian by A. B. Sosinskiĭ. MR874337

[9] John K. Hunter and Ralph Saxton, *Dynamics of director fields*, SIAM J. Appl. Math. **51** (1991), no. 6, 1498–1521, DOI 10.1137/0151075. MR1135995

[10] John K. Hunter and Yu Xi Zheng, *On a completely integrable nonlinear hyperbolic variational equation*, Phys. D **79** (1994), no. 2-4, 361–386, DOI 10.1016/0167-2789(94)90093-0. MR1306466

[11] Victor G. Kac, *Infinite-dimensional Lie algebras*, 3rd ed., Cambridge University Press, Cambridge, 1990, DOI 10.1017/CBO9780511626234. MR1104219

[12] B. G. Konopel′chenko, *Nonlinear integrable equations*, Lecture Notes in Physics, vol. 270, Springer-Verlag, Berlin, 1987. Recursion operators, group-theoretical and Hamiltonian structures of soliton equations, DOI 10.1007/3-540-17567-9. MR882726

[13] Joseph Krasil'shchik and Alexander Verbovetsky, *Geometry of jet spaces and integrable systems*, J. Geom. Phys. **61** (2011), no. 9, 1633–1674, DOI 10.1016/j.geomphys.2010.10.012. MR2813504

[14] I. S. Krasil'shchik, A. M. Verbovetsky, and R. Vitolo, *A unified approach to computation of integrable structures*, Acta Appl. Math. **120** (2012), 199–218, DOI 10.1007/s10440-012-9699-x. MR2945634

[15] Joseph Krasil'shchik, Alexander Verbovetsky, and Raffaele Vitolo, *The symbolic computation of integrability structures for partial differential equations*, Texts and Monographs in Symbolic Computation, Springer, Cham, 2017, DOI 10.1007/978-3-319-71655-8. MR3753682

[16] I. S. Krasil′shchik and A. M. Vinogradov, *Nonlocal symmetries and the theory of coverings: an addendum to Vinogradov's "Local symmetries and conservation laws" [Acta Appl. Math. **2** (1984), no. 1, 21–78; MR0736872 (85m:58192)]*, Acta Appl. Math. **2** (1984), no. 1, 79–96, DOI 10.1007/BF01405492. MR736873

[17] I. S. Krasil′shchik and A. M. Vinogradov, *Nonlocal trends in the geometry of differential equations: symmetries, conservation laws, and Bäcklund transformations*, Acta Appl. Math. **15** (1989), no. 1-2, 161–209, DOI 10.1007/BF00131935. Symmetries of partial differential equations, Part I. MR1007347

[18] Boris Kruglikov and Oleg Morozov, SDiff(2) *and uniqueness of the Plebański equation*, J. Math. Phys. **53** (2012), no. 8, 083506, 11, DOI 10.1063/1.4739749. MR3012650

[19] Boris Kruglikov and Oleg Morozov, *Integrable dispersionless PDEs in 4D, their symmetry pseudogroups and deformations*, Lett. Math. Phys. **105** (2015), no. 12, 1703–1723, DOI 10.1007/s11005-015-0800-z. MR3420595

[20] V. B. Matveev and M. A. Salle, *Darboux transformations and solitons*, Springer Series in Nonlinear Dynamics, Springer-Verlag, Berlin, 1991, DOI 10.1007/978-3-662-00922-2. MR1146435

[21] Oleg I. Morozov, *Contact integrable extensions of symmetry pseudo-groups and coverings of* $(2+1)$ *dispersionless integrable equations*, J. Geom. Phys. **59** (2009), no. 11, 1461–1475, DOI 10.1016/j.geomphys.2009.07.009. MR2569267

[22] Oleg I. Morozov, *Deformed cohomologies of symmetry pseudo-groups and coverings of differential equations*, J. Geom. Phys. **113** (2017), 215–225, DOI 10.1016/j.geomphys.2016.09.010. MR3603767

[23] Oleg I. Morozov, *Deformations of infinite-dimensional Lie algebras, exotic cohomology, and integrable nonlinear partial differential equations*, J. Geom. Phys. **128** (2018), 20–31, DOI 10.1016/j.geomphys.2018.02.007. MR3786183

[24] Oleg I. Morozov, *Lax representations with non-removable parameters and integrable hierarchies of PDEs via exotic cohomology of symmetry algebras*, J. Geom. Phys. **143** (2019), 150–163, DOI 10.1016/j.geomphys.2019.05.001. MR3959217

[25] Oleg I. Morozov, *Nonlinear nonisospectral differential coverings for the hyper-CR equation of Einstein-Weyl structures and the Gibbons-Tsarev equation*, Differential Geom. Appl. **75** (2021), Paper No. 101740, 12, DOI 10.1016/j.difgeo.2021.101740. MR4221517

[26] Oleg I. Morozov, *Isospectral deformation of the reduced quasi-classical self-dual Yang-Mills equation*, Differential Geom. Appl. **76** (2021), Paper No. 101742, 14, DOI 10.1016/j.difgeo.2021.101742. MR4230510

[27] Oleg I. Morozov, *Integrability structures of the generalized Hunter-Saxton equation*, Anal. Math. Phys. **11** (2021), no. 2, Paper No. 50, 21, DOI 10.1007/s13324-021-00490-8. MR4213759

[28] Oleg I. Morozov, *Integrable partial differential equations and Lie-Rinehart algebras*, J. Geom. Phys. **181** (2022), Paper No. 104661, 11, DOI 10.1016/j.geomphys.2022.104661. MR4477948

[29] S. Novikov, S. V. Manakov, L. P. Pitaevskiĭ, and V. E. Zakharov, *Theory of solitons*, Contemporary Soviet Mathematics, Consultants Bureau [Plenum], New York, 1984. The inverse scattering method; Translated from the Russian. MR779467

[30] S.P. Novikov. On exotic De-Rham cohomology. Perturbation theory as a spectral sequence. `arXiv:math-ph/0201019`, 2002

[31] S. P. Novikov, *On metric-independent exotic homology* (Russian, with Russian summary), Tr. Mat. Inst. Steklova **251** (2005), no. Nelineĭn. Din., 215–222; English transl., Proc. Steklov Inst. Math. **4(251)** (2005), 206–212. MR2234383

[32] Peter J. Olver, *Applications of Lie groups to differential equations*, 2nd ed., Graduate Texts in Mathematics, vol. 107, Springer-Verlag, New York, 1993, DOI 10.1007/978-1-4612-4350-2. MR1240056

[33] J. F. Plebański, *Some solutions of complex Einstein equations*, J. Mathematical Phys. **16** (1975), no. 12, 2395–2402, DOI 10.1063/1.522505. MR386609

[34] C. Rogers and W. F. Shadwick, *Bäcklund transformations and their applications*, Mathematics in Science and Engineering, vol. 161, Academic Press, Inc. [Harcourt Brace Jovanovich, Publishers], New York-London, 1982. MR658491

[35] A. V. Bocharov, V. N. Chetverikov, S. V. Duzhin, N. G. Khor′kova, I. S. Krasil′shchik, A. V. Samokhin, Yu. N. Torkhov, A. M. Verbovetsky, and A. M. Vinogradov, *Symmetries and conservation laws for differential equations of mathematical physics*, Translations of Mathematical Monographs, vol. 182, American Mathematical Society, Providence, RI, 1999. Edited and with a preface by Krasil′shchik and Vinogradov; Translated from the 1997 Russian original by Verbovetsky [A. M. Verbovetskiĭ] and Krasil′shchik, DOI 10.1090/mmono/182. MR1670044

[36] H. D. Wahlquist and F. B. Estabrook, *Prolongation structures of nonlinear evolution equations*, J. Mathematical Phys. **16** (1975), 1–7, DOI 10.1063/1.522396. MR358111

[37] V. E. Zakharov, *Integrable systems in multidimensional spaces*, Mathematical problems in theoretical physics (Berlin, 1981), Lecture Notes in Phys., vol. 153, Springer, Berlin, 1982, pp. 190–216, DOI 10.1007/3-540-11192-1_38. MR723902

[38] Pasha Zusmanovich, *Deformations of* $W_1(n) \otimes A$ *and modular semisimple Lie algebras with a solvable maximal subalgebra*, J. Algebra **268** (2003), no. 2, 603–635, DOI 10.1016/S0021-8693(03)00295-3. MR2009323

Trapeznikov Institute of Control Sciences, 65 Profsoyuznaya Street, Moscow 117997, Russia
Email address: oimorozov@gmail.com

Contemporary Mathematics
Volume **788**, 2023
https://doi.org/10.1090/conm/788/15829

Energy functionals and normal forms of knots and plane curves

A. B. Sossinsky

Dedicated to the memory of Aleksandr Vinogradov

ABSTRACT. In this paper, a new theory, called theory of thickened knots, is described. Roughly speaking, a thickened knot is a tube smoothly embedded in Euclidean space considered up to isotopy. The theory depends on a parameter $\varepsilon \in (0,1]$, which is equal to the diameter of the two-dimensional disk that sweeps out the tube. It is shown that for each fixed ε, there is only a finite number of nonisotopic knots. An approach to the solution of the classification problem of thickened knots, based on the idea of knot energy, is suggested. Concrete energy functionals for thickened knots are proposed, their local minima yield certain positions of the given knot which are called normal forms. Experiments with wire models of thickened knots are described, and they are in agreement with the theory. A number of promising conjectures are formulated.

This note is based on the talk that I gave at the Alexandre Vinogradov Memorial Conference, or rather the talk was a summary of a first draft of this note. The note is about knot theory, which is a topic that always interested Aleksandr Vinogradov: he translated the book by Ralph Fox on the subject and wrote an important supplement to the book.

In this note, we develop the theory of thickened knots. A thickened knot is a thin tubular neighborhood of a smooth ordinary knot in $\mathbb{R}^3$ (the exact definition is given below), or rather an isotopy class of such objects. The theory is developed for each fixed $\varepsilon \in [0,1]$, so there are infinitely many such theories. The theory becomes interesting for small (i.e., close to 0) values of ε (whereas there is only one thickened knot for $\varepsilon = 1$). Thickened knots should not be confused with so-called "fat knots" and ideal knots in the sense of [**6**], where the word "thickness" has a special meaning unrelated to the size of the cross section of the tube.

Why study thickened knots? Because thickened knots are a more adequate model of real-life knots (e.g. made out of rope or wire) than smooth curves, and for each $\varepsilon > 0$, there are finitely many thickened knots, so that the main problems of knot theory (their classification, the unknotting problem, and the search for a complete system of invariants) appear to be considerably more accessible than in

2020 *Mathematics Subject Classification.* Primary 57K10.
Key words and phrases. Thickened knots, plane curves, energy functionals.
This research was partially supported by the Jim Simons Foundation.

©2023 American Mathematical Society

classical knot theory, which may be regarded as the limit of the theory of ε-thickened knots as ε tends to 0.

1. Main definitions

Let us fix some value of ε in $(0,1]$. If $\varepsilon > 0$, we define a *thickened knot* K (or *ε-thickened knot*, if we wish to specify ε, or *ε-knot* for short) any embedding $K : \mathbb{S}^1 \times \operatorname{Int} \mathbb{D}^2 \to \mathbb{R}^3$ such that the length of the curve $K(\mathbb{S}^1 \times \{0\}) =: \heartsuit(K)$ is 2π and the image of K can be represented as the union of all disks of diameter ε centered at points of $\heartsuit(K)$:

$$\operatorname{Im} K = \bigcup_{\varphi \in [0,2\pi]} K(\{\varphi\} \times \operatorname{Int} \mathbb{D}^2).$$

The set of all ε-thickened knots, where $\varepsilon \in (0,1]$, is denoted by $\mathcal{K}_\varepsilon$.

The curve $\heartsuit(K)$, which is an ordinary smooth knot, will be called the *midline* of the thickened knot K. It follows from definitions that the curvature of the midline of any thickened knot is bounded away from zero.

For any fixed $\varepsilon \in [0,1]$, we say that two ε-thickened knots K and K' are *isotopic* if there exists a smooth isotopy of $\mathbb{R}^3$, $h_t : \mathbb{R}^3 \to \mathbb{R}^3$ that begins with the identity $h_0 = \operatorname{id}$ and ends by taking K to K', i.e., $h_1(K) = K'$. The relation of being isotopic is an equivalence relation, thus $\mathcal{K}_\varepsilon$ splits into equivalence (*isotopy*) classes. In what follows, by the abuse of notation customary in knot theory, we will use the symbol K to stand for the the whole equivalence class of K, for the map defining the concrete thickened knot K itself, and for the image $\operatorname{Im} K$ of that map. When we want to emphasize that we mean $\operatorname{Im} K$, we refer to K as the *position of* K.

2. Basic properties

Thickened knots possess the following important simple properties, which we state as theorems.

Theorem 1. *For any ordinary smooth knot $k : \mathbb{S}^1 \to \mathbb{R}^3$, there is a positive ε for which there exists a ε-knot K whose midline is k: $\heartsuit(K) = k$.*

The assertion of the theorem follows from the fact that the curvature of $\heartsuit(K)$ is bounded away from zero.

Note that if $\heartsuit(K)$ is very complicated (i.e., has a large crossing number) and ε is not very small, there is no ε-knot with midline $\heartsuit(K)$. For example, for $\varepsilon = 1/2$, there is no ε-knot K for which $\heartsuit(K)$ is the eight knot 4_1.

Theorem 2. *For any positive ε, the number of (isotopy classes of) ε-thickened knots is finite.*

This can be proved by showing that the volume of an ε-knot whose midline has "too many" crossing points is greater than it should be $(2\pi^2\varepsilon^2)$.

Theorem 3. *For any positive ε, the invariants of classical knot theory (knot polynomials, Vassiliev knot invariants, Khovanov homology, etc.) are also invariants of ε-thickened knots.*

We define an invariant of a thickened knot as the corresponding invariant of its midline. Hence the theorem follows from the fact that an isotopy of thickened knots defines an isotopy of their midlines.

3. Two reasonable conjectures

The two following conjectures claim that for any $\varepsilon \in (0,1]$, ε-knots are classified by a complete system of invariants (namely the Vassiliev invariants), and the recognition (=classification) problem is algorithmically decidable by a computer implementable algorithm. Unlike the same conjectures for classical knot theory, they do not appear to be hopeless.

Conjecture 1. *For any postive ε, Vassiliev knot invariants constitute a complete system of invariants for ε-thickened knots. Moreover, there is a practically implementable algorithm that, for any positive ε, indicates the order of Vassiliev invariants that form a complete system of invariants for ε-thickened knots.*

Conjecture 2. *There exists a computer implementable algorithm that, for any positive ε, decides whether or not two given ε-thickened knots are isotopic. In particular, there exists an algorithm that, for any positive ε, decides if a given ε-thickened knot is or isn't the ε-unknot.*

4. Energy functionals for thickened knots and normal forms

We propose to attack the classification problem by means of the knot energy approach due to H.K. Moffat [**4**], developed by V.I. Arnold [**5**], and already successfully used in a different context in my own work and joint work with O. Karpenkov and S. Avvakumov (see the papers [**1**]-[**5**]). The idea is to supply ε-thickened knots with a functional and construct an algorithm minimizing the functional that would bring the given knot K to a position corresponding to a local minimum of the functional. We call such positions normal forms of the thickened knot. The formal definition is as follows.

By the *normal form* of an ε-thickened knot corresponding to a real-valued functional Φ defined on the space of midlines of ε-thickened knots, we mean a position of the knot at which Φ has a local minimum. The motivation here was that effectively finding normal forms would yield a solution of the classification problem for knots (with a fixed crossing number).

In the first theme of our paper [**3**], devoted to the classificattion of regular curves in the plane up to regular homotopy, we worked with the functional equal to the integral of the squared curvature along the curve $k : \mathbb{S}^1 \to \mathbb{R}^3$, i.e.,

$$\Phi_1(k) = \int_0^{2\pi} \big[\kappa(k(s))\big]^2 ds, \tag{4.1}$$

where s is the natural arc length parameter on k, $\kappa(k(s))$ is the curvature of k at the point $k(s)$, and obtained a series of normal forms bijectively corresponding, as expected, to the winding number $w(k) \in \{0,1,2,3,\dots\}$ of the given curve, thereby giving a new proof of the classical Whitney–Graustein Theorem based on the calculus of variations.

As a by-product, we solved the *elasticae problem* set by J. Bernoulli, which was left unsolved by Euler in the particular case of closed curves (for the details, see [**3**]).

In the second topic of [**3**], our aim was to study the evolution of knots under the minimization of a functional equal to the sum of two functionals. The first one, is Φ_1 (see formula (1)), the second one, Φ_2, prohibits the self touching and self crossings of the given knot k as $\Phi_2(k)$ evolves to its local minimum (for the exact definition, see [**3**]). However, this could not be done by variational calculus methods, and so we discretized the problem.

Instead of smooth knots $k(\mathbb{S}^1)$, we worked with *admissible* closed broken lines, i.e., closed polygonal lines of fixed length and small local curvature with a fixed (large) number of edges of fixed (very small) length, and used discrete approximations of the two functionals. The algorithm works discretely step by step by small shifts along the gradient, producing a sequence of positions that stabilises near a fixed position – the normal form.

To our surprise, in all our computer experiments, the algorithm converged to normal forms, some of which were unexpected. In particular, we obtained a normal form for the eight knot (4_1 in knot tables) that looks exactly like the wire knot in the right-hand side of Figure 2 below. The fact that the algorithm always converged is not the consequence of a mathematical theorem, it is due to the fact the probability that our algorithm will produce a non-admissible polygonal curve is tiny.

What is more, in all our simulations (carried out for randomly chosen positions of prime knots with seven crossings or less), we obtained the same *unique normal form* for each of these knot types. In fact, there were three types of normal forms, that we call *almost flat*, *circular*, and *spherical*. In the almost flat normal form, the knot lies on a (say, horizontal) plane, and only at the crossing points do small intervals rise above the plane. For an example of an almost flat normal form, look at the second photograph of the wire model in Fig. 1 below. In the circular normal form, the knot also lies on a horizontal plane and winds around a round circle. For an example, look at the first photo in Fig. 1. In the spherical normal form, the knot touches the horizontal plane at three points, it is "very three-dimensional" and appears to be inscribed (in a certain approximate sense) in a 2-sphere. For an example, look at the second photo in Fig. 2.

The results obtained by mathematical simulation are in complete agreement with physical experiments described in the next section. We will return to a summary of the simulation results in Sec. 7.

5. Physical experiments

Figure 1 shows photos of a wire model of a thickened knot placed on a table, namely the thickened knot K whose midline $\heartsuit(K)$ is the eight knot (4_1 in knot tables). The knot is shown in two equilibrium positions, the one on the left photo has the form of a closed braid (i.e., is circular), while the one on the right has a different, more familiar almost flat shape.

The circular equilibrium position of the wire on the left-hand side of Fig. 1 is unstable – if moved slightly off that equilibrium, the wire jumps up off the table and acquires a stable position in which it lies flat on the table (shown in the photo on the right).

Figure 2 shows photos of a different wire model of a thickened knot placed on a table, whose midline is also the eight knot. The knot is shown in two equilibrium positions, the one on the left photo is circular, while the one on the right has an unexpected "very three-dimensional" spherical shape.

The equilibrium position of the wire on the left-hand side of Fig. 2 is unstable – if moved slightly off that equilibrium, the wire jumps up off the table and acquires the position shown on the photo on the right, touching the table at three points,

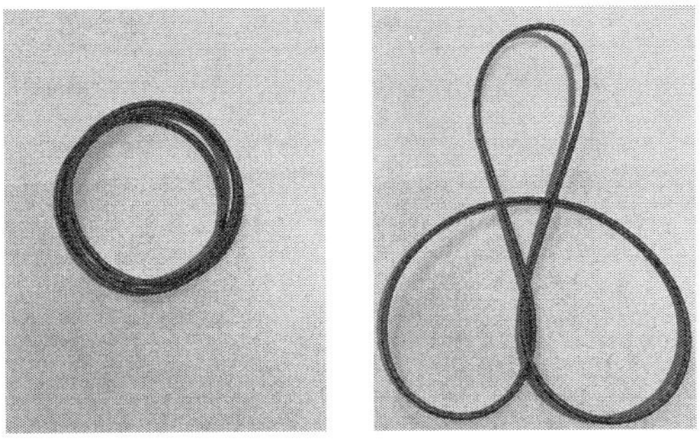

FIGURE 1. Two equilibrium positions of a model of a knot K with ♡(K) the eight knot

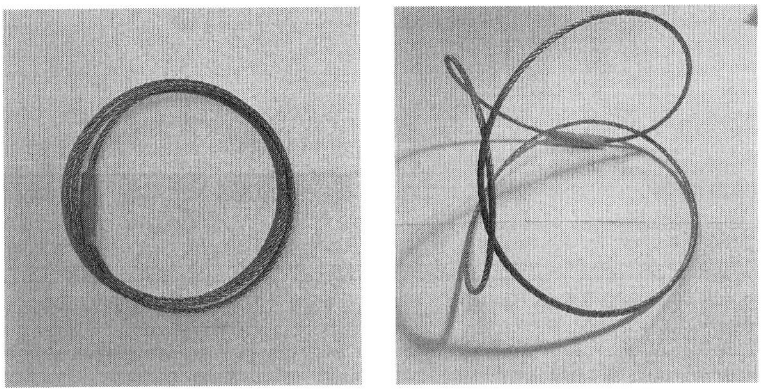

FIGURE 2. Equilibrium positions of another model of a knot K with ♡(K) the eight knot

and this position is stable. Note that in that position the wire has four "double points", i.e., points where different parts of the wire touch each other.

The different behavior of the two wire models is due to the difference in the mechanical properties of the wires from which they were made. In the second model, the straightening out force was much stronger than in the first one, and the weight of the second model was much less than that of the first. The wire appearing in Fig. 1, being very heavy, the model acquires its almost flat shape because of the gravitational force that pushes it vertically down to the horizontal table. By lifting parts of it carefully off the table, I was able to see that it can acquire a 3D shape just like that of the second wire. That shape (which I was unable to photograph) would clearly be a stable equilibrium position in the absence of gravity.

It should also be noted that some of the normal forms (in particular almost flat ones) obtained by simulation are very similar, but not identical to the corresponding ones in the wire models. This is because the frictional forces acting in the models are not taken into account in the simulations.

Recently I have begun a series of experiments of new models of thickened knots made of different types of thin wire cables that can be twisted before they are locked to form a closed tube, thus ensuring that besides the straightening out force, there is a substantial torque force acting on them. The first expriments show that the normal forms obtained differ drastically from those obtained from the the wire models without twisting force. Hopefully, the results of those experiments and the mathematical theory behind them will be the subject of a future publication.

6. Circular and almost flat normal forms

The knots 3_1, 5_1, and 7_1 evolved to circular normal forms, the knots 4_1, 6_1 to spherical normal forms, and the knots $6_2 - 6_3$, $7_2 - 7_7$ to almost flat normal forms. Figure 3 schematically shows the almost flat normal forms of all knots with crossing number ≤ 7, except the knot 6_1, which has no almost flat normal form.

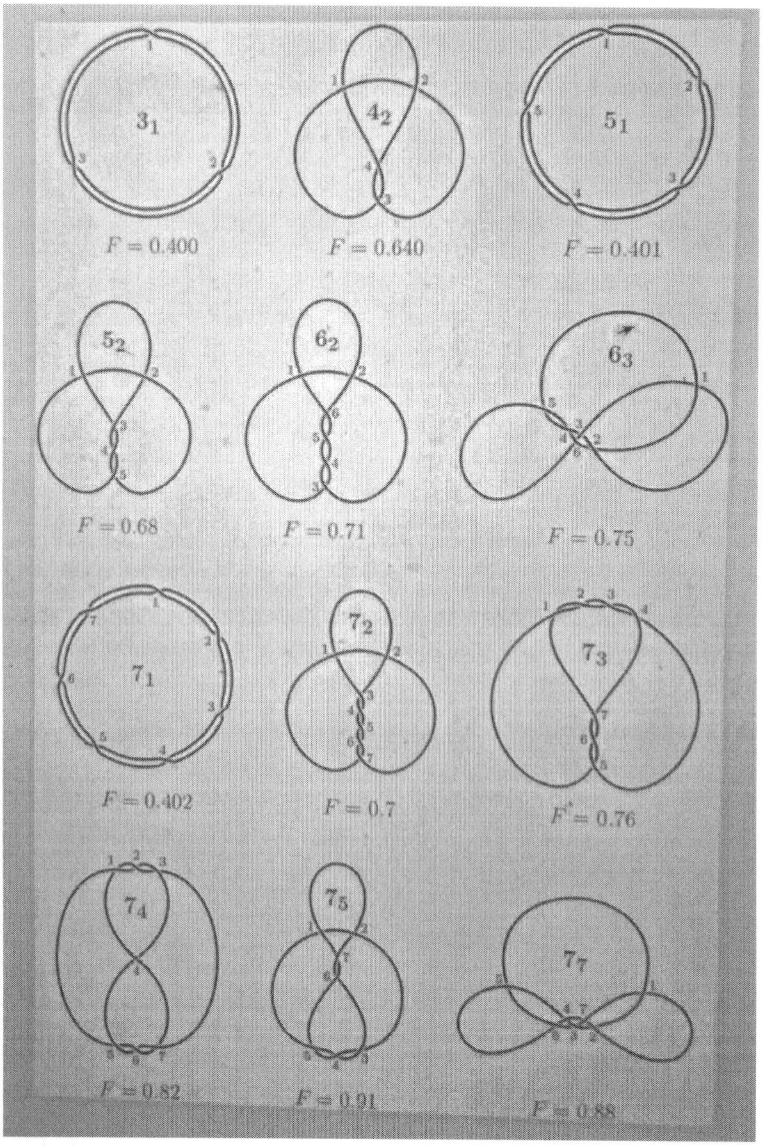

FIGURE 3. Almost flat and circular normal forms of certain knots

In the figure, F stands for the locally minimal value of the functional corresponding to the given knot. We see that *the value of our functional classifies knots with crossings number* ≤ 7. The knot denoted by 4_2 in the figure is actually the eight knot 4_1 in a position which isn't a normal form; we have included it here to show that the corresponding value of our functional ($F = 6.5$) is much larger than its values corresponding to the spherical and circular normal forms of the eight knot 4_1.

7. Normal forms vs. ideal knots

I have noted in the introduction that the idea of studying tubular knots rather than ("infintely thin") classical knots is not new ([**6**], [**7**]). The authors of those papers have defined a certain shape of a tubular knot, which they call ideal knot. In their theory, ideal knots plays a role similar to that of the normal forms in our theory. The motivation and the mathematical definition of the notion of ideal knot is very convincing. I cannot say which of the two notions (ideal knot, normal form) is better from the mathematical and the practical viewpoint, I only want to mention that normal forms are more pleasing to the eye than ideal knots.

In the case of classical knots, a knot energy functional similar to ours was described in the 1980-ies by Jun O'Hara (see [**8**]).

Ackowledgements

The author is grateful to Valery Balikoev for constructing the physical models of knots used in this study and for many stimulating discussions.

References

[1] A. B. Sossinsky, *Mechanical normal forms of knots and flat knots*, Russ. J. Math. Phys. **18** (2011), no. 2, 216–226, DOI 10.1134/S1061920811020105. MR2810996

[2] A. B. Sossinsky, *Normal forms of twisted wire knots*, Russ. J. Math. Phys. **19** (2012), no. 3, 394–400, DOI 10.1134/S1061920812030119. MR2965554

[3] S. Avvakumov, O. Karpenkov, and A. Sossinsky, *Euler elasticae in the plane and the Whitney-Graustein theorem*, Russ. J. Math. Phys. **20** (2013), no. 3, 257–267, DOI 10.1134/S1061920813030011. MR3096373

[4] S. Avvakumov and A. Sossinsky, *Bringing closed polygonal curves in the plane to normal form via local moves*, Math. Notes **103** (2018), no. 3-4, 466–473, DOI 10.1134/S0001434618030124. MR3789104

[5] A. B. Sossinsky, *Normal forms of unknotted ribbons and DNA*, Russ. J. Math. Phys. **25** (2018), no. 2, 241–247, DOI 10.1134/S1061920818020097. MR3813023

[6] Oscar Gonzalez and John H. Maddocks, *Global curvature, thickness, and the ideal shapes of knots*, Proc. Natl. Acad. Sci. USA **96** (1999), no. 9, 4769–4773, DOI 10.1073/pnas.96.9.4769. MR1692638

[7] R. A. Litherland, J. Simon, O. Durumeric, and E. Rawdon, *Thickness of knots*, Topology Appl. **91** (1999), no. 3, 233–244, DOI 10.1016/S0166-8641(97)00210-1. MR1666649

[8] Jun O'Hara, *Energy of a knot*, Topology **30** (1991), no. 2, 241–247, DOI 10.1016/0040-9383(91)90010-2. MR1098918

Independent University of Moscow, 119002, Bolshoy Vlasyevskiy Pereulok 11, Moscow, Russia

SELECTED PUBLISHED TITLES IN THIS SERIES

788 **I. S. Krasil'shchik, A. B. Sossinsky, and A. M. Verbovetsky, Editors,** The Diverse World of PDEs: Geometry and Mathematical Physics, 2023

785 **A. Leroy and S. K. Jain, Editors,** Algebra and Coding Theory, 2023

784 **Dinh-Liem Nguyen, Loc Hoang Nguyen, and Thi-Phong Nguyen, Editors,** Advances in Inverse Problems for Partial Differential Equations, 2023

783 **Alla Detinko, Michael Kapovich, Alex Kontorovich, Peter Sarnak, and Richard Schwartz, Editors,** Computational Aspects of Discrete Subgroups of Lie Groups, 2023

782 **Galina Filipuk, Alberto Lastra, and Sławomir Michalik, Editors,** Recent Trends in Formal and Analytic Solutions of Diff. Equations, 2023

781 **Emanuel Indrei, Diego Marcon, and Levon Nurbekyan, Editors,** Geometric and Functional Inequalities and Recent Topics in Nonlinear PDEs, 2023

780 **Erik Koelink, Stefan Kolb, Nicolai Reshetikhin, and Bart Vlaar, Editors,** Hypergeometry, Integrability and Lie Theory, 2022

779 **Samuele Anni, Valentijn Karemaker, and Elisa Lorenzo García, Editors,** Arithmetic, Geometry, Cryptography, and Coding Theory 2021, 2022

778 **Carlos Galindo, Alejandro Melle Hernández, Julio José Moyano-Fernández, and Wilson A. Zúñiga-Galindo, Editors,** p-Adic Analysis, Arithmetic and Singularities, 2022

777 **Bang-Yen Chen, Nicholas D. Brubaker, Takashi Sakai, Bogdan D. Suceavă, Makiko Sumi Tanaka, Hiroshi Tamaru, and Mihaela B. Vajiac, Editors,** Differential Geometry and Global Analysis, 2022

776 **Aaron Wootton, S. Allen Broughton, and Jennifer Paulhus, Editors,** Automorphisms of Riemann Surfaces, Subgroups of Mapping Class Groups and Related Topics, 2022

775 **Fernando Galaz-García, Cecilia González-Tokman, and Juan Carlos Pardo Millán, Editors,** Mexican Mathematicians in the World, 2021

774 **Randall J. Swift, Alan Krinik, Jennifer M. Switkes, and Jason H. Park, Editors,** Stochastic Processes and Functional Analysis, 2021

773 **Nicholas R. Baeth, Thiago H. Freitas, Graham J. Leuschke, and Victor H. Jorge Pérez, Editors,** Commutative Algebra, 2021

772 **Anatoly M. Vershik, Victor M. Buchstaber, and Andrey V. Malyutin, Editors,** Topology, Geometry, and Dynamics, 2021

771 **Nicolás Andruskiewitsch, Gongxiang Liu, Susan Montgomery, and Yinhuo Zhang, Editors,** Hopf Algebras, Tensor Categories and Related Topics, 2021

770 **Stéphane Ballet, Gaetan Bisson, and Irene Bouw, Editors,** Arithmetic, Geometry, Cryptography and Coding Theory, 2021

769 **Kiyoshi Igusa, Alex Martsinkovsky, and Gordana Todorov, Editors,** Representations of Algebras, Geometry and Physics, 2021

768 **Dražen Adamović, Andrej Dujella, Antun Milas, and Pavle Pandžić, Editors,** Lie Groups, Number Theory, and Vertex Algebras, 2021

767 **Moshe Jarden and Tony Shaska, Editors,** Abelian Varieties and Number Theory, 2021

766 **Paola Comparin, Eduardo Esteves, Herbert Lange, Sebastián Reyes-Carocca, and Rubí E. Rodríguez, Editors,** Geometry at the Frontier, 2021

765 **Michael Aschbacher,** Quaternion Fusion Packets, 2021

764 **Gabriel Cunningham, Mark Mixer, and Egon Schulte, Editors,** Polytopes and Discrete Geometry, 2021

763 **Tyler J. Jarvis and Nathan Priddis, Editors,** Singularities, Mirror Symmetry, and the Gauged Linear Sigma Model, 2021

For a complete list of titles in this series, visit the
AMS Bookstore at **www.ams.org/bookstore/conmseries/**.